# Environment and Sustainability in a Globalizing World

From the *Foundations in Global Studies* series, this student-friendly text follows a two-pronged approach to help readers understand the globalizing processes relating to environment and sustainability, which are examined in a range of disciplines, including environmental studies, geography, global studies, political science, international affairs, comparative politics, and other disciplines. First, it presents foundational material that gives students the conceptual underpinnings required for rigorous analysis. Following the Introduction and Overview, Part One presents a brief historical overview of the concerns revolving around environmental sustainability in the modern era. The text then covers key concepts and theoretical constructs that define the global context for sustainable environmental practices, such as the key thinkers and theories pertaining to sustainable environmental practice, and the key international agencies and treaties involved in global discussions. The first part then explores the various models and ways to measure sustainability, the range of environmental domains at play in the sustainability dialogue, and the controversies surrounding them. Part Two employs case studies to examine theory and practice at work in particular situations. The case studies have been selected with an eye toward comprehensiveness of coverage across disciplines and across regions.

**Andrea J. Nightingale** is a Professor in the Department of Sociology and Human Geography at the University of Oslo. She has been Chair of Rural Development in the Global South at the Swedish University of Agricultural Sciences and a Professor II at Noragric, Norwegian University of Life Sciences.

**Foundations in Global Studies**
Series Editor: *Valerie Tomaselli, MTM Publishing*

**Recent Books in the Series**

**Political and Economic Foundations in Global Studies**
*Editors: Michael Anderson and Stephanie Holmsten*
**Social and Cultural Foundations in Global Studies**
*Authors: Eve Stoddard and John Collins, St. Lawrence University*
**Environment and Sustainability in a Globalizing World**
*Editor: Andrea Nightingale, Swedish University of Agricultural Sciences*

**East Asia in the World: An Introduction**
*Editor: Anne Prescott, Five College Center for East Asian Studies at Smith College*
**Latin America in the World: An Introduction**
*Editors: Daniel J. Greenberg with Antonia Garcia-Rodriguez, Pace University*
**The Middle East in the World: An Introduction**
*Editor: Lucia Volk, San Francisco State University*
**South Asia in the World: An Introduction**
*Editor: Susan Snow Wadley, Syracuse University*

# Environment and Sustainability in a Globalizing World

Edited by
Andrea J. Nightingale

Associate Editors: Tom Böhler,
Ben Campbell, and Linus Karlsson

Routledge
Taylor & Francis Group
NEW YORK AND LONDON

First published 2019
by Routledge
52 Vanderbilt Avenue, New York, NY 10017

and by Routledge
2 Park Square, Milton Park, Abingdon, Oxon, OX14 4RN

*Routledge is an imprint of the Taylor & Francis Group, an informa business*



*Library of Congress Cataloging-in-Publication Data*
Names: Nightingale, Andrea J., editor.
Title: Environment and sustainability in a globalizing world / edited by Andrea J. Nightingale.
Description: New York, NY : Routledge, 2019. | Series: Foundations in global studies series | Includes bibliographical references and index.
Identifiers: LCCN 2018052989 (print) | LCCN 2019004014 (ebook) | ISBN 9781315714714 (eBook) | ISBN 9780765646439 (hbk) | ISBN 9780765646446 (pbk) | ISBN 9781138363779 (ebk)
Subjects: LCSH: Sustainability. | Sustainable development. | Globalization—Environmental aspects. | Sustainability—Case studies. | Sustainable development—Case studies. | Globalization—Environmental aspects—Case studies.
Classification: LCC HC79.E5 (ebook) | LCC HC79.E5 E5695 2019 (print) | DDC 338.9/27—dc23
LC record available at https://lccn.loc.gov/2018052989

ISBN: 978-0-7656-4643-9 (hbk)
ISBN: 978-0-7656-4644-6 (pbk)
ISBN: 978-1-315-71471-4 (ebk)

Typeset in Times New Roman
by Apex CoVantage, LLC

# Contents

# About This Book and Series

From the *Foundations in Global Studies* series, *Environment and Sustainability in a Globalizing World* provides an incisive introduction to the idea and practice of sustainability in the context of the global forces shaping the world today. The book aims to expose students to the conceptual tools fundamental to how scholars, activists, and policymakers understand problems relating to "sustainability"—an idea itself that is considered critically throughout the book. This basic methodology—interrogating underlying assumptions and accepted precepts that shape sustainability discussions—applies analytical rigor to varying viewpoints and interpretations that sustainability comprises. Throughout the text, the authors apply the fundamental questions of: *sustainability of what* and *sustainability for whom?* The imperative to act in the face of what is viewed as a world of limited resources is great, but what exactly are the problems to be addressed and what are the specific ways to address them across various local contexts?

## Outline of the Book

As in the rest of Routledge's *Foundations in Global Studies* series, *Environment and Sustainability in a Globalizing World* employs a two-part strategy. Conceptual and theoretical underpinnings discussed in the first part are explored in practice through case studies in the second part.

Part One: Background, Theory, and Context opens with an Introduction and Overview that explores the conceptual framework at the root of sustainability thinking and practice. Chapter 2 offers an overview of the history of sustainability, and its global context, especially insofar as its contested nature has affected approaches to it. Chapter 3 turns toward key concepts and processes *vis-à-vis* various, often conflicting, narratives of sustainability revolving around how the relationship between society and environment is conceptualized. Chapter 4 continues the exploration of key concepts and processes by investigating how these different narratives are "enacted" on the ground. Scale—one of the most critical aspects of how sustainability is implemented—is the subject of Chapter 5, and in a concluding foundational discussion, Chapter 6 explores the range of concerns that are addressed by sustainability scholars and policymakers. Part Two, then, moves to specifics. It presents eight case studies—transition towns in the UK, for example, or community forestry in Nepal—that help to illuminate the conceptual underpinnings presented in Part One.

## Special Features of This Innovative Series

Several tools are used throughout the book—and the series—that augment coverage:

- Text boxes are employed to expand and emphasize specific material: they are used to open up the coverage to related topics or to call attention to especially critical material, such as historical milestones or key vocabulary.
- Resource boxes, set in a grey background, offer links that point readers to sources—mostly online—on the topics discussed. The links, which are live in the ebook version, include connections to timely data, reports on recent events, official sites, local and country-based media, and visual material. Taken together, they establish a rich archive of additional material for readers to draw on. The URLs included are known to be current as of February 1, 2019, and in the case of expired URLs, enough information has been provided for the reader to locate the same, or similarly useful, resources.
- At the same time, back-of-chapter References and Further Research lists help students to trace the material used by authors or to follow more general leads relating to the topics covered in the chapters.
- Images highlight specific details of the case studies, helping to bring the subjects alive. The captions and source notes for the photos will also help readers seek additional resources on the case studies as their interests dictate.

These special features—along with the simple two-part strategy of employing case studies to explore concepts and theory—aim to chart an innovative path of high-interest and enhanced utility for readers. They are the cornerstones of the entire *Foundations in Global Studies* series.

PART ONE

# Background, Theory, and Context

# 1

# Introduction and Overview

Andrea J. Nightingale, Tom Böhler, and Ben Campbell

Sustainability has been on the global agenda since 1972 when the idea was first debated at the United Nations Conference on the Human Environment held in Stockholm, Sweden. At that conference, sustainability was defined as "an economy in equilibrium with its basic ecological support system." This definition echoed concurrent thinking in ecological science that was concerned with absolute limits, or "carrying capacity," for populations within ecosystems. Researchers working in sustainability translated this thinking to human societies and economies, investigating how to control population growth and extractive practices (that is, economic activities that draw on the earth's finite natural resources) to avoid overexploitation. Since the 1972 Stockholm conference, the idea of sustainability has undergone several transformations, but the core goal remains to understand how human societies can live in managed balance with their ecosystems.

This book is an introduction to environment and sustainability in a globalizing world taking two core principles as starting points. First, "sustainability" is not an environmental state that can be known or achieved. Rather, as a concept that has been proposed in various ways to regulate perceived damaging uses of the environment, sustainability requires critical scrutiny into its range of meanings and especially with respect to the different parties attempting to define it. Second, given that it is a concept (rather than an achievable state or goal), sustainability must be explored within the framework of globalization to understand the different ways that sustainable practices have been proposed or implemented around the world. How and why do ideas of sustainability show up in particular places, at specific moments in time? Who takes up the idea, and for what purposes? What are the consequences socially, politically, and ecologically? At core, then, this book is concerned with how sustainability is fundamentally a politically and historically contingent concept that can never be neutral. This lack of neutrality does not imply that a given initiative calling itself "sustainable" is necessarily flawed or illegitimate; the point of this book is that all such initiatives require us to think carefully about sustainability *of what* and *for whom*.

Researchers working on sustainability have tended to accept the *three pillars of economies, ecologies, and societies* as the main components shaping how humans exist within their environments. Precisely how these components operate together is hotly debated, and most definitions hold each one as conceptually separated. Sustainability is closely related to the practice of sustainable development, which can be understood as attempts to operationalize, or put into effect, ideas of sustainability. Sustainable development programs seek to govern or change relations among economies, societies, and ecologies in order to achieve sustainability. Sustainability is also often used to signal a commitment to economic growth, assuming that social and ecological sustainability will follow from that. Tensions between which of the three pillars are more important, whether they capture the full range of sustainability dilemmas, and how to integrate them expose some of the contradictions within the sustainability field. This text critically probes sustainability and sustainable development as concepts, and explores the implications of how these concepts are defined and the contradictions that exist between different applications of them.

The book performs these functions in a systematic way. Part One: Background, Theory, and Context lays out, in six chapters, the foundations needed to explore the environment and sustainability in a globalizing world. This chapter, Introduction and Overview, discusses the conceptual framework underlying many approaches to sustainability; it explores how the concepts of natural limits, the importance of cultural diversity and biodiversity, and ecosystems' durability or fragility have affected specific local environment–society dynamics and the work of sustainable development projects. Chapter 2 explores the historical development of sustainability and its underlying conceptual, political, and contested nature. The rest of Part One discusses the general shape of the sustainability field on the ground. Chapter 3 focuses on the narratives that are shaped by theorists and practitioners, and the ideas that underlie them; Chapter 4 explores how those narratives are put into practice, or "enacted," with a strong focus on how sustainability is measured; Chapter 5 turns toward "scale" as one of the main issues in enacting sustainable solutions; and Chapter 6, concluding Part One, highlights the range of environmental concerns that theorists and practitioners of sustainability address.

Part Two: Case Studies presents different contexts wherein questions of sustainability arise in on-the-ground projects. Some of the chapters look specifically at sustainable development projects and trace the resulting changes in both societies and environments. Other case studies explore technologies that raise sustainability questions, either positively or negatively, while the remainder are concerned with alternatives that seek to radically reframe what a sustainable society might look like. Throughout the book we return to the core questions: sustainability *of what?* and *for whom?* It is our conviction that the nature of and solutions for environmental problems change depending on the place and the scale at which the problem is viewed.

## Aim of the Book

The main aims of this book are therefore twofold. The first is to explore how sustainability and sustainable development have been defined historically, and to assess the consequences of these definitions (and their transformations) over time and across different communities, including ordinary people, and those engaged in various development practices, businesses, and social movements. We probe how ideas of sustainability and sustainable development in different parts of the world both reflect and manifest within society–environment dynamics. Perhaps most importantly, we attend to what these differences mean for understanding the work that sustainability does in different parts of the world.

The second aim is to critically examine projects in different parts of the world that claim to create sustainable society–environment interactions. We suggest that one of the reasons so many sustainability projects have failed is because society–environment interactions have been inadequately conceptualized. The phrase "society–environment interactions" implies a separation between society and its ecosystems. Yet many scholars and activists have argued that it is precisely this way of thinking that has led to the problems of sustainability that many fear today. In Chapter 3: Narratives of Sustainability, we discuss in more detail the idea of "socionatures" and the importance of thinking about society–environment interactions in a fundamentally different manner. Socionatures is a concept used in academia to insist on the inseparability of society from its environment. The concept helps us think about how any change in society necessarily entails simultaneous changes in environments and ecosystems and vice versa, begging the question of whether the three pillars of sustainability are really the right kind of framing for transforming nature–society dynamics in a manner that promotes both social and environmental welfare. In Chapter 2: Background and History, we also introduce new movements that do not articulate their sustainability aims necessarily in these terms, but are similarly redefining how the conceptual and everyday flows and boundaries between society and environment are drawn. We explore the lessons that can be gleaned from efforts to live differently. Space, place, and scale are important. Some relatively small-scale projects seem to be more successful at fundamentally changing socionatures in a direction that appears to integrate society and environment in a more holistic way. Yet, we also see problems with the way that these ventures themselves serve to (re)define who has access to what resources and services from what spatial scales. How are the needs of local people balanced against the needs of the global community, for example? As such, we go back to our core questions: sustainability of what, and for whom?

## Making Sense of Sustainability: Sustainability of What? Sustainability for Whom?

Throughout this book we make a distinction between the concepts "sustainability" and "sustainable development." The term "sustainability" (or "sustainable") is used to depict certain kinds of relationships between society and the biophysical environment. Precisely what constitutes sustainability has changed over time to the extent that some have argued it is meaningless as a concept. We argue the opposite. Large numbers of people and significant financial resources have been mobilized in the name of sustainability, all of which serve to shape society–environment dynamics, often with lasting consequences. So while sustainability remains for us a contested concept, it nevertheless deserves careful scrutiny.

When introduced for the first time in 1974 at a conference convened by the World Council of Churches, a "sustainable society" was defined as a society where scarce resources are equally distributed; where all people can participate in social decisions; and where the global capacity to supply food will exceed the need for food worldwide. As such, the concept evokes goals and desirable conditions that most people can stand behind, which helps to explain why this definition of the term is in such widespread use. It is important to recognize, however, that ideas about what sustainability or a sustainable society entails have changed over time, and today it is used to refer to everything from recycling campaigns to economic growth targets for businesses. Thus, in this book we therefore suggest that sustainability in itself is meaningless conceptually without defining more carefully our core questions: sustainability of what, and for whom?

The concept of "sustainable development," on the other hand, has to do with the realization of sustainability and usually involves investment into projects intended to improve both societies and their environments. This desire to improve indicates the way that sustainability and sustainable development are normative and political; they are not neutral or value-free concepts but rather involve competing claims over what improvement entails. As such, these values and political dimensions—who defines them, who promotes them, and with what consequences for whom—cannot be removed from the ideas themselves. While the most common priorities are framed around justice, environment, and economy, there are many ways of realizing these goals. How different dimensions of sustainability are prioritized across projects and among interest groups is closely tied to how sustainability, progress, and development themselves are defined and measured.

What is less frequently discussed is who is included and who is excluded from any given sustainable development program and how these processes of inclusion and exclusion are generated both locally and globally. As some of the case studies in the book show, sustainable development projects require planners to make assumptions about benefits, inclusion, and needs that cannot be extended to larger scales and inevitably exclude some people. Local projects aimed at realizing sustainable development, for example, often depend on waste disposal or resource exploitation far away from the local project area. Sustainability plans will always create opportunities for some people in one place, and close down possibilities for others elsewhere.

This analysis implies that, while there are many possible visions for sustainable development, any credible vision must consider global spatial implications as well as questions about justice. Different countries or groups of people may, owing to ideology, cultural preferences, or natural conditions, want to follow different strategies in their struggle for a sustainable society. Understanding sustainable development in this way, the terms "developed" and "developing countries" become problematic. These terms indirectly imply that all societies are on the same development path, and thus close down our imagination of alternative development visions.

Developed countries, normally associated with fossil fuel–based, highly industrialized economies, at the moment tend to define sustainable development. Indeed, the concept of development itself has its roots in post–World War II attempts to rebuild the ruined economies, cities, and landscapes of Europe. Yet, considering the global environmental problems caused by "developed" countries, it has become clear that alternative environment–society interactions are required. This way of thinking has special importance when it comes to issues regarding sustainable development and globalization. A fundamental question remains about whether economic globalization can be "ecologized," that is, made ecologically benign, as at the moment, sustainable development is either practiced on a very small scale, or diverted towards economic sustainability as the main aim. A frequently occurring argument within the current global sustainability narrative is that some kind of global governance is needed to overcome all of our global environmental problems. But questions of justice recur when we consider which mentality or ideology should underpin that global governance. Should it be based on the material growth of the most highly industrialized countries of today, or should it be based on the various ideas of living well (*buen vivir*), "resilience," "de-growth," and/or "antiglobalization"? This book does not seek to reconcile these questions; our goal is to hold them in creative tension. We argue that it is impossible for one group of people to resolve questions of who decides what pathway to take towards sustainability. Open and lively debate is required.

## The Conceptual Framework of the Book

As stated above, this book begins from the premise that sustainability is not a state or something knowable with definable criteria, but rather a contested concept that organizes society–nature interactions in particular (and varying) ways. Throughout the book, we pay particular attention to the use of *concepts*, each of which is in some sense problematic in how it consciously and unconsciously influences our way of looking at the world. As all concepts are simplified interpretations of the world, they are also political; they represent one out of many possible perspectives. Yet some concepts stand out as less political or less ideological than others, particularly those that are framed as technical definitions or solutions. In this book we highlight this imbalance in terms of whether a concept is perceived as political or not. To be clear, all interpretations and solutions regarding sustainability are political because they are underpinned by particular commitments to ordering the world, whether they are framed as technical or not.

In order to avoid separating "societies, politics, and economies" from "ecologies," we invoke the term "socionatures." From this perspective, interconnections between different social, ecological, political, and economic practices and concepts are given special attention. In other words, *understandings* of processes or interrelations across borders or categories are as important as the borders and the categories themselves. Any analysis of them requires us to make arbitrary choices in time and space. In order to think in a holistic way about these interconnections and contingent outcomes, we use the terms "co-production" and "mutual becoming" to signal the dynamic qualities of socionatures, demonstrating how relationships between different dimensions are never stable and are constantly influencing each other. These ideas make our analytical task more complex, but also serve to reflect life in a more realistic manner. In Chapters 2 and 3 we elaborate on how these different concepts have shaped sustainability thinking and practice.

From this discussion, four core themes emerge that run throughout the book:

1. Narratives of sustainability and how these constructions succeed in guiding society–nature interactions.
2. Justice and how normative concerns are always part of sustainability claims.
3. Governance and the ways that sustainability motivates and organizes particular forms of participation in environmental governance.
4. Globalization, or how sustainability concerns serve to both connect and separate parts of the world and to raise serious questions about the current political economic order.

These themes are not distinct from each other, but help us to organize the discussion and to highlight their importance.

## Narratives of Sustainability

Narratives refer to the languages and stories used to describe sustainable, as well as unsustainable, socionatures. Social scientists take narratives very seriously as important world-making tools that shape how people think and act. They define what is possible and desirable, questions that are central to sustainability issues. Here we introduce three core narratives that derive originally from environmental science:

1. Balance
2. Limits
3. Diversity

These three concepts underpin most meanings of sustainability.

The idea that nature is characterized by *balance*, or stabilizing relationships, is central within environmental sciences, especially ecology. For example, ecosystems are normally understood as sets of relationships that over time have evolved into so-called stable domains that will last, provided external disturbance is not too severe. Many sustainable development projects thus seek to retain relationships and conditions as they are, or to return them to a state in the past that was deemed in balance.

The second concept, *limits*, is closely linked to "carrying capacity," or the idea that a given ecosystem can provide a finite amount of food and other necessary resources. Populations can outgrow the carrying capacity of their ecosystem, causing their numbers to decline or even crash. This concept first came from predator–prey studies, where the number of predators in a given area was assumed to reach a peak, after which the population would decline because its prey would become too scarce. As predators declined, the prey species would increase, allowing more predators to thrive, thus beginning the cycle again. Very few empirical studies, however, have been able to show this relationship, in part because the concept of carrying capacity assumes that only pressure from predators controls prey populations and vice versa. In reality, many other factors influence population numbers, including disease. Today, a new related concept, "planetary boundaries," has emerged from the natural sciences, which attempts to define a wide set of different biophysical domains with which one can identify the "safe operating space" for humans. Planetary boundaries are presumed to govern the stability of the global biosphere, and if these boundaries are crossed, major transformations will occur. With climate change, there are increasing fears that planetary limits will be reached and the biosphere as we know it will collapse.

Sustainability has further been linked to the idea of *diversity*. Within ecology, diversity has to do with the number of species present within a community as well as the relative abundance of each species. Diversity, or at least the conservation of diversity within a specific ecosystem, is considered intrinsically desirable and necessary for sustainability. As a result, the conservation of biodiversity has become a goal in itself. Over the last couple of centuries, many species have gone extinct, nearly all of them due to human activities. Yet how to protect vulnerable species is not necessarily straightforward. Attempts to save animal species by breeding them in zoos have seen only limited success, for example. And many sustainability advocates argue that the health of entire ecosystems should be the focus, not the preservation of specific species. Diversity, seen from this perspective, is simply a symptom of a healthy ecosystem, rather than a goal in itself. We probe these narratives in more detail in Chapters 3 and 4.

## Justice and the Normative Aspect of Sustainability

Narratives of sustainability are tied to issues of justice. They can bring into view, or obscure, particular inequalities and unfair practices. Using the three concepts we discussed above—balance, limits, diversity—here we introduce some of the quandaries concerning justice attached to them. The first concept, stability or balance, raises questions about what states are desirable and what forms of disturbance are unacceptable. People often take current ecosystem states as "natural" and fail to probe how human interactions are crucial to the production and maintenance of that state. Perhaps most importantly from a justice perspective, we need to ask who decides which states are desirable in which places and at what times and for what purposes. While many sustainability projects appeal to a common-sense understanding of a "good ecosystem," from a socionature perspective, no clear guidelines for what constitutes a good ecosystem exist. It is a bit easier to define "bad ecosystems"—for example, those that are polluted with toxic chemicals—while we struggle to so clearly define a "good" one.

Similarly, the idea of planetary boundaries has been embraced as common sense, but many critique this natural science–driven framework. From a justice perspective, concerns exist that an overemphasis on limits masks distributional issues. A long tradition of scholarship extending back to at least the 1980s has shown how access to resources for particular populations is driven by political economies of distribution rather than absolute limits. Within the wider debate, there are tensions between those who focus on defining "planetary boundaries" (which themselves are contested within natural science) in opposition to those who argue for more efficient use of resources through new technologies and those who place more emphasis on just systems of distribution. In this book, we aim to consider the consequences of these tensions. Within the popular debate, "planetary boundaries" has been taken up very quickly as a defining concept for environmental governance. Yet the concept does little to narrow down core sustainability problems, not least because measuring and defining planetary boundaries are more or less impossible. Issues like these highlight the necessity to both understand and reflect upon the positive and negative effects of taking physical limits seriously. What are the limits, and in what ways and contexts are they highlighted? And most importantly perhaps, who or what is affected by these limits and how? Is it the limits themselves at issue, or is it the way certain limits are prioritized over others?

The third concept, diversity, raises important normative questions about how humans use the environment in ways that affect other species. Diversity is assumed to produce stability, but this is not well substantiated empirically. Ecosystems in the tropics normally have greater diversity than ecosystems around the poles, but they are not necessarily more stable or durable. When protecting or increasing diversity becomes an aim in itself, crucial questions emerge about what kind of diversity, for which species, for whom, and how this will be achieved. For example, environmentalists often use biological diversity as an argument to stop other humans from using "nature" as their livelihood or resource base. But in many ecosystems, the harvesting of plants and animals by humans, such as cutting of trees within a forest, is important for promoting and maintaining diversity. While too much forest harvesting can be destructive, some scientists characterize a healthy forest ecosystem as containing open areas, or gaps, that allow for a greater diversity of species to flourish. Perhaps most significantly, which dimensions of biodiversity are important is most often defined by environmental scientists, yet the people who depend on healthy biodiversity for their livelihoods often have their own knowledge of its value as well as their own systems for producing and protecting biodiverse environments. Embracing such alternative knowledges can be crucial for how we conceive of and communicate the concept of sustainability. These insights draw attention to the need to continually ask: sustainability of what? and sustainability for whom?

## Governance

Very closely related to sustainability narratives and the normative questions they raise are issues of governance. All sustainability debates are infused with the underlying question of "who decides"? As a result, a wide number of local, regional, national, and global bodies have the responsibility for promoting and governing sustainability. Today many local governments are given an explicit mandate to promote sustainability, yet many government employees struggle to know what precisely that implies in terms of action and priorities. The environmental sciences are often used as a neutral referee to determine whether particular practices, resource uses, or ecosystems are sustainable. But as we have seen above, it is impossible for the environmental sciences to be a neutral referee. The natural sciences themselves are infused with competing values and normative ideas of sustainability, and they are also not the only way to understand the world around us. Many social movements and indigenous communities are insisting that their knowledge systems be taken seriously when sustainability questions emerge. While valuable and important, scientific knowledge alone cannot resolve questions of sustainability for whom and of what.

Which organizations are authorized to promote and regulate sustainability is therefore also a political and normative question. In many democratic countries, the nation-state is assumed to be best placed to define laws that address different sustainability issues and to ensure they are enforced. Yet increasingly, nation-states have been criticized for failing to live up to sustainability standards. The United States' actions on climate change are one example. Its refusal to sign the Kyoto Protocol and later pulling out of the Paris Agreement—both aimed at addressing climate change through reducing carbon emissions—shows how one nation-state's failure to live up to its sustainability obligations can derail global initiatives. In other countries, people see the state as being one of the prime perpetrators of unsustainable practices as the case study of Ecuadorian oil exploration in Chapter 13 outlines. Also of significance in that case, the question of sustainability of what continually re-emerges to change how indigenous people perceived the practices of oil exploration by the government.

The Kyoto Protocol was developed under the United Nations Framework Convention on Climate Change (UNFCCC) and first adopted in Japan on December 11, 1997. The Protocol did not come into force until February 16, 2005. A summary of the Kyoto Protocol along with the Marrakesh Accords, which outlined how the Kyoto Protocol was to be implemented, can be found at the UNFCCC website:

http://unfccc.int/kyoto_protocol/items/2830.php.

Further information on the 2015 Paris Agreement to limit emissions can be found at the European Commission website:

http://ec.europa.eu/clima/policies/international/negotiations/paris/index_en.htm.

The full Paris Agreement can be downloaded from the UNFCCC website:

https://unfccc.int/process-and-meetings/the-paris-agreement/the-paris-agreement.

As sustainability issues are increasingly framed in global terms (for example, global climate change), calls are increasing for either existing international bodies to take on a greater governance role, or for the formation of new governance bodies that would have authority over nation-states. A supranational environmental governance body could help to ensure that all nation-states are held to global sustainability standards. For example, in 2015, the United Nations proposed a set of Sustainable Development Goals (SDGs) as a refinement of its earlier Millennium Development Goals that were used as benchmarks for developing countries and programs to assist them.

The Sustainable Development Goals were adopted in September 2015 and seek to expand and refine the original Millennium Development Goals. More information can be found at the United Nations website:

https://sustainabledevelopment.un.org/?menu=1300.

At the same time, many groups of people are dissatisfied with centralized control over questions that have profound implications for their local environments. Various social movements seek to promote their own understandings of sustainability and to hold states or other governance bodies accountable to not only legal standards, but also to new standards based upon alternative logics of good living (*buen vivir*). Who should govern our pathways towards sustainability and at what scale are therefore also central questions of this book. The case studies in Part Two illustrate some of the tensions that emerge when competing narratives of sustainability encounter competing governance arrangements.

## Globalization and Local–Global Connections

Calls for global governance of the environment represent one dimension of globalization within sustainability debates. Other dimensions include the global economy and questions of local–global connections. The global economy is often singled out as either the biggest threat to sustainability or, conversely, the best way to promote sustainability. In today's global economy, resources extracted in one place are transported over long distances to be used in other places, and the waste from these processes often ends up in a third place. This spatial redistribution of resources makes it difficult to understand the full impact of their use and redistribution. It also masks the many lifestyles in the Global North that are dependent on resources not locally available and the deposition of waste in the Global South or China. The concept of "ecological footprints" that accounts for how much land a community or nation needs to maintain its standard of living (discussed in Chapter 4) is one attempt at highlighting such relationships. Others see this kind of global redistribution of resources as a positive development for sustainability. Many definitions of sustainable development, predicated on the notion of capitalist growth, insist that a sustainable economy is one that is growing. This is supported by the fact that as economies in the Global North have become wealthier, they have also reduced their environmental impact, at least within their own nation-state boundaries. The concept of ecological modernization (also discussed in Chapter 4) is an example of a narrative that embraces economic growth as a necessary quality of a sustainable society.

Like concerns over global governance, many social movements oppose the ongoing commodification of life. They argue that sustainability cannot be achieved under the present capitalist system and the environmental values attached to modern life. In the United Kingdom and the United States, the idea of "transition towns" has emerged wherein people in a community make a commitment to radically reshape their use of resources and overall carbon emissions. Chapter 8: Alternative Food Movements and Transition Towns in the United Kingdom provides a look at one such endeavor. Indigenous movements have for the last several decades organized around the idea that their lifeways are under threat and that they are not interested in living by the terms dictated through modern capitalist societies. It is crucial, however, to view these movements as part of globalization. Indeed, the most successful movements are those that connect people from disparate localities in order to leverage more political power. In addition, attempts are increasing to link localities in the Global South with localities in the Global North to redistribute resources and make direct connections between people living very differently (for more, see Chapter 12 on permaculture initiatives in Zimbabwe).

Finally, thinking about sustainability in a globalizing world draws attention to scale. As discussed above, our present economic system in some sense makes the world a smaller place and demands that we think at multiple scales when considering sustainability issues. Global climate change, for example, has made us profoundly aware of how we cannot hide in environmentally pristine places because overexploitation of resources and polluting activities end up impacting those places anyway. The interconnections of the globe are not necessarily all negative, however. They also offer the possibility to think more creatively about how to address resource shortages in particular localities and how to engage on multiple scales at once. Chapter 5 discusses in more detail questions of scale and how sustainability issues often serve to redefine the scale at which problems are identified and addressed.

In summary, this book provides a concept-oriented introduction to the wide terrain of sustainability and sustainable development. Throughout the book we probe the underlying ideas of environment–society relations within different framings of sustainability and think through the consequences of these conceptual underpinnings for action on the ground. Our aim is to provide readers with the tools to think critically about the work that sustainability achieves around the world. For while it is certainly not *one* concept, and while we do not believe a universally agreed-upon principle for sustainability is possible, there is no question that sustainability as an idea is able to motivate large numbers of people to engage in a huge diversity of projects intended to make our world a better place. This book is dedicated to thinking through who defines what "better" means, who is authorized to guide and regulate our pathways towards "better," and what the consequences might be for those who have alternative visions of "better." Before delving into a critical exploration of sustainability of what and for whom, Chapter 2 presents a brief history of the idea of sustainability as a way to understand how current debates are informed and framed.

## References and Further Research

Bell, Simon, and Stephen Morse. 2013. *Measuring Sustainability: Learning From Doing*. New York: Earthscan.

Cunningham, William P., and Mary Ann Cunningham. 2010. *Environmental Science: A Global Concern*. Volume 11. Boston: McGraw-Hill Higher Education.

Dresner, Simon. 2008. *The Principles of Sustainability*. Second ed. New York: Earthscan.

Mol, Arthur P. J., David Allan Sonnenfeld, and Gert Spaargaren. 2009. *The Ecological Modernisation Reader: Environmental Reform in Theory and Practice*. London: Routledge.

Newell, Peter. 2012. *Globalization and the Environment: Capitalism, Ecology and Power*. Cambridge: Polity Press.

Robertson, Margaret. 2014. *Sustainability Principles and Practice*. New York: Earthscan.

# 2

# Background and History of Sustainability

Andrea J. Nightingale, Tom Böhler, Ben Campbell, and Linus Karlsson

The principles that underpin sustainability, outlined in Chapter 1, are certainly not new. The historical record shows that concerns over human impact on the earth extend back well over 2,000 years, and are characterized by a variety of different explanations for why human habitats seem to decline. In this chapter, we explore how the idea of sustainability has changed over time and within different regions. While the idea of sustainability came into wide circulation in the 1970s, modern understandings of human–environment relations have emerged from a long history of contested ideas over the best way to live with nature.

## History of Environment and Sustainability

The word *sustainability* first came into the English language in the mid-twentieth century, but the ideas that underpin it have a much longer history. The notion in European thought that the earth needs protection from humans and that humans are exploitative of their environments extends back at least to the early Greeks. The historical record from across the world is checkered with narratives of humans "falling from grace" or the "decline from Eden."

### From Ancient Philosophy to the Enlightenment

While many of these narratives are particularly associated with Judeo-Christian thought, as Clarence Glacken shows in his 1967 book, *Traces on the Rhodian Shore*, other traditions have similar narratives of environmental decline associated with human activities. The Greeks, for example, attributed environmental degradation to moral and political decline within a society that failed to live in harmony with its landscape. Others in the Roman Empire explained the decrease in society's agricultural output as the result of an aging planet. These explanations of how the earth necessarily declines over time are very persistent within many human societies and continue to shape how sustainability is understood today.

Another important influence on modern sustainability thinking is the notion that the earth is alive. Carolyn Merchant shows in her 1980 book, *The Death of Nature: Women, Ecology and the Scientific Revolution*, that before the seventeenth century European thought was dominated by ideas that the earth was a living organism that had power over physical and spiritual aspects of human life. Current proponents of sustainability often explicitly look back to the time when such holistic understandings of the earth predominated. The concept of Gaia, or framing the earth as a living being, popularized by James Lovelock in the early 1970s, is one modern manifestation of this looking back. Yet, as explored in more detail in Chapter 3, one irony of current notions of sustainability is the extent to which they continue to separate society from nature, rather than seeing the two as part of one living whole.

James Lovelock's Gaia hypothesis has been criticized from several different perspectives and recently has been challenged as inaccurate. The hypothesis nevertheless has had an important impact on how science has been done and how many people think about the earth today. Read more about Gaia and criticisms of it at this New Scientist web page:

www.newscientist.com/round-up/gaia/.

The Enlightenment—an intellectual movement in several European countries primarily during the eighteenth century—marked a distinct shift in how society–environment dynamics were conceptualized. Key Enlightenment thinkers argued that religious dogmatism and superstition distorted our understanding of physical reality. Instead, they proposed a mechanistic conceptualization of the earth that was closely related to the rise of science as a dominant way to understand the world. This was also the period when science as a particular mode of inquiry emerged, leading to an emphasis on observable, empirical knowledge. The scientific method, which creates generalized knowledge by showing that the results of an experiment are repeatable, is a product of this turn towards isolating observations of one variable from outside influences in order to understand them. The Enlightenment worldview also proposed that humans possess the power to change and improve their own life conditions. This new way of looking at the world tended to frame the environment as a resource without intrinsic values, and changed the focus of environmental thought from the past to the future as people increasingly sought progress.

The Enlightenment was not only a time of change in environmental philosophies, it was also concurrent with rapid socioeconomic change, particularly the development of mechanized industries known as the Industrial Revolution. The Industrial Revolution was closely tied to the exploitation of natural resources from countries in the Global South, colonized by countries in the Global North. While intraregional trade has very deep historical roots that extend back at least 3,000 years, mercantile colonialism that dominated world trade beginning in the sixteenth century and reaching its peak in the eighteenth and nineteenth centuries, linked parts of the world at a scale and rate that were significantly greater than in previous centuries. Colonialism also helped to establish capitalism as an economic system in many parts of the world.

### The Industrial Revolution and Sustainability

The Industrial Revolution has been credited with producing many of the technologies we depend upon in the Global North for our lifestyles: electricity, telephones, combustion engines that power cars, boats, and airplanes, along with machinery for the mass production of consumer goods. Mechanization also allows fewer people to produce more food, such that more people can engage in nonagricultural activities. Yet today, many are questioning whether these innovations are positive. At the core of many sustainability debates is a fundamental tension between those who turn towards more technological progress to address sustainability challenges, and those who argue for a return to historical lifestyles that depended far less on machines. Throughout this book, you will find those tensions evident in competing conceptions and practices of sustainability.

The Industrial Revolution along with colonialism brought about widespread changes in economies and environments upon which the emerging global economy depended. Capitalism is based upon the selling of labor power for wages, and the continual circulation of capital (money as well as assets, or land, equipment, natural resources, etc.) to generate growth. Growth is essential to the healthy functioning of a capitalist system. And perhaps most importantly, the global economy that emerged from this so-called "mercantile capitalism" established a world order wherein natural resources were extracted in the colonies and moved to the industrial centers of Europe and North America for processing into goods that were then sold for significant profits. In other words, a global capitalist economic system translates value from resources and human labor into money and other forms of wealth, and redistributes that wealth to other places. As such, capitalism is highly dependent on the (cheap) extraction of resources in one place, and the generation of wealth (often) in another place. This colonial, capitalist economy caused widespread movement of people, resources, and other species, and produced exploitative relationships among different countries, societies, and environments, legacies that persist today. There are many who argue that the unequal relations between the Global South and the Global North were established during the era of mercantile–colonial capitalism. While a comprehensive accounting of how this time period impacted environments and societies is outside the scope of this book, it is crucial to recognize that many of today's present sustainability challenges have their roots in colonialism as well as in our current form of global economy.

For an overview of the history of international trade, see "International Trade in Historical Perspective" by Elhanan Helpman, the Onasis Prize Lecture, under the auspices of Harvard University and the Canadian Institute for Advanced Research, September 1, 2012:

www.cass.city.ac.uk/__data/assets/pdf_file/0004/141295/E-Helpman-Presentation.pdf.

The expansion of the Industrial Revolution in the eighteenth and nineteenth centuries also brought environmental concerns firmly onto the political agenda and established many of the environmental–social–economic relations that continue to shape sustainability challenges today. As part of establishing a global economy based upon highly unequal relations across space and societies, capitalist economic relations were in the process of more or less subsuming older, barter and feudal-oriented economies in many parts of the world. Adam Smith (1723–1790), considered the founder of classical economics, was one of the first to argue that society needed to reduce its dependence on resource extraction. Yet his argument was not derived from a concern for the planet or for exploited colonial societies. Rather, he was a strong advocate of economic growth, competitive commodity production, and free markets.

Smith assumed, as neoliberal economics does today, that an economy organized in this manner will solve problems of uneven accumulation and overexploitation of labor and resources. As we return to in Chapter 3, there is a connection between the need for growth within capitalist economies and the generation of wealth in one place at the expense of other regions. These issues of distributive justice are at the heart of many social debates over sustainability today.

Adam Smith was an extremely influential scholar during his lifetime and many of his ideas, including those in his book *An Inquiry into the Nature and Causes of the Wealth of Nations*, usually shortened to *The Wealth of Nations*, underpin different strands of current economic thought. Read more about Adam Smith's life and his ideas at this Library of Economics and Liberty website:

www.econlib.org/library/Enc/bios/Smith.html.

An electronic version of *The Wealth of Nations* is available at this web page on the Library of Economics and Liberty website:

www.econlib.org/library/Smith/smWN.html.

While Smith was a respected economist during his lifetime, it was not until the end of the eighteenth century that ideas about limits to human consumption and economic growth first became popular. Thomas R. Malthus (1766–1834), an economist and Anglican priest, published *An Essay on the Principle of Population* in 1798, arguing that a constantly increasing human population would overrun the earth's ability to produce enough food. His ideas ran counter to popular beliefs at the time, which instead were focused on the abilities of technology and economic growth to overcome agricultural limitations. While Malthus maintained that population growth would eventually lead to a crisis of resources, he simultaneously believed that technological progress could postpone the inevitable need to reduce demands on resources.

**Malthusianism**

Malthusianism is a school of ideas derived from the writings of the British economist and Anglican pastor Thomas Malthus (1766–1834). In his *An Essay on the Principles of Population* (1798), Malthus argued that a population grows at a faster rate than its means of subsistence, and unless it is controlled or restrained, poverty and environmental degradation inevitably follow. He identified two "checks" that reduce populations to sustainable levels: "preventive checks" performed by individuals (abstinence), and "positive checks" such as wars, famines, and diseases that return populations to a lower level (known as Malthusian catastrophes). Malthus' work was motivated by his concern over the rate of reproduction of the poor in England. He imagined a future in which "civilized" people would be overrun by masses of the undereducated and uncivilized poor. Neo-Malthusianism refers to a line of thought advocating population control to ensure resource availability for current and future populations.

Smith and Malthus in these ways laid an important foundation for probing the link between economic activities and environmental degradation. Smith highlighted economic limitations within a resource-heavy economy, while Malthus was more concerned about the relationships between environment and food production, or what is known today as carrying

Adam Smith, considered the founder of classical economics.

*Source:* John Kay/Wikimedia; https://commons.wikimedia.org/wiki/File:AdamSmith1790.jpg.

Thomas Malthus, one of the first Western thinkers to raise concerns about population growth, whose ideas continue to be controversial today.

*Source:* Wellcome Images/Wikimedia; https://commons.wikimedia.org/wiki/File:Thomas_Robert_Malthus_Wellcome_L0069037.jpg.

capacity. Following them, more explicit critiques of capitalist growth appeared in the second half of the nineteenth century. Economists like John Stuart Mill (1806–1873) argued that since constant growth in society's production will lead to environmental degradation, transition to an economic system that regulated growth, or a steady-state economy, was desirable. This was also the period when some of the environmental consequences of the Industrial Revolution and merchant capitalism became more evident. These consequences included severe air pollution, a major decline in wood resources in Europe, and severe ecological decline in some island colonies. George Perkins Marsh's book *Man and Nature* (1864) was the first in this period to put forward the idea that human activities could permanently damage the earth. He advocated for a precautionary approach to human–environment relations but many of his arguments were not taken up until much later. Ideas of limits to economic growth, the need for moderating human exploitation of resources, and the desirability of steady states thus have a long history. Their influence has grown over time and now underpin much of modern environmentalist and sustainability thought.

**Steady States**

The Merriam-Webster online dictionary defines steady state as:

> a state or condition of a system or process (as one of the energy states of an atom) that does not change in time; *broadly*: a condition that changes only negligibly over a specified time.

Steady state is used in the environmental field to indicate an ecosystem, organism, or place that is in balance. It does not change radically in its overall attributes over time, despite smaller changes that occur constantly. For example, a mature oak forest remains an oak forest even though there are continual processes of growth and decay within it. This is also referred to as "dynamic equilibrium." Beginning in the early part of the twentieth century, ecologists believed that healthy ecosystems are in balance, or in a steady state. It was not until the 1970s that ecologists began seriously questioning whether dynamic equilibrium was in fact a characteristic of healthy ecosystems or if a model of more dynamic change was required.

As a result of Enlightenment thinking, concepts such as "resources," and isolating one variable from others to understand natural systems, emerged. Today, these ideas are so entrenched in how we think and do science it is hard to imagine that they are products of a particular conceptualization of the earth. Yet, it is important to understand them as historically situated ideas. During the scientific and Industrial Revolution, improvement, human development, and progress became powerful, ideological, guiding concepts. Notions of how to best maximize resource utilization were also heavily influenced by John Locke's stance on landownership (1690). Locke argued that private property rights would encourage landowners to invest in labor to improve land over its natural state, and this idea paved the way for privatization of common lands. The hegemonic power of these concepts prevailed until the environmental impact of industrial society became more visible, giving rise to ideas of environmentalism and sustainability, which in the 1960s and 1970s began to challenge some of the bases of Enlightenment logic.

For more on dynamic equilibrium and steady states see the web page, with video links, "Equilibrium and Steady State" at this MIT's Teaching + Learning Lab website:

http://tll.mit.edu/help/equilibrium-vs-steady-state.

## The Origins of Sustainability

As shown above, modern notions of sustainability can be traced back to ancient times and draw from a variety of historical traditions. The word "sustainability" in English is believed to have derived from the eighteenth-century German forestry term "sustained yield." Sustained yield was based upon Malthus' concepts, seeking to strike a balance between extraction and forest management that would give the highest yields over a period of time. It emerged out of the application of scientific ideas to resource extraction, particularly forests, which were adopted to meet the challenges posed by declining forest resources in Europe. Many ancient forests in Europe were harvested to build sailing ships for transporting goods, slaves, and convicts, in addition to immigrants fleeing from the potato famine and dispossession of their lands due to privatization of the commons and rising capitalist economic relations in Europe. A series of wars among the major European powers and later with the Americans (1690–1815) also consumed timber resources for shipbuilding.

Sustained-yield forest management was quickly internationalized through colonial landholdings. The extraction of resources in the colonies of European countries followed a "western" logic, focused on the production of trees as timber and often backed up by the scientific language of "sustainable-yield management." Traditional practices found in Africa, Asia, and

Latin America, such as slash-and-burn (or "swidden") agriculture, were deemed primitive and destructive. Sustained-yield, in contrast, was consistent with contemporary ideas of economic growth, and efficient and rational use of natural resources. Sustained-yield is thus an early example of how environmental narratives and practices have been globalized.

In the United States, the influence of nineteenth-century U.S. environmental thinkers such as Henry David Thoreau, Ralph Waldo Emerson, and George Perkins Marsh began to contribute to the growing interest in environmental degradation. While not strictly pondering on the concept of sustainability or the German notion of "sustained yield," these authors drew on personal experience and romantic ideas of nature to frame the environment as a source of inspiration, peace, and connection. They were writing at a time when the U.S. frontier was rapidly disappearing and many areas that were perceived as pristine by early colonists had been brought under agriculture or industrial uses. While today the claim that North American environments were free from human use (i.e., pristine) before European colonization has been largely rejected, there is no question that colonial activities and industrial expansion transformed the U.S. landscape significantly. It was these transformations that environmental thinkers began to criticize.

Concerns over the impact of industrial economies on the environment increased in the mid-twentieth century, both within science by people like Rachel Carson and her book on

Rachel Carson, author of the groundbreaking work, *The Silent Spring*, in the field with Bob Hines during her tenure at the U.S. Fish and Wildlife Service. The book is often considered a catalyst for the modern environmental movement.

*Source:* U.S. Fish and Wildlife Service/Wikimedia; https://commons.wikimedia.org/wiki/File:Rachel_Carson_Conducts_Marine_Biology_Research_with_Bob_Hines.jpg.

the dangers of pesticides, *Silent Spring* (1962), as well as within society at large. Carson's book was groundbreaking in emphasizing invisible threats to the environment. Until that point, pesticides and new chemical fertilizers were embraced as miracle products that could eradicate hunger and free most people from agriculture through industrial production of food. Sustainability concerns were mainly voiced in relation to environmental decline and nostalgic calls for a return to simpler lifestyles that disrupted the environment less. The 1960s also saw a massive popular critique of affluence, consumerism, and pollution, issues that were often connected to other political movements such as civil and women's rights. Alternatives were being spoken of and advocated for at the international scale by scientists and groups that were outside government and industry, as discussed later in this chapter. These movements were also catalyzed by the catastrophic effects on the environment of nuclear bombs in places where they were tested, and most catastrophically, where nuclear weapons were used against a population in Japan at the end of World War II.

The twentieth century also saw a shift in academia from relatively holistic disciplines like geography, which sought to understand human–environment relations from the scale of the individual to that of society, to the rise of quantitative science and the splintering of disciplines into more narrow specializations, like the separation of macro- from microeconomics or human from physical geography. The ecological sciences focused on the dynamics among plants within a particular place, seeking to explain how they developed over time into what was considered the "climax" or "steady state." This model of ecology assumes that a healthy ecosystem is one where change is minimal and relatively slower growing, but longer lived species ("climax species") dominate. Ecosystems go through a process of succession wherein quick growing, colonizing species are gradually replaced by slower growing vegetation until the climax species come to dominate. Ecosystems today continue to be conceptualized within systems thinking as webs of interactions, with multiple relationships among plants, animals, and the environment within which they grow, although the idea of climax has been rejected. These relationships are largely assumed to exist outside of human influence, and indeed, it is when human activities intervene, that the balance of nature is disturbed. These ideas of stability have been somewhat overturned since the 1970s by resilience thinking, which uses systems theory to think about how socio-environmental systems are characterized by constant change, but they continue to underpin most sustainability science.

> The concept of climax ecosystems has been largely rejected within ecological science, but it dominated ecology and, importantly, natural resource management until very recently. Many environmental managers today continue to manage for stability and prioritize mature vegetation types over vegetation that is quick to colonize disturbed areas. For more on climax vegetation and environmental management see this article: "Climax Theories and a Recommendation for Vegetation Classification: A Viewpoint" in the *Journal for Range Management* (https://journals.uair.arizona.edu/index.php/jrm/article/viewFile/7757/7369).

It is perhaps as a reaction to the increasing specialization and narrowing of disciplines that modern environmentalism has sought to restore holistic thinking. Certainly, that is an explicit point of contention for many grassroots and indigenous environmental movements, to which we return at the end of the chapter. Yet there is also no question that a lot of sustainability science has its origins firmly in ecology and the natural sciences, traditions that generally follow the scientific method and seek to separate out components in order to study them in isolation from other components. And most importantly, these traditions understand stability over time and space to be desirable attributes of an ecosystem. As we probe more fully throughout this book, stability is not necessarily desirable; what is stable can depend on which attributes of a system are valued most, and at what scale stability is evaluated.

## Sustainability as a Global Order

So far we have shown how sustainability emerged from recurring concerns over human impacts on the earth. As such, sustainability needs to be conceptualized both as something that *guides* societies as well as something that is *part of* how societies understand their interconnections with the environment or nature. Sustainability as a formal global political order, however, with national and international bodies creating communities of practice and policies around it, first emerged in the 1970s. The United Nations Conference on the Human Environment held in Stockholm, Sweden, in 1972 is often named as an important watershed during which environmental sustainability was first put forward as a theme that required a global response. Attended by then secretary general of the United Nations U Thant of Burma, as well as representatives of several UN departments, specialized agencies, observers from intergovernmental organizations, representatives of international nongovernmental organizations, and 113 invited states, the conference framed sustainability as "an economy in equilibrium with basic ecological support systems." The purpose of the conference was to discuss "problems of the human environment" and "identify those aspects of it that can only, or best be solved through international cooperation and agreement." It was, in other words, an attempt to internationalize environmental protection. And indeed, one of the outcomes of the Stockholm Conference was the creation of the United Nations Environment Programme (UNEP), a first move towards international cooperation for tackling the environmental challenges created by an industrial, global economy. Although UNEP did not challenge the idea that nation-states are ultimately responsible for environmental governance, the program sought to foster better cooperation.

This institutionalization of sustainability and global environmental governance, however, reflected wider trends within society in many respects, some dating back to the 1940s, when development as a global political order first emerged. The world wars of the twentieth century left the economies of Europe in ruin. World leaders met in 1944 and signed the Bretton Woods agreement, which promised technological and financial support to rebuild Europe. In his inaugural address in 1949, president of the United States Harry Truman promoted an internationalization of this new approach to supporting struggling economies. He said: "we must embark on a bold new program for making the benefits of our scientific advances and industrial progress available for the improvement and growth of underdeveloped areas." These ideas of progress, growth, and the need for some communities to support others socioeconomically became consolidated in development practice and, more recently, sustainable development.

As the development model became rooted in international relations and global policy initiatives, other trends were emerging. During the 1960s and 1970s, warnings were sounded concerning the limitations of our natural resource base, including finite resources such as oil and copper and other ores, as well as renewable resources like fish and water. Researchers and activists urged consideration of the global population as well as its resource uses. Books like Paul R. Ehrlich's *The Population Bomb* (1968) and Donella H. Meadows and colleagues' *The Limits to Growth* (1972) brought into mainstream policy discussions a debate over whether a capitalist economy was really the best way to organize society's use of resources. In the last section of this chapter, we discuss some of the popular movements relating to agriculture that emerged out of these critiques.

The UN continues to hold a major international gathering at regular intervals to debate the state of the environment and forge agreements on how to respond to perceived threats (see also the text box on Major UN Conferences and Agreements). Each of these conferences has had a distinctive title and focus. Today, many of them are considered watersheds in defining sustainability debates, albeit in different ways. A UN resolution created, for example, the 1983 World Commission on Environment and Development (WCED), an independent body tasked with compiling a report on the state of the world's environment. The resulting publication, *Our Common Future* (1987), commonly known as the Brundtland report (after former prime minister of Norway, Gro Harlem Bruntland, who oversaw the conference and the production of the report), put forward a collective statement about the importance of "meeting the needs of the present without compromising the ability of future generations to meet their own needs." This continues to be one of the most commonly used definitions of sustainability today.

The Brundtland definition of sustainability represented a number of significant shifts. First, it explicitly linked sustainability to both an increasing population and the attendant strain on available resources. "Needs" encompasses a wider set of concerns than simply resource capacity. Second, this definition also added a longer time dimension. By invoking the importance of maintaining the environment for future generations, it established a moral obligation to those not yet born. Perhaps most importantly, the Brundtland definition does not attempt to define "needs," nor is it prescriptive about how sustainability is evaluated. It makes no attempt to outline the criteria through which actions will be assessed for their immediate and long-term impacts, nor does it give authority to any particular institution or set of actors to monitor and enforce whether this goal is being met. So while the Brundtland definition continues to be the touchstone upon which people seek to be more specific, in many respects it did little other than to open up the idea of sustainability to encompass more than the economic system–environmental resources dynamic, and to insist on the consideration of future generations.

Because many environmental problems transcended national borders, transnational corporations, nongovernmental organizations, and various other international institutions emerged in the 1970s and 1980s hoping to influence the debate about sustainability. The globalization of sustainability concerns led many such organizations to embrace the view that nation-states, while remaining the appropriate bodies for regulating environmental governance, should act in the interest of the planet, and not for the nation only. Consequently, a number of joint reports and agreements were signed in the 1980s by the world's countries. For instance, the International Union for Conservation of Nature (IUCN), United Nations Environment Programme (UNEP), and World Wide Fund for Nature (WWF) released the *World Conservation Strategy* in 1980, focusing on conservation of nature and population pressure on resources. Other international agreements include the 1985 Vienna Convention for the Protection of the Ozone Layer; the 1987 Montreal Protocol on Substances that Deplete the Ozone Layer; and the 1989 Basel Convention on Transboundary Movement of Hazardous Wastes and Their Disposal. These agreements, among others, supported the view that international cooperation is both feasible and the appropriate mechanism for regulating the environment.

The second conference in the UN environment series, the 1992 UN Conference on Environment and Development, held in Rio de Janeiro, Brazil, and better known as the Rio Earth Summit, is often considered to be the most important event in which world political, industrial, and civil society leaders signed on to the environmental agenda. One of the central aims of the conference was to identify principles for operationalizing the ideas in the Brundtland report. An action plan, Agenda 21 (referring to the twenty-first century), was launched together with the establishment of a *Commission on Sustainable Development* to monitor and facilitate the implementation of this action plan. Agenda 21 outlined how sustainability would be achieved in economic and social terms, with respect to resources and conservation of resources for development, through strengthening the role of major groups (ranging from women to global institutions like the UN), and by establishing the means for implementation, particularly finance mechanisms.

**Major UN Conferences and Agreements Related to Sustainability, 1972 to the Present**

1972 United Nations Conference on the Human Environment is convened in Stockholm. Stockholm Declaration is adopted.
1973 United Nations Environment Programme (UNEP) is established.
1985 Vienna Convention for the Protection of the Ozone Layer is adopted. Entered into force in 1988.
1987 Montreal Protocol on Substances that Deplete the Ozone Layer is adopted. Entered into force in 1989.
1987 The Brundtland report is published.
1992 United Nations Conference on Environment and Development is convened in Rio de Janeiro. Agenda 21 and Rio Declaration are adopted.
1992 United Nations Convention on Biological Diversity is adopted. Entered into force in 1993.
1992 United Nations Commission on Sustainable Development (CSD) is established.
1997 Kyoto Protocol on Climate Change is adopted.
2000 Millennium Summit is held in New York City. The Millennium Declaration is adopted and the Millennium Development Goals (MDGs) are announced.
2002 World Summit on Sustainable Development, also known as Rio+10, is held in Johannesburg. Johannesburg Declaration and Plan of Implementation are adopted.
2012 United Nations Conference on Sustainable Development, also known as Rio+20, is held in Rio de Janeiro. The Sustainable Development Goals (SDGs) are proposed.

In many respects the Rio Earth Summit was an important collective agreement to tackle the effects of capitalist economies as part of one large problem, rather than as separate issues. Emphasizing the linkages between society and the environment, the summit represented a departure from the view that development is detrimental to environmental goals. The lack of development in many parts of the world was seen as driving environmental decline. Impoverished peasants were assumed to overexploit their resources due to their poverty, whereas in industrialized countries, problems of industrial pollution and a fossil fuel economy were blamed as the main culprits of environmental decline. Agenda 21 sought to overcome these contradictions by employing more integrated thinking to development in both the Global North and the Global South.

The Rio Earth Summit itself did not accomplish this shift in how the global community approached environmental questions, however. Rather, Rio represented a culmination of multiple influences that were changing the collective commitment to the environment, of which three are particularly important to mention. First, there was growing evidence that industrial pollution was affecting parts of the world far from the smoke stacks and factories that emitted toxic chemicals into the air and water. Acid rain, for example, was causing trees and fish to die in locations far from the sources of pollution, as had become apparent not just to scientists but to people living in Scandinavia, Canada, and remote, nonindustrial areas of the United States.

Second, there was increasing concern over the hazards of nuclear waste both in the short and long term. Nuclear power was creating large stockpiles of highly radioactive waste, toxic to humans and animals, with no good solutions emerging for how to store it safely for the thousands of years necessary until it becomes less hazardous to life, and no surefire safeguards against catastrophic failures. The meltdown of the Chernobyl nuclear power plant in 1986 highlighted the latter in stark terms. The meltdown at Chernobyl spread radioactive material into the atmosphere, which was carried by wind and rain across large parts of Europe and Russia. Chernobyl, in the Ukraine, was part of the former Soviet Union at the time and thus behind the Iron Curtain, a symbolic barrier during the Cold War marking the social and political division between the communist Soviet Union (and its Eastern European satellite states) and the capitalist-oriented countries of Western Europe. The spread of nuclear radiation after the accident through water and the air highlighted how the Iron Curtain was simply a political, symbolic barrier rather than a physical one. While the flow of people, goods, and news across the East–West divide was severely limited, the Chernobyl disaster showed that ecological boundaries are nonexistent.

Third, changing commitments to environmental problems came about after the fall of the Berlin wall in 1989. The Berlin wall, a physical barrier erected through the center of Berlin to separate the "Global East" and the "Global West," was a powerful symbol of the Cold War. In 1989, the wall was torn down by protestors on both sides, catalyzing a profound shift in global geopolitics. On the one hand, the end of the Cold War meant that the threats to global society now seemed to be even more focused on industrial and poverty-related environmental issues rather than nuclear war. On the other hand, it opened up more political space for alternative voices to be heard at the Rio Earth Summit. Those voices demanded attention to everyday threats to livelihoods rather than financial and political attention focused on a global order defined largely around political tensions linked to Europe, North America, and the former Soviet Union—that is, the Global North. The end of the Cold War focused much more attention on developing countries' needs, although, in practice, some critics have argued this was not always the case.

Agenda 21 therefore was influenced by concerns over poverty, unequal development, and the lack of basic needs across many countries in the Global South. It represented a profound shift in thinking about environmental issues from concerns with ecological limits and environmental decline, to the social and political causes of overexploitation. As a result, Agenda 21 in many respects is a commitment to particular kinds of development as the main pathway towards environmental sustainability, although, as we explore in Chapter 3, these pathways were not consistent and very quickly diverged. Most importantly, it marked a split in sustainability thinking between those concerned with sustainability and those concerned with sustainable development. Sustainable development came to be considered the tools through which sustainability can be achieved. But at the same time, sustainable development took on the idea of social sustainability, while the concept of sustainability alone continues to be more narrowly focused on environmental concerns and, as we discuss in detail in Chapter 4, on economic sustainability.

## Sustainability from Rio to the Present

After the Rio Earth Summit, sustainability and sustainable development as concepts began to proliferate and were taken up in a wide variety of fields. A number of subsequent global agreements and conferences sought to refine the concepts as well as apply them to ever widening domains. Many critics have suggested that sustainability and sustainable development are meaningless as a result, that they can be whatever one wants them to be. In this book, we agree with the need to probe carefully how sustainability and sustainable development are conceptualized and applied, but we otherwise profoundly disagree with this critique. Rather, we see that a lot is done all over the world in the name of sustainability and sustainable development. Millions of dollars have been invested, thousands and thousands of projects, both big and small, have been implemented, and significant networks of people have rallied together under the banner of sustainability. We therefore place our emphasis on what sustainability does as it travels around the world. How has it changed as different groups of actors take up the idea, and what kind of results can we observe in its name?

At the global scale, a number of international agreements that followed the Rio Earth Summit continued to place sustainability as a central guiding principle. Agenda 21 helped to shape development priorities that were focused on the fulfillment of basic needs. In the twenty-five years after Rio, development projects across a range of sectors, from water resources to women's development, promoted local participation and addressed people's needs for food, shelter, water, and a clean environment. Programs that had the twin goals of poverty alleviation and producing healthy forests, such as community forestry, were launched. It is only since about 2010 that we again see a significant conceptual shift in development practice from focusing on sustainability as a condition, or state, that can be achieved in specific localities, to sustainability as emerging from particular kinds of investments and networks.

The hopes and expectations that Rio created quickly cooled off, however, in part because of neoliberal economics. Neoliberalism refers to an economic and political logic that takes Adam Smith's original ideas about the "invisible hand of the market" as the right tool for regulating social and environmental relations to advocate for free markets, deregulated banking systems, and international investment. With the breakup of the former Soviet Union, capitalist economic systems came to dominate most of the world in the 1990s, helping to expand intensive trade, investment, and capital markets globally. This kind of economic growth was in contrast to Agenda 21's focus on creating sustainability through a global partnership including national governments, and international, national, regional, and local level organizations, as well as nongovernmental organizations and civil society. Such a global partnership would meet the basic needs of populations, improve living standards for all, protect ecosystems, and lead to a prosperous future. In contrast, the so-called neoliberal Washington Consensus guided economic activity for the world's superpowers, advocating fiscal and monetary soundness, deregulation of markets and financial institutions, and privatization of public services. Sustainability therefore increasingly became synonymous with sustained economic growth, rather than sustained global ecosystems. As we explore below, many of the social movements that have emerged under the banner of sustainability engage in an explicit critique of this kind of global economy and corresponding narratives of sustainability.

### Neoliberalism Pros and Cons

Neoliberalism as a political and economic project has attracted both staunch supporters and critics alike. Some argue that neoliberal policies are necessary to ensure healthy domestic and global economies, while others blame these same policies for creating one of the biggest gaps in wealth ever recorded in history. The figures vary depending on how they are calculated, but the *Washington Times* reported in 2014 that the top 1 percent of U.S. citizens owned 40 percent of the nation's wealth, while the bottom 80 percent of the population owned just 7 percent. Regardless of whether these figures are the best way to calculate the wealth gap, most observers agree that economic inequality has grown radically since the 1950s.

For more on neoliberalism see this Global Issues primer on neoliberalism:

www.globalissues.org/article/39/a-primer-on-neoliberalism.

For more on wealth inequalities see Inequality.org, a project of the Institute for Policy Studies (http://inequality.org/wealth-inequality/), and the *Washington Post* blogpost "If You Thought Inequality Was Bad, Get a Load of Wealth Inequality":

www.washingtonpost.com/news/wonk/wp/2015/05/21/the-top-10-of-americans-own-76-of-the-stuff-and-its-dragging-our-economy-down/.

Another factor emerging in the post-Brundtland era was the scientific understanding of nature as the basis for ensuring sustainability. The Intergovernmental Panel on Climate Change (IPCC) was established in 1988 and has since received a great deal of attention in the media, serving to raise public awareness of climate change and ecosystem degradation. Recently, concerns with environmental sustainability have been largely subsumed under worries over global climate change, and today climate change continues to be the primary arena about which questions of interlinked environmental and economic sustainability are raised. Social sustainability has been almost entirely neglected by the climate change agenda until the past few years (see also Chapter 15). The post-Brundtland era is thus characterized by increasing emphasis on scientific knowledge and documentation of climate change and biophysical processes, and their implications on human well-being and planetary health. This emphasis on scientific knowledge, produced within formal institutions such as universities and government-funded research institutes, serves to further globalize environmental governance. The physical sciences have given more attention to technical solutions and making nature knowable and thus governable. This has led to ideas of environmental or geo-engineering, which rest on assumptions that ecological limits and balances can be modified or regulated artificially.

In this manner, climate change increasingly dominated formal global environmental sustainability governance, and concurrently, neoliberal economic policies were clearly failing to cause the "trickle-down" effects that advocates of free markets and deregulation promised. It was at this time that the UN launched its eight Millennium Development Goals (MDGs) in 2000. The MDGs were a direct acknowledgment that despite decades of significant development investment, few improvements were made in economic inequality and sustainability and therefore more concrete targets and timelines were needed for achieving Agenda 21. The MDGs were therefore in some respects a refinement of Agenda 21 and focused on: (1) eradication of extreme poverty and hunger; (2) achieving universal primary education; (3) promoting gender equality; (4) reducing child mortality; (5) improving maternal health; (6) combating HIV, malaria, and other diseases; (7) ensuring environmental sustainability; and (8) developing a global partnership for development. Of all the goals only MDG7 is explicitly about sustainable development, but the others are linked to key dimensions of sustainability that were identified in Agenda 21. It is noteworthy that the MDGs are directed at developing countries and do not explicitly recognize the monetary and moral responsibility of industrialized countries. As such, they do not say much about global justice and the ways that the global economy produces inequality, even if these goals are an attempt at addressing such inequalities within specific places.

While the MDGs represent a renewed commitment to what are often considered fundamental human needs, the means to accomplish them were seen to lie within neoliberal economic approaches and, in particular, the engagement of the private sector. This shift from a basic needs to a neoliberal economic approach to development became obvious at the World Summit on Sustainable Development in Johannesburg, in 2002 (also known as Rio+10). At this UN conference, the interdependence of economic, social, and environmental development was acknowledged, and one aim was to revitalize and accelerate the whole process of sustainable development, which was seen to have stagnated. An outcome of the conference was the development of the new *type II* partnerships initiatives, where private and civil actors were included in the work and management of sustainable development. These type II partnerships were intended to bring more momentum into the implementation of the UN's Agenda 21 and its Millennium Development Goals, particularly at subnational local levels, than the existing state- and government-centered *type I* partnership initiatives.

The next in the UN conference series, Rio+20, in Rio de Janeiro in 2012, was a major stocktaking moment, where the MDGs were revisited along with Agenda 21 to set the global environmental agenda for the next twenty years. The conference also highlighted how different global networks had begun to diverge substantially in their conceptualizations of and interests in sustainability. By this time, climate change had begun to dominate the sustainability agenda of formal, global institutions, such that one World Bank report, compiled as a background paper for the meeting, complained that climate change was the only sector in which sustainability continued to be a guiding principle. But while the UN hoped to establish a roadmap for a new green global economy, developing countries wanted to discuss new sustainability goals that better protect the environment, secure food production, alleviate poverty, and give poor people more power. In short, competing narratives of what sustainable development might mean and how to achieve it were further apart than they were at the Rio Earth Summit.

The 1992 Rio Earth Summit, and many of the documents resulting from it, including links to the three volumes of the "Report of the United Nations Conference on Environment and Development," are presented here at the Sustainable Development Goals Knowledge Platform.

www.un.org/wcm/webdav/site/climatechange/shared/gsp/docs/GSP1-6_Background%20on%20Sustainable%20Devt.pdf.

The document produced from Rio+20, "The Future We Want" can be found here:

www.un.org/disabilities/documents/rio20_outcome_document_complete.pdf.

The main outcome of Rio+20 became a nonbinding document, "The Future We Want," in which governments renewed their political commitments to sustainable development. But the document also contains supportive statements regarding development of new so-called Sustainable Development Goals (SDGs) in 2015 that address criticisms of the earlier Millennium Development Goals (MDGs) regarding the environmental and justice aspects of development. For example, nations agreed to explore alternatives to Gross Domestic

Product indices that can take both environmental and social factors into account. It was also recognized that changes in societies' consumption and production patterns were necessary to achieve global sustainability. In Chapter 4 we look in more depth at the SDGs.

Looking back at the decades since the first UN Conference on the Human Environment in Stockholm in 1972, we can see many changes in the conceptualization of environmental problems. In the 1970s the focus was on limits to growth rather than sustainable development. Economic growth was seen as something that could threaten our survival. The Brundtland report, however, served to open up this framing of human–environment dynamics. Environmental problems were not only considered from an ecological and scientific perspective but also from economic, social, and political perspectives as well. The three last environmental summits, 1992 (Rio), 2002 (Johannesburg), and 2012 (Rio), have all produced and reaffirmed the view that development is inextricable from environmental issues. For example at the 2012 summit in Rio, it was stated that eradicating poverty was the greatest global challenge facing the world today and an indispensable requirement for sustainable development. After 2015 we are beginning to see a shift away from this kind of narrative, however. Limits are back on the agenda with ideas of "planetary boundaries" and "de-growth." We return to these ideas in Chapter 4 to probe how they are shaping the understanding of sustainability today.

## Social Movements and Critiques of Sustainable Development

As discussed above, many of the political agendas around sustainability have been closely tied to growing concerns over the consequences of the resource-intensive, industrial, global economy that has led to widespread redistribution of wealth and resources around the globe. Criticisms of these forms of sustainability have focused on precisely these dimensions.

Agriculture has been one area where activists have been particularly strong. During the twentieth century, agriculture and agribusiness expanded through intensification of production and increasingly capitalized technical inputs (mainly machinery and chemicals). The global West pushed a technocratic version of modern agriculture, epitomized by the so-called Green Revolution (see also the accompanying text box). The Green Revolution promoted the use of chemical fertilizers and pesticides in order to maximize crop yields. It was assumed that more agricultural output would lessen, and possibly eradicate, poverty and famine from those countries that adopted intensification. Subsequently, the Green Revolution has been heavily criticized on environmental and social grounds. In particular, problems of famine are widely accepted to be produced as much by an unequal *distribution* of food as they are by *lack* of food. And in part because of these problems, the promotion of Green Revolution ideas was paralleled by a counter-tendency among a range of movements and organizations advancing concerns for soil health, biodiversity, and the value of local ecological knowledge. In this section we take a closer look at some of those movements.

### The Green Revolution

The Green Revolution refers to the development of high-yielding varieties of grain crops in the 1940s and their widespread promotion across the Global South in the 1950s and 1960s. Largely credited with launching this "revolution," the American scientist Norman Borlaug began research in Mexico in the 1940s to develop new disease-resistant high-yield varieties of wheat. He was awarded a Nobel Prize in 1970.

Wheat, rice, and corn were especially targeted for crop development after Borlaug's initial experiments in Mexico. Farmers and scientists have bred food crops for millennia by cross-fertilization—the techniques used for the original Green Revolution—so in this sense, it was not new. What was new was the emphasis on new seeds in combination with mechanized production and the introduction of chemical fertilizers and pesticides to increase grain yields radically over historical rates. These new techniques and seeds were sold by seed and chemical companies and promoted through governments and international development aid across the Global South, particularly India, Mexico, and the Philippines. Their adoption was widespread and did serve to increase production, at least in some places, for some time.

Quickly, however, the miracle of the Green Revolution began to be questioned. Over time, chemical fertilizers strip soils of their naturally occurring nutrients and pesticide resistance in insects, fungus, and other targeted pests became major problems. Yields declined accordingly and farmers found themselves struggling with debt incurred from needing to buy seeds and inputs rather than using locally sourced materials.

Today there are renewed calls for a Green Revolution to meet food security demands generated by the confluence of a growing population and a changing climate. Africa is particularly being targeted for improved varieties, both genetically modified and those bred through cross-fertilization techniques, supported by wealthy private foundations. These calls are being met with resistance within the academic and activist communities alike. Nevertheless, so-called improved seeds and inputs are a cornerstone of the U.S. Agency for International Development's agricultural development policies in the Global South.

For a critique of the Green Revolution by renowned environmental activist Vandana Shiva, see this Living Heritage Trust web page:

http://livingheritage.org/green-revolution.htm.

The new calls for a Green Revolution in Africa are elaborated at these web pages:

"A New Green Revolution for Africa?" at Grain.org: www.grain.org/article/entries/74-a-new-green-revolution-for-africa, and

"Alliance for a Green Revolution in Africa" at the Bill and Melinda Gates Foundation website: www.gatesfoundation.org/How-We-Work/Resources/Grantee-Profiles/Grantee-Profile-Alliance-for-a-Green-Revolution-in-Africa-AGRA.

Wendell Berry, a passionate writer on the problems with industrial agriculture, was an early voice in the movement for sustainability. His book, *The Unsettling of America* (1977), develops a historical critique of the profit-driven expansion of the American frontier to show the importance of local knowledge in small-scale farming practices and community integration in enabling sustainable food cultures. Scientific research played a part in these alternative movements. Berry was in close conversation with agricultural scientists including Wes Jackson of The Land Institute (TLI) in Salina, Kansas. This institute explores the possibility of developing perennial grains and growing them in polycultures, such as prairie ecosystems, that do not require the use of fertilizers or pesticides.

Similarly, interest was revitalized in the 1911 book *Farmers of Forty Centuries*, which highlighted historical practices of permanent agriculture from East Asia that could maintain agricultural systems over millennia. Composting systems such as the Indore method that was developed for maintaining soil health in tropical conditions on colonial plantations were put forward as alternatives to chemical intensive agriculture. There was a broad appeal for such systems among communities of growers and members of social movements who ascribed to a worldview that rejected technocratic modernism. This new wave of self-described sustainability social movements was literally built up from the grassroots, as activists compiled examples of successful soil, plant, and animal health interactions conducive to human well-being as a basis for their ideologies.

For more info on the Indore composting method, among other methods, see these sources:

- "Small-scale composting," at this Food and Agricultural Association web page: www.fao.org/3/y5104e/y5104e06.htm, and
- "Sir Albert Howard and the Indore Process," and other helpful articles on this Permaculture Noosa web page: https://permaculturenoosa.com.au/sir-albert-howards-indore-compost-method/.

Another good example of organizations promoting sustainability thinking include the UK-based Soil Association, which pioneered the certification of foods that were produced from organically grown, nonchemical methods. They worked alongside the Organic Gardening organization, to lead the way in encouraging people to take charge of their own food systems and use crop rotation to ensure healthy food supplies from even small garden patches. Beginning in the 1970s, the Alternative Technology Centre, built on a barren former quarry in west Wales, promoted food growing, renewable energy technologies, self-build architectural systems, and educational outreach as interlinked parts of sustainable ways of living. Meanwhile in Australia, the permaculture movement developed a robust conceptual approach to ecological location (especially in arid places) that worked with local plants and rainfall constraints. This approach emphasizes mutually advantageous relationships and self-organizing systems, such as "vertical" growing along forest edges. These kinds of integrated agricultural, social, and forested systems are a profound contrast to the intensive agricultural practices reliant on chemical inputs and large-scale production of monoculture crops that is epitomized in early Green Revolution technologies. Even those practices are now beginning to incorporate lessons from organic agriculture, such as the use of earthworms to maintain soil health, but they are nevertheless still primarily based upon chemical inputs.

The emancipatory and social justice dimensions of these land-based movements were quite varied, and in some cases even reactionary. In the United States, some activists in the Earth First movement resorted to acts of vandalism, like filling the gas tanks of bulldozers with sand, to try to stop industrial developments they opposed. There were also, however, many movements that emerged out of working-class communities who organized around rights to healthy landscapes. For example, in the United States, the United Farm Workers organized grape and other food boycotts to raise awareness of how farm workers were being exposed to dangerous levels of pesticides. In the United Kingdom, there were strong working-class pressures for the public right to access land for leisure among urban communities in the north of England. It was the industrial working classes who led mass trespass movements to secure rights to walk in the Peak District, which became the UK's first national park before World War II. The working-class tradition of keeping allotment gardens also provided important supplements to family budgets and a basis for community autonomy, family life, and handmade art forms.

The United Farm Workers have been important in raising awareness not only of the dangers of pesticides, but also of their immediate health impacts on farm workers. Yet the union has also been heavily criticized. For more information about their history and some of the boycotts they have instigated, see their own website at http://ufw.org/ and the article "The Rise and Fall of the United Farm Workers" at the *Monthly Review* website: http://monthlyreview.org/2010/05/01/the-rise-and-fall-of-the-united-farm-workers/.

However, it could be said that for sustainability to become the force that mobilized the post–Cold War global community in 1992, a significant confluence of political movements needed to take place. This included feminist concerns about women's exclusion from many aspects of development in urban and rural settings, and a recognition that cultural minorities and indigenous peoples not only should be allowed to participate in development benefits, but that they have significant contributions to make in terms of actively defending knowledge of biodiversity, food, medicines, and traditions of how to live well.

A number of celebrated cases drew worldwide attention to popular, grassroots movements. These movements demonstrated that caring for and protecting biodiversity in conjunction with livelihoods did not mean having to choose between intensive agriculture and environmental protection. Both were often done at the expense of local people's access to land and traditional resources. Perhaps one of the best known is the Chipko movement that began in the Indian Himalayas in the 1970s. Groups of villagers hugged the trees and chained themselves to equipment to prevent commercial logging by companies that had been given licenses for the work by the government. The villagers' resistance successfully led to a ban on harvesting timber from forests above an altitude of 3,280 feet (1,000 meters), and initiated extensive studies and projects in support of people's rights to nontimber forest products. Influenced by the power of movements from below to protect forests, people in central Nepal subsequently opposed the creation of a national park that was to be patrolled by the army, and came up with an alternative plan for a community-based conservation area, the Annapurna Conservation Area, which today is one of Nepal's most popular trekking areas. Other examples of local people defending livelihoods and knowledge systems against government and commercial encroachments on lands and forests include the Brazilian Rubber tappers group led by Chico Mendes, murdered for his activism 1988; the Penan people's refusal to abandon the rainforest in Borneo to logging companies; and Rigoberta Menchu's struggle against racist plantation landlords' genocidal and environmentally destructive regime in Guatemala. These are some of the better known examples of how sustainability narratives have been taken up and redefined by social movements. Note how these movements are very locally based, and yet their successes have brought them into globalized networks. In some cases, these movements have been dependent, for their survival, on developing these widespread supportive networks of people who share their vision for sustainability.

These examples of peoples' resistance movements against unsustainable development processes were given new legitimacy from Agenda 21. The mid-1990s witnessed large-scale movements, including opposition to big dam building projects in central India on the Narmada River. These movements brought together an alliance of indigenous people (Adivasis) and peasant farmers. They were so widespread and tenacious that the World Bank was persuaded the dams would cause as many, if not more, problems for greater numbers of the poor than benefits. World Bank funding for the Narmada river dams was therefore withdrawn.

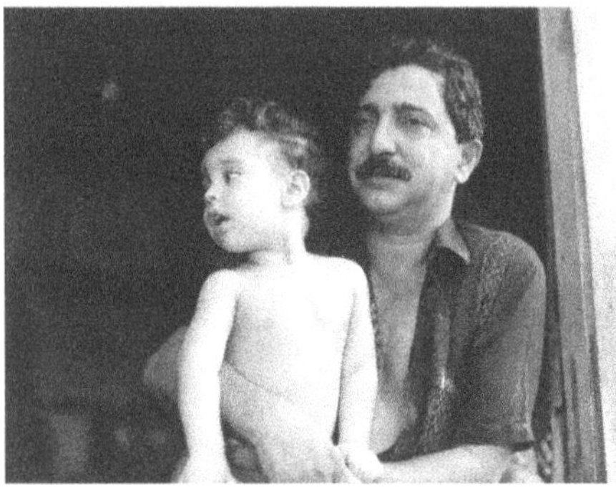

Chico Mendes with his son Sandino in 1988, five weeks before his assassination.

*Source:* Miranda Smith, Miranda Productions/Wikimedia; https://commons.wikimedia.org/wiki/File:Chico_Mendes_with_Sandino_2.png.

Guatemalan activist Rigoberta Menchu, left, with the Ecuadorian ambassador in 2012.

*Source:* Cancillería Ecuador/Wikimedia; https://commons.wikimedia.org/wiki/File:Llegada_de_Rigoberta_Mench%C3%BA_al_Ecuador_(6982895712).jpg.

This type of resistance in the Global South had a cultural impact back in the Global North. There is no question it helped challenge large-scale development interventions like big dams, and fueled a groundswell of popular dissatisfaction with mass consumer society and the lack of democratic participation in techno-centric economic growth models. The North had become increasingly postindustrial as most manufacturing industries migrated to Asia and many communities in the former manufacturing belts of the United States and the UK found themselves out of work or having to radically change their lifestyles. In the UK the cause of defending ecology against the Conservative government policies in the 1980s and 1990s led to protest movements against roads that crossed rural areas. For example, the battle of Twyford Down took place in 1991 to oppose new road building through iconic southern English "chalk down" woodland; the fight was led by people who had pursued alternative lifestyles that opposed the values and consumption patterns advocated by the mainstream.

From the mid-1990s in the Unites States and Europe, the focus of critical popular opinion and protest against unsustainable political and economic trends shifted towards foods and genetically modified (GM) crops as well as globalization. Spearheaded by organizations such as Greenpeace and Friends of the Earth, European ports were blockaded in opposition to the introduction of GM content in the food chain from U.S. maize and soya. The 1999 Battle for Seattle witnessed mass protests against corporate domination of the world food system and advocated rights to autonomous food cultures in contrast to fast-food expansion. This huge popular protest against the World Trade Organization meeting in Seattle included targeted vandalism against big multinational food chains. The French leader of Confédération Paysanne termed such chains "mal-bouffe" or "junk food." Meanwhile his artisanal crafted Roquefort cheeses made from the unpasteurized milk of sheep raised on hillsides in France's Auvergne region were denied import licenses to the United States. These movements brought into widespread popular consciousness the problems of commercial agriculture, globalization, and the perils of a consumer culture with growth as its only objective. They also showed the power of mass protests to block corporate- and government-sponsored development. As the case studies in Part Two of this book show, such gains are sometimes only short-lived. Indeed, one of the main challenges for sustainability movements today is how to keep momentum strong and to avoid simply pushing the problems into new places. The European ban on GM crops, for example, has been repeatedly challenged by multinational corporations keen to penetrate EU markets.

The World Trade Organization's Ministerial Conference in Seattle in 1999 was met with large protests by anti-globalization activists. This article in *The Atlantic*—"The Dark Side of Globalization"—discusses the issues that triggered the protests and argues that from the perspective of 2014, their demands were justified. Read it here:

www.theatlantic.com/business/archive/2014/01/the-dark-side-of-globalization-why-seattles-1999-protesters-were-right/282831/.

## Conclusion

This chapter has shown how ideas of sustainability have deep roots in human concerns over environmental decline—decline that has been attributed to radically different causes (natural processes vs. human activities). The solutions put forward similarly reflect very different conceptualizations of what attaining sustainability might mean. Sustainability as a global order, with formal institutions, and scientific and development projects, as well as organized social movements networked around the idea of sustainability, emerged in the 1970s. The sustainability debate as well as the global governance of sustainability has been shaped by science, popular movements, and formal global networks, albeit not always working with precisely the same narratives of sustainability. As a result, sustainability needs to be understood as a spectrum of concepts rather than something that is knowable or achievable. It is crucial to understand which kinds of networks have emerged in the name of sustainability, what they seek to achieve, and which dimensions of sustainability they privilege. In the next chapter, we take a systematic look at current notions of sustainability in order to understand better how sustainability travels around the world.

## References and Further Research

Berry, Wendell. 1977. *The Unsettling of America: Culture & Agriculture*. San Francisco: Sierra Club Books.

Carson, Rachel. 1962. *Silent Spring*. Boston: Houghton Mifflin.

Ehrlich, Paul R. 1968. *The Population Bomb*. New York: Ballantine Books.

Glacken, Clarence J. 1967. *Traces on the Rhodian Shore: Nature and Culture in Western Thought from Ancient Times to the End of the Eighteenth Century*. Berkeley: University of California Press.

Grove, Richard H. 1995. *Green Imperialism: Colonial Expansion, Tropical Island Edens and the Origins of Environmentalism, 1600–1860*. New York: Cambridge University Press.

IUCN, UNEP, and WWF. 1980. *World Conservation Strategy: Living Resource Conservation for Sustainable Development*. Gland, Switzerland: International Union for Conservation of Nature (IUCN).

King, Franklin Hiram. 1911. *Farmers of Forty Centuries: Or, Permanent Agriculture in China, Korea and Japan*. Mineola, NY: Dover Publications.

Locke, John. 1690. *Two Treatises of Government: In the Former, the False Principles and Foundation of Sir Robert Filmer and His Followers are Detected and Overthrown, the Latter is an Essay Concerning the True original, Extent, and End of Civil Government*. London: Printed for Awnsham Churchill.

Malthus, Thomas R. 1798. *An Essay on the Principle of Population, as it Affects the Future Improvement of Society*. London: Printed for Joseph Johnson.

Marsh, George Perkins. 1864. *Man and Nature: Or, Physical Geography as Modified by Human Action*. New York: Scribner.

Meadows, Donella H., Dennis L. Meadows, Jørgen Randers, and William W. Behrens, III. 1972. *The Limits to Growth: A Report for The Club of Rome's Project on the Predicament of Mankind.* London: Earth Island.

Merchant, Carolyn. 1980. *The Death of Nature: Women, Ecology and the Scientific Revolution.* New York: Harper & Row.

Mill, John Stuart. 1848. *Principles of Political Economy: With Some of Their Applications to Social Philosophy.* London: J.W. Parker.

Sitarz, Dan. 1993. *Agenda 21: The Earth Summit Strategy to Save Our Planet.* Hauppauge, NY: Nova Science Publishers.

Smith, Adam. 1776. *An Inquiry into the Nature and Causes of the Wealth of Nations.* London: Printed for W. Strahan and T. Cadell.

UN World Commission on Environment and Development. 1987. *Our Common Future. (The Brundtland Report).* New York: Oxford University Press.

# 3

# Narratives of Sustainability

## Key Concepts and Issues

Andrea J. Nightingale, Linus Karlsson, Tom Böhler, and Ben Campbell

As shown in the previous two chapters, sustainability is a concept that has traveled far. It is invoked in a remarkably wide number of contexts and taken up by formal governments and international agencies as well as grassroots social movements. In this chapter, we review some of the key concepts and issues in sustainability debates in order to better understand how sustainability can be invoked for such a wide range of purposes. By scrutinizing the narratives of environment and society that underpin sustainability, a wide range of conflicting conceptualizations becomes evident. Awareness of conceptual conflicts helps us to understand how sustainability can mean so many contradictory things, and assess why certain ideas about the world prevail and others do not. This evaluation helps show why sustainability and sustainable development are essentially deeply contested concepts, rather than concepts that are knowable or achievable with more science.

We begin with a discussion about the contrasting conceptualizations that underpin different narratives of sustainability. This is followed by a more detailed elaboration of the distinction between sustainability and sustainable development that we introduced in Chapter 2. Talking about sustainable development leads us to the "three pillars" of sustainable development (environment, society/culture, and economy) and related concepts on limits to growth and measurement. We attempt to outline the range of conceptual underpinnings to sustainability through eight different narratives and show how they are linked differently to questions of justice, governance, and globalization.

### Sustainability and Socionatures

One way to approach the different narratives of sustainability is to unpack how we conceptualize the relationship between environment and society. Currently, many people assume that human societies interact with their environments, or that societies affect the environment. This commonsense understanding seems reasonable enough, but it implies that society is a domain separate from the environment or nature. This assumed divide between society and environment is fundamentally a product of Enlightenment thought as explained in Chapter 2. Earlier conceptualizations of how humans and nonhumans interact were based on metaphors that assumed a more entangled relationship, one where there were less clear boundaries between them. More recently, a number of reconceptualizations of society–environment relations have sought to overcome Enlightenment binary thought and reimagine the world as an integrated place where the boundaries between society and environment are rejected. (The idea of "socionatures" that we introduced in Chapter 1 and elaborate in the next section is a good example.)

We can identify a continuum upon which most conceptualizations of environment–society interrelations fall. On one end of the continuum is the idea that society is a distinct realm from nature or environment. On the other end is the idea that the entire world is simultaneously both social and natural so that it is impossible to separate the two conceptually, metaphorically, and practically. In between these two poles most current ideas about society–environment lie.

Most sustainability narratives are predicated upon some separation of society from environment. Indeed, only by holding humans and human societies separate from environment can we think about how humans affect nature, a core concern of sustainability advocates. Within the biological sciences, most people accept the idea that human actions and environments (i.e., cities, even the insides of our houses) are fundamentally different than natural environments. In natural environments, other species are able to live as they are supposed to live—for example mice live in fields as opposed to in houses. However, a quick look at insights from biology suggests that this distinction between natural environments and human environments is a bit problematic. Since the 1970s some ecologists have become interested in how humans have been crucial to the formation of the very ecosystems they are interested in protecting from human activities. In the Midwest of the United States, for example, oak savanna landscapes are now recognized as a product of regular burning by Native American societies. As farmland took over and fire suppression became standard practice within the United States, the remaining oak savannas began to transform into a forested ecosystem. Some land managers have thus returned to regular burning in order to protect this valued landscape. William Cronon in his 1983 book, *Changes in the Land*, illustrates several other cases of ecosystems maintained by Native Americans with the use of fire.

For more information on oak savannas in the Midwestern United States and the importance of fire, see:

Tallgrass Prairie and Oak Savanna Fire Science Consortium, A Joint Fire Science Program (JFSP) Knowledge Exchange Consortium.

www.tposfirescience.org/.

"Iowa Prairie Conference, July 2013." Tallgrass Prairie and Oak Savanna Fire Science Consortium, A Joint Fire Science Program (JFSP) Knowledge Exchange Consortium.

www.tposfirescience.org/videos.

"Managing Fire in Oak Savannas." Savanna Oak Foundation, Inc.

http://oaksavannas.org/fire-fuel.html.

The oak savanna example shows how human activities are part of ecosystems, and if humans are part of ecosystems, then perhaps we cannot think about human impacts as unnatural. Rather, definitions of human impacts are crucial to the kinds of transformations people accept. The example given here shows that land managers first wanted to exclude human-induced forest fires because they thought they were unnatural. But over time, they realized that fire was important to the ecosystems they valued, and therefore had to rethink the role of fire (and by extension, human impacts) in those ecosystems.

The biological sciences have also provided a platform for conceptualizing societies and environments as part of one, integrated whole. Humans are classified as a species like any other, and evolutionary biology as well as genetics shows our close relationships to other animal species. From this point of view, it is difficult to understand how human actions can be "unnatural." It is ironic, therefore, that while the biological sciences accept humans as a species related to other animals, conservation biologists rarely question the "humans impacting their environments" approach to sustainability. Conservation biologists tend to see nature or environment as a domain that can be contaminated or ruined by humans and they advocate for conservation practices that limit human use of particular places, as with fire suppression in the oak savannas just discussed. This conceptualization retains a clear distinction between society and environment, a worldview that can be a barrier to rethinking society–environment relations in new ways. If humans are an integral part of their environments, both as a product of and producers of those environments, it requires more complex management strategies, ones that continually review how relationships between human activities and nonhuman processes are evolving *together*. At present, most of our conservation strategies are aimed at stabilizing or fixing human–nonhuman relationships, but if we radically change our perspective, and start by assuming that we are part of a big, ever changing world, then perhaps we need more dynamic management strategies.

Further support for questioning the idea that society and environment can be analyzed separately can be found in the social sciences. To begin with, many social scientists suggest that holding environment and society separate fails to recognize the complex ways in which humans are a product of nonhuman life and vice versa. In several different social science disciplines, scholars have therefore tried to conceptualize new ways of understanding environment–society relations. These attempts can be broadly classified into those that seek to link the two through our representations (epistemologically, or how we can know nature–society), and those who seek to link them fundamentally (ontologically, or how we assume nature–society is).

For more explanations of ontology and epistemology, see:

"What Is the Difference Between 'Ontology' and 'Epistemology'?" Quora. www.quora.com/What-is-the-difference-between-ontology-and-epistemology.
"Difference Between Ontology and Epistemology." *DifferenceBetween.net*. www.differencebetween.net/science/health/difference-between-ontology-and-epistemology/.

Epistemological attempts at an integrated approach to sustainability questions retain the categories, environment and society, as useful analytical tools. The emphasis is on representing how societies are embedded in nature, and how nature is embedded in society, in order to question the clean boundaries that have historically been assumed to exist. Some of the terms used to describe this conceptual approach are "dialectical relations" and "mutual constitution," words that try to capture the interdependencies between society and nature. Research of this type looks for ways in which any change in society necessarily leads to a change in nature and vice versa. A good example is the mutual constitution of forests and harvesting practices. Where there are big trees, people harvest them for timber, but as they cut more, there are fewer big trees, changing the composition of the forest. But, if there are no big trees to harvest, then people do not attempt to extract timber. More complex studies have looked at how forest harvesting practices can be important in creating distinctions within societies, based on gender, race, or class, and these distinctions in turn can shape how harvesting is limited. For example, logging for timber is considered a high-status activity in some places and a very low-status activity in others. Therefore, engaging in logging can change how you are viewed by society (see the analyses by Nightingale 2005 and Peluso 2009), and social hierarchies can literally shape forests. These complex interactions make it important to recognize that forests and societies are not distinct from each other. A mutual constitution approach acknowledges their interrelation, and yet analytically, keeps two categories as useful for evaluating what kinds of changes are taking place.

Socio-ecological systems (or social-ecological systems; SES) is another example of epistemologically linking society and environment. SES was developed in the natural sciences and links different components of a system through feedback loops. Conceptually, SES approaches work with methodological objects that fall fairly easily into either a social or environmental domain. Researchers then seek to understand the feedback loops through which a change in one domain causes changes in the other. These loops can be very complex. Many studies of climate change, for example, use SES frameworks to try to understand the multiple ways in which changes in societies cause changes in the atmosphere, which in turn cause changes in society. To take a simple example, industries emit carbon dioxide, which has been shown to cause global warming. Concerns about global warming have led to limitations on the amount of carbon dioxide that can be emitted by individual industries in Europe and to a more limited extent the United States, requiring industries to change some of their technical processes to comply with those limits. These technological changes have, in turn, lowered the amount of carbon dioxide entering the atmosphere from those industries, which (hopefully) will help slow down climate change over time. These kinds of feedback loops (global warming, climate policy) can be represented in a SES and their relative impacts on different systems components (industries, atmosphere, technology) can be evaluated. The advantage of representing society and environment as part of one big system using SES or dialectical approaches (joining society and environment epistemologically) is that they are amenable to using established techniques for understanding processes in both social and environmental domains. It makes data collection easier, even if relating the results to each other requires more innovation (see Nightingale 2015).

For more in-depth discussions of dialectics and mutual constitution see:

"Hegel's Dialectics." *Stanford Encyclopedia*. Stanford University Center for the Study of Language and Information. http://plato.stanford.edu/entries/hegel-dialectics/.

For more in-depth discussion of SES see Chapter 4 as well as these resources:

McGinnis, M. D., and E. Ostrom. 2014. "Social-Ecological System Framework: Initial Changes and Continuing Challenges." www.ecologyandsociety.org/vol19/iss2/art30/.

"Critically Reflecting on Social-Ecological Systems Research." Resilience Science. http://rs.resalliance.org/2015/02/11/critically-reflecting-on-social-ecological-systems-research/comment-page-1/.

While SES have been enthusiastically embraced by interdisciplinary scientists, others have argued that a far more profound shift in our thinking is required. When the link between societies and environments is tackled epistemologically, it fails to really challenge the cognitive divide that some have argued is at the root of unsustainability today. Rather, they argue that a more fundamental, ontological reimagining of the world is required, one that understands all of human societies and actions as inherently composed of both human and nonhuman dimensions. For example, harvesting trees is not simply a human act. It enrolls nonhuman entities such as the logging equipment, and the trees themselves. A change in any one of these actors would change the entire operation.

Making this ontological shift is accompanied by new vocabularies that avoid separating nature from society. Some of the terms used to describe this kind of conceptualization include "vibrant ecologies," "hybrid natures," and "socionatures." Vibrant ecologies emphasize aliveness; all entities are living, according to Jane Bennett, and as such, are able to exert a living force on the world. Hybrid natures, according to Sarah Whatmore, emphasize how everything in the world is at the same time a product of how we understand social and environmental processes, thus the word "hybrid." Socionatures, the term we prefer because it is somewhat self-explanatory, similarly emphasizes that we cannot separate ourselves, our representations, and our actions from environment or nature. So while differences in emphasis exist across these narratives, they attempt to avoid clear distinctions between society and nature or environment.

Linking society and environment ontologically through a socionatures conceptualization recognizes that entities like forests, grasslands, human bodies, even the atmosphere, are socio-natural entities. They are created and transformed by processes we happen to call "social" and "environmental" by convention, but these separations are not intrinsic to the processes and relationships *themselves*. Above, we briefly discussed the oak savanna

Oak savanna in Michigan, in early summer.

*Source:* Brent Kryda; http://americanvoyages.blogspot.com/2012/05/oak-savannas-of-southeastern-michigan.html.

example. Oak savannas in the Midwestern United States cannot exist without human intervention (fires), making it inherently socio-natural. The atmosphere may be a less obvious example, but we would argue it is also ontologically socio-natural. The atmosphere is conceptualized as gases that occur on earth and are circulated by winds, which in turn are influenced by the earth's rotation and gravity as well as the incoming energy from the sun. All of this sounds natural and seems to involve environmental processes only. However, as global climate change makes evident, the composition of the atmosphere is heavily influenced by social processes, as discussed in many sources including a 2014 report by the International Panel on Climate Change (IPCC). Industry produces greenhouse gases at alarming rates. Large-scale animal farming systems concentrate animals into small spaces to minimize human labor (see accompanying text box), which creates large amounts of methane emissions, a potent greenhouse gas source. But large-scale farms are not a biological necessity. They are a product of particular social and economic relationships that are currently considered more efficient. But that efficiency comes at the cost of greenhouse gas emissions. These emissions are changing the composition of the atmosphere, perhaps irreversibly. And this in turn means that temperatures on earth change, which will change which emissions we produce (i.e., more energy used for air conditioning) and how humans are able to farm, perpetuating the cycle.

### Large-Scale Farming

Factory farming, or large-scale farming, is a system of raising livestock or fish where high numbers of animals are kept in a relatively small space, referred to as high stocking density. This system of farming is associated with industrialization, and has allowed for the production of large quantities of meat, fish, and eggs with the input of less labor than in other kinds of farming systems. It is considered efficient as measured by the ratio of labor to food-calories produced. Factory farming can also refer to intensive, highly mechanized production of monoculture crops such as corn, soybeans, wheat, and rapeseed (canola).

A number of efficiency and ethical questions have been leveled at factory farming, however. From an efficiency perspective, if all ecological costs are taken into account, it is questionable whether factory farming really is more efficient than less dense stocking approaches. Factory farming produces a lot of manure in a small area, requiring careful handling to avoid producing methane emissions (a very potent greenhouse gas), and contamination of water sources with bacteria that can be harmful to humans and other species. Animal waste can further pollute waterways and endanger fish, amphibians, and aquatic insect species with nitrogen, which causes overgrowth of micro- and macro-plants in a process called eutrophication. Once these environmental costs are factored in, factory farming is not always more efficient than lower density stocking. For grain crops, there is evidence that the heavy use of chemical fertilizers, herbicides, and pesticides has wide ranging environmental consequences that are not factored in to the cost of farming in that manner.

Ethically, animal rights activists have challenged the conditions under which factory farmed animals live. Beef cattle, for example, are kept confined in barns, unable to move significantly for their entire lives. Chickens are similarly kept in very small cages. Even free-range chickens are kept in barns with such high stocking densities that most of the animals never venture outside even though they can. Because of high stocking densities, the risk of disease increases radically, requiring farmers to use antibiotics and other medicines on the entire herd, leading to public-health concerns about antibiotic resistance.

For more information on factory farming see:

Fraser, David. 2005. "Animal Welfare and the Intensification of Animal Production: An Alternative Interpretation." Rome: Food and Agriculture Organization:

www.fao.org/docrep/009/a0158e/a0158e00.HTM.

Attempts at linking society and environment ontologically aim to foundationally change our worldview. Despite their importance conceptually, these narratives still suffer from a lack of good methodological tools for research. To examine forest harvesting as a socionatural entity, for example, available methods are primarily developed with plants, or human practices, in mind, but not the two together. There are no research methods that see human practices and plants as *one thing*. Most studies that begin from a socionatures perspective rely on observation as the main research method and, when they do incorporate other kinds of data, they generally have to fall back into an analytical distinction between society and environment. So while the need for an ontological shift is pressing, there is also a need to think more creatively about how we construct research objects and methodologies in order to put these ontological shifts into practice.

## Anthropocentrism and Ecocentrism

Two other narratives that are particularly important for understanding sustainability are anthropocentrism and ecocentrism (sometimes also known as biocentrism). These two narratives cut across the conceptualizations of society and environment discussed above. Just because a conceptualization tries to rethink the divide between society and environment, it is not automatically ecocentric, for example. Both anthropocentric and ecocentric positions have been powerful justifications for advocating different kinds of sustainability. It is therefore important to recognize these influences when evaluating sustainability narratives.

Anthropocentrism is based on human values, and the welfare of the human population is a valid justification for any action or intervention. There is usually an assumption that humans are exceptional compared to the rest of the world based on privileged traits of language, symbolic representation, and technology, and as a result, humans should take precedence over other species. In this worldview, nature has value only when it directly or indirectly serves human interests. Anthropocentrism advocates often promote technical and managerial approaches to solve environmental crises. For example, anthropocentric approaches to climate change have focused on how a changing climate will affect society. This underlying assumption drives attempts to limit greenhouse gas emissions or to remove some of the carbon in the atmosphere in order to minimize the damage for human societies. While the impacts on other species are not entirely ignored, these impacts are considered significant because of how they will influence the ability of humans to live as we do now. One of the criticisms of anthropocentrism is that it is overly utilitarian in its consideration of other species. As explored in the following sections, this perspective underpins most approaches to sustainable development.

In contrast to anthropocentrism, ecocentrism argues that nature has intrinsic, autonomous value. As a result, the needs of all species together are of foremost concern, above the needs of a single species, including humans. Sustainability in this view can only be achieved by emphasizing ecosystems above social systems. From an ecocentric perspective, many sustainable development approaches are problematic because they, largely, attempt to use neoliberal market economy solutions for problems that the market created in the first place. (See Chapter 2 for more on how sustainable development approaches have changed over time.) Going further, many ecocentric proponents reject the idea that modernity has produced a good society. Rather, they advocate for a return to simpler lifestyles that embrace the diversity and importance of other species in everyday life. Ecocentric perspectives therefore argue that technological fixes fail to prioritize the needs of nonhumans and should be rejected as a result. For example, ecocentric approaches to climate change focus on the impact of atmospheric change on the planet as a whole. While some attention is paid to how human societies will have to adjust, their primary concern is whether nonhumans can adjust to climate change. They tend to focus on lowering greenhouse gas emissions through major lifestyle transformations, rather than technological advances that can allow current lifestyles to continue relatively unchanged. One of the criticisms of ecocentrism is that we, as humans, cannot know with any certainty what nonhumans need. Furthermore, ecocentrist narratives have often been extreme in placing the needs of nonhumans over those of humans, leading some critics to brand them "ecofascist."

## Sustainability and Sustainable Development

Clarifying the differences between sustainability and sustainable development, which are often used as synonyms, is a prerequisite to exploring current conceptualizations of sustainability. As we have emphasized in this book, we use sustainability as an overarching concept that can encompass sustainable development within it. The two ideas should not be collapsed into each other, however, because they can have quite different ideological bases. Sustainability is a specific vision of society in which prosperity is based on concern for environmental, societal, cultural, and economic well-being over time, a goal that many people can accept. Indeed, it is quite unusual to hear someone suggest that they are in favor of an unsustainable future. But note how environmental, societal, cultural, and economic well-being remain undefined, which is part of why a sustainability agenda is so readily embraced by most people. Sustainable development, in contrast, refers to the many possible, but politically different, processes and pathways to achieve the goal of sustainability. In other words, sustainable development operationalizes the idea of sustainability in specific ways.

How one ought to go about achieving sustainability is highly contested, and cannot be reduced to sustainable development alone. As the case studies in Part Two of the book demonstrate, competing visions for a sustainable future society exist. The most widely accepted visions of sustainability are anthropocentric; nearly all formulations of sustainability and sustainable development are concerned primarily with the perpetuation of human societies and lifestyles. Indeed, two of the three pillars of sustainable development—economy and society—are rooted in human systems. However, many alternative visions are ecocentric, and other, non-anthropocentric narratives of sustainable development "decenter" the human subject, to avoid simply replacing nonhumans as the central concern of sustainability. Some of the case studies in Part Two provide excellent examples of communities trying to put such ideas into effect. As the anthropologist Kay Milton described in her 1996 book, *Environmentalism and Cultural Theory*, non-anthropocentric narratives of socionatures have formed the basis of many societies in the Global South. Today, social movements seeking to articulate competing visions of sustainability often draw upon these historically rooted value systems and ways of conceptualizing socionatures to move beyond anthropocentrism and ecocentrism.

Sustainable development therefore is also a contested narrative with quite radically different implications depending on the ontological and political stance of the people mobilizing it. Many proponents of sustainable development, including the World Bank, talk about it as "visionary development," or development that seeks to transform the world. We therefore find it contradictory that within such aspirations sustainability is often used as a justification to continue with present economic practices. Sustainable development can, on the one hand, be understood as developing sustainability as a concept for fundamental transformations of human societies. Many social movements, scientists, and local communities who articulate narratives of sustainability have aspirations for such transformational change. On the other hand, sustainable development can be simply a form of greenwashing—a somewhat derogatory term that highlights how naming something as sustainable, perhaps with a few token dimensions that address environmental concerns, can serve to mask the ways that an activity simply carries on with business as usual. In this sense, sustainable development has been criticized for legitimatizing a reformist approach to highly unequal global social and economic practices. Some of the case studies show specific examples of such transformations of societies as well as greenwashing.

In the next section we explore in more detail key concepts that underpin ideas of sustainability to illustrate how these two positions (socio-natural and non-anthropocentric approaches on the one hand; greenwashing and business as usual on the other) mark the two ends of the sustainability spectrum. In key concepts, clues are found for how both agendas—maintaining the status quo and profound transformations in societies and lifestyles—can be justified under the banner of sustainable development. The slogans of "visionary development" and "meeting the needs of the present without compromising the ability of future generations to meet their own needs" (as expressed in the Brundtland report) do not in themselves provide us with the right pathway for achieving them. Rather we need to return to the central questions of this book and ask more probing questions about sustainability of what and for whom and at what cost.

## The Three Pillars of Sustainable Development

Sustainable development is often discussed, as previously mentioned, in terms of three pillars. These were first put forward at the Rio Earth Summit in 1992 and are encompassed within Agenda 21. These pillars are:

- Environmental protection (ecology).
- Social welfare and cultural integrity (equity).
- Economic development/prosperity.

The three pillars are often portrayed as a triangle, or as overlapping domains of environmental, social, and economic sustainability. Note how the pillars inherently reproduce an ontological divide between society and environment, something that is quite important for attempts to measure sustainability. The three pillars also underpin many anthropocentric narratives of sustainability. For example, increasing the social welfare of the human population is assumed to conflict with environmental constraints and the protection of the environment. Therefore, central to these narratives is the idea of balancing the three pillars, or the need for trade-offs between them in order to achieve sustainability.

Together, the three pillars are believed to encompass the domains that are foundational to promoting long-term sustainability. Environmental protection is often assumed to be the most important starting point since that is the domain from which concerns over sustainability emerged. Here, advocates seek to understand the processes and relations of ecosystems that are necessary for their long-term viability. These conceptions of ecological sustainability span a broad spectrum: from the anthropocentric—ensuring that human needs are met; to the ecocentric—ensuring that ecosystems function for all species.

The social sustainability pillar has a strong justice dimension to it, in part because of the deep concerns over development that characterized the 1992 Rio Earth Summit. The spectrum here varies from those who argue that sustainability means the perpetuation of our ways of life as they exist now and the maintenance of basic needs for all humans, to the need for a complete transformation of our present lifestyles, particularly those within industrialized economies. Attending to social sustainability therefore does not necessarily imply creating equality. Rather, it can simply mean that the needs of all people are fulfilled, but that inequality in terms of access to resources, wealth, and overall well-being may continue. Other sustainability advocates, however, insist that without addressing underlying inequalities, social sustainability cannot be achieved. Most approaches to sustainability sit somewhere in between these two poles. The social sustainability pillar usually encompasses culture, but some conceptualizations insist on separating that domain into a fourth pillar. Because modernity has threatened many traditional ways of life, numerous sustainability advocates have insisted on the need for protection, respect, and diversity of cultural differences as part of the sustainability formula.

If environmental protection is the starting point for sustainability, the economic pillar is often considered the main pathway, or means, to sustainability. Many argue that economic sustainability is a requirement for sustainability, and this stance typically implies the importance of economic growth. Agenda 21 placed a strong emphasis on the generation of wealth for achieving sustainability, largely because poverty was assumed to be a significant cause of environmental degradation. If poverty causes overexploitation of resources, then environmental conservation will follow economic improvement. While such a simplistic relationship between poverty and environment has long been rejected by a group of scholars known as political ecologists, many people today continue to focus on economic growth as a necessary condition for achieving sustainability. In contrast, other sustainability advocates argue that unsustainability lies precisely within current models of economic growth. Global capitalism is blamed for producing extreme inequalities in wealth, access to and uses of resources, and overexploitation of ecosystems. The need for continuous growth within capitalism, in order to avoid recession and economic hardship, is considered fundamentally unsustainable. As a result, many sustainability thinkers and stakeholders advocate for alternative economic relations. Most recently, there have been calls for "de-growth," an idea discussed below.

Of concern to us in this book are two issues. First, how the sustainability triad is conceptualized—which pillar is considered most important and how they relate to each other—has a major impact on the kind of economic practices that are assumed to be necessary. Second, the economic pillar is most often used in attempts at greenwashing, or in trying to justify questionable practices in the name of perpetuating economic growth. Regardless of how it is conceptualized, today the economic pillar is overwhelmingly given the most consideration in sustainable development policy. Indeed, across the spectrum of sustainability studies and activists, economic systems—wealth, comfortable lifestyles, and stable economies—are considered foundational to whether sustainability is achievable or not.

The core concepts explored in this chapter, including the three pillars, and the measures of sustainability explained below (and in Chapter 4), all highlight a fundamental concern in analyzing sustainability and sustainable development: the concepts are so vague they can mean almost anything. While certainly definitions of sustainable development exist along with targets for achieving them—the UN's Millennium Development Goals, along with the new Sustainable Development Goals, are a good case in point—the concepts themselves do not ensure transformation from the status quo. At the same time, many proponents of sustainability are committed to fundamentally changing how society is organized economically and socially. Interestingly, across the spectrum of sustainability advocates, from greenwashing to profound transformation of current lifestyles, few question the three pillars themselves. Rather, the relationship between social processes, economic practices, and ecologies are assumed to underpin sustainability.

A lot of effort has therefore been made to integrate the three pillars into an all-encompassing sustainability narrative. Some have tried to include secondary pillars in addition to the core three, like cultural protection noted previously, in order to capture more dimensions of sustainability. Yet, such efforts may be missing the point. To approach sustainable development holistically has proven to be very difficult because no consensus exists on what is to be sustained, what is to be developed, for what purpose, and with what means. The three pillars in themselves cannot resolve such issues, and adding more pillars simply serves to complicate the question. Ontologically, the pillars are rooted in a separation between environment and society and, as a result, continue to treat human societies as separate from ecological systems, providing justification for thinking of human actions in isolation from environmental processes. By treating the three pillars as distinct entities, sustainability practices are characterized by trade-offs and substitutions, rather than synergies between society and economy on the one hand, and environment on the other. Retaining an ontological divide between nature and society perpetuates a sense that the environment is here for the service of humans. Politically this divide raises questions over who defines sustainability and for what purposes. These questions, we believe, need to be foundational to any approach to sustainability.

## Weak and Strong Sustainability

In addition to the three pillars, a distinction is often made between weak and strong sustainability. These conceptions differ in the extent to which trade-offs between the economy and environment are tolerated. These narratives are an attempt to make more explicit the contrasting ways that sustainability can be conceived, even under the rubric of the three pillars. Profoundly anthropocentric, weak sustainability is usually defined as conserving economic capital into the future and allowing unlimited substitution between natural and economic capital. "Capital" is a term often used in sustainability debates to signal economic, natural, and economic resources that are important for long-term well-being. The goal in weak sustainability is to sustain human lifestyles, with little consideration of other species. Protection of the environment is therefore important because of its role in sustaining humans. Under a weak sustainability view, it is possible to substitute economic gains for environmental ones in order to achieve sustainability. Ecotourism, for example, is often used to provide income and economic stability in areas of natural beauty that might otherwise be used for resource extraction. But ecotourism itself is not necessarily environmentally benign, as desirable destinations are subjected to overuse by tourists. The environmental costs of ecotourism are justified because the economic gains help balance the three pillars. Of course this requires moving resources from one place to another—banning resource harvesting in beautiful places and replacing them with resources bought from elsewhere. Weak sustainability thus assumes that gains in one domain can compensate for losses in another. It maintains a very clear ontological separation of environment and society and takes a utilitarian approach to sustainability questions. As long as current economic conditions are maintained, then trade-offs with the other domains are acceptable. To come back to the core questions of this book, however, we need to ask: for whom are economic conditions maintained and over what time period?

Strong sustainability, in contrast, is based on the principle that the intrinsic processes in the world limit substitutions between the three pillars. Ecosystems provide some functions that are not replaceable by human processes. Examples include endangered species and vital ecosystems, but they can also include other functions such as culturally valued landscapes and species, and spiritual relationships with particular places. These cannot be replaced by economic development. The emphasis on intrinsic values makes strong sustainability less anthropocentric than weak sustainability, but it is possible to maintain a strong sustainability stance and still put human needs first. We explore in more detail the idea of limits below, but here it is important to note how strong sustainability is founded upon the idea of limits, particularly ecological limits. Some versions of strong sustainability also insist on considering economic *use* values, rather than the unpredictability of monetary and market fluctuations as a measure of economic sustainability. Strong sustainability is generally considered to be closer to the original spirit of the Brundtland report and the intentions of Agenda 21. While people have criticized those two political documents as allowing a revisionist approach to our current economic system, they are nevertheless informed by an understanding that human activities need to be contained or limited in order to sustain environments over time.

Neither weak nor strong sustainability questions the assumption that human needs are the starting point for evaluating sustainability and as such are profoundly anthropocentric. While environmental protection is given more consideration in strong sustainability, proponents generally stop short of placing the needs of nonhumans as the most important priority (an ecocentric perspective). Similarly, neither stance opens up the possibility to think about socio-natural processes and objects, seeing the world as a socio-natural system rather than one composed of society and environment. Both thus reflect an ontological divide between society and environment.

If one takes a socionature approach, the three pillars themselves are part of creating sustainability problems because they recreate analytical categories that allow thinking about trade-offs and substitutions. They preclude reimagining our environmental problems in a radically different manner. Rather, if we conceptualize our present sustainability dilemmas as socio-natural, wherein economics, social relations, and environmental transformations cannot be separated from each other, then the pillars are not the right starting points. Balancing economy, society, and environment is not the goal, but rather a more profound reframing of how we live in the world is needed. Such a reframing would include embedding a consideration of other species into all dimensions of our lives. The idea of limits to substitution propounded by strong sustainability narratives serves to entrench, rather than critically question, the idea that society and economy are separate from each other, as well as being separate from environment.

## Different Narratives of Sustainable Development

The three pillars are nearly ubiquitous in framings of sustainable development, yet in themselves, they do not clarify precisely what sustainable development ought to be. As a result, a number of different, often conflicting, conceptualizations have emerged. Here, eight narratives are discussed, each informed by different conceptual approaches to achieving sustainability. Note that these perspectives are ideal types and nearly all field practices are a blend of more than one. Many individuals, networks, and organizations at times act within one or more of these ideals. It is, therefore, perhaps better to view these as lying on a spectrum.

1. **The Development as Growth** narrative is underpinned by the premise that economic growth is essential for human welfare and sustainable practice. The foundational principle is that affluence provides incentives for protecting the environment. *The Economist* magazine explained in a 1999 article that in rich countries where economic growth has been strongest, the environment has also improved, whereas in poor countries, declining environmental conditions are increasingly a hazard to ecosystem health. In many respects the idea that development means economic growth informs the vast majority of approaches to sustainable development—and indeed, development in general. This approach does not fundamentally challenge how our societies and economies are organized today, but rather advocates for an extension of Global North practices to more parts of the world. It is particularly favored by analysts who advocate for a neoliberal approach to economics, leading them to suggest that free markets are the best pathway towards affluence and, ultimately, sustainability. As a result, many proponents of sustainable development, including the World Bank and USAID, argue for the removal of trade barriers in the name of sustainable development.
2. **Development as Ecological Modernization** is an attempt to retain a commitment to growth without sacrificing the environment. Ecological modernization theory argues that economic growth and a sustainable environment do not have to oppose each other. On the contrary, with the development of new and greener technologies, it is possible to decouple, or dematerialize, the economy from its natural resource base. Rather than posing limitations, environmental problems and their solutions are an important engine for growth as it stimulates the need for new innovations and markets, like renewable energy. This argument is frequently seen in climate change arenas where calls for technologies to help sequester carbon are endorsed as growth industries that can simultaneously address climate change and promote economic prosperity.

   However, many environmental researchers, such as Nicolas Kosoy and colleagues, have highlighted the limited extent to which it is possible to delink increases in energy and material use from economic growth. Developing more resource-efficient technologies is important, but at the same time energy and material efficiency make consumption cheaper, leading to overall increases in resource use. Such increases would cancel out any sustainability gains made by more efficient technologies. Critics of ecological modernization have questioned the technological optimism embedded in this perspective, arguing that the ability of technology to overcome ecological limits is restricted. There is also a strong Global North bias in ecological modernization with the assumption that strong, growing economies will be green economies, whereas this is not necessarily the case when we look at some rapidly growing economies in the Global South. Perhaps most importantly, no mention is made of structural inequalities that maintain globally uneven resource use. Finally, as many of the following narratives highlight, ecological modernization lacks reflection on how contemporary societies are alienated from nature and what that means in terms of society–environment dynamics. The formal sustainability strategies of most highly industrialized countries (e.g., United States, United Kingdom, Japan, Germany, Netherlands, Sweden) are based upon ecological modernization.
3. **Development as De-Growth**, a reaction against ecological modernization, argues explicitly against economic growth as a desirable goal for sustainability. Rather, proponents argue that our economies need to shrink and find an equilibrium wherein stable lifestyles are not dependent on a growing economy. In 2010, the Second Conference on Economic De-Growth for Ecological Sustainability and Social Equity was held in Barcelona, Spain. More than five hundred participants from forty countries gathered to discuss how to shrink the global economy. According to one of the movement's

leading thinkers, Serge Latouche, de-growth should not be confused with economic decline, and nor should it imply suffering. Rather, de-growth is described as a "diet for better health." Achieving de-growth will require dramatic shifts in individual and collective consumption patterns, as people's main ecological impacts come from food production and consumption, housing, and transportation. We need to reimagine how to live well in a manner that values practices like reusing goods, reducing transportation, and owning less. Proponents of de-growth argue that people in highly industrialized countries have to live more simply, travel less, and eat lower on the food chain.

4. **The Development as Freedom** narrative, advocated by Amartya Sen in his 1999 book by that name, argues that development must be predicated upon political, human, and individual rights; equity; and economic opportunities. The emphasis on human well-being makes this a fundamentally anthropocentric perspective; in fact, environmental concerns are almost absent. A key difference between this approach and the development as growth approach is the emphasis on economic opportunities rather than simply growth. This approach does not necessarily reject the need for economic growth, but it probes who benefits from growth, rather than growth itself being an indicator of development. Those who advocate for this perspective suggest that sustainability initiatives tend to neglect the poor and their acute vulnerability to environmental degradation. Even programs targeted specifically at sustainable development, these analysts argue, usually include efforts to bring poor people in the Global South into food and commodity chains that allow most of the profit to be accumulated by those who are already well off. For example, growing cash crops can help provide necessary income and increase overall household well-being, but most of the profits from food sold in commercial markets is captured by merchants rather than growers. Reliance on cash crops can also lead to overuse of land in order to make more income. In this sense, the neoliberal global capitalist system that presently dominates the world is often singled out as a cause of both environmental degradation and inequality and therefore in need of radical transformation.
5. **The Deep Ecology (or Bioenvironmental)** perspective, first promoted by Arne Naess and colleagues, begins by challenging the human-centered, anthropocentric assumptions embedded within sustainable development. In particular, "deep ecologists" argue that capitalism, economic globalization, and consumer culture are not adequately challenged by sustainable development formulations. In this sense, deep ecologists share a stance with the "development as freedom" perspective albeit without the emphasis on human well-being. They fundamentally reject the idea that economic growth in developing countries will bring about enhanced levels of ecological conservation. Rather, they argue for placing the needs of nonhumans as the central frame of reference around which human societies and economies need to be organized. Of great concern to them is the need to expand one's sense of self to embrace more than the human body, and extend it to other dimensions of our surroundings. For example, deep ecologists have embraced North American First Nations' understandings of human–animal relationships that consider bears as brothers rather than simply animals. They often refer to the idea of limits to biophysical systems as a justification for needing to halt growth and overturn the current capitalist economic system.
6. **Ecological Economists** join the "development as freedom" and "deep ecology" thinkers in criticizing the current capitalist system, but stop short of arguing against growth. They begin from the stance that the planet has finite resources and that actions taken on smaller scales will ultimately have impacts on the planetary scale. As a result, they seek to incorporate ecological, social, and ethical concerns into the "development as growth" approach. They have taken up the three pillars and tried to translate them

into measurable quantities that they can use in ecological models. This results in treating the economy as a closed system, such that the ecological costs of resource depletion and environmental damage are included in the price of commodities and services. For example, when we drive cars, the total cost of extracting and refining oil and the emissions that occur at each step of this process are not factored into the prices paid. Emissions are dangerous not only for climate change, but also for the health of humans and nonhumans. Oil and gas contain toxic substances that are breathed in or absorbed through the skin and can cause cancer and other illnesses. Ecological economists argue that the cost of dealing with these negative effects should be included in the price of fuel at the pump. This sits in contrast to neoclassical economics (see accompanying text box) wherein the costs of goods and services (exchange value) does not include ecological costs; this is more or less how the cost of oil is factored today, except for a handful of countries who impose heavy taxes in acknowledgment of ecological costs. Ecological economists have also tried to contest competition as a foundation for capitalist economies, and advocate instead for more cooperation, alliance-building, and stable relationships among different actors in the economic system. In this manner, ecological economists have tried to ascribe value to ecological, social, and ethical concerns, but of course how this should be done is debated.

**Neoclassical Economics**

Neoclassical economics refers to a conception in economic thinking that promotes the market as the prime mediator of economic transactions. When people engage in competitive market transactions, supply and demand thus determine the fair price of goods. It assumes that people are rational individuals who engage in market transactions in order to maximize utility, while businesses seek to maximize profits. Most economic thinking today is underpinned by at least some of the assumptions and approaches to neoclassical economics, especially microeconomics. Neoliberal economic ideas have emerged from neoclassical economics and privilege the market as the prime controller of the economy.

For more on neoclassical economics see: "Neo-Classical Economics." *Business Dictionary*. WebFinance Inc. www.businessdictionary.com/definition/neo-classical-economics.html.

7. **Political Ecology** is an approach that emerged in geography and anthropology and is closely related to the "development as freedom" perspective, but with somewhat more emphasis on understanding how society and environments are mutually constituted. Political ecologists focus on how discourses, societal structures, and power relations shape the way that human–environment dynamics unfold and the possibilities for social justice as a result. They have traced how the actions of landowners in specific localities are shaped by political economic relations at larger scales, meaning that larger scale structures shape the life possibilities of individuals in localities. For example, the global market for cut flowers is important in motivating farmers in Africa and Latin America to grow flowers on their land rather than crops they can eat. Some of their farming practices can be damaging to soils and water from chemical inputs or intensive farming. But to blame the consequences of flower cultivation only on small farmers is

not really fair, according to political ecologists. Without markets for cut flowers in the Global North, African and Latin American farmers would not devote so much of their land (if any) to flower cultivation. So while political ecologists might not dispute some of the effects of land degradation evident in highly impoverished areas, they do question whether people on the ground are really responsible. Rather, they argue that global capitalism and the way it extracts natural resources and capital from one place and translates it into consumption and wealth in another place is at the root of ecological degradation. Importantly, it is not always poverty that causes overexploitation of resources. Political ecologists have also shown how affluence at different scales can cause environmental decline as the overexploitation of resources for rich consumers can lead to destruction of environments in other places.

8. **Development as Living Well (*Buen Vivir*)** narratives refer to notions of environmental justice and development that are embedded in other people's worldviews, such as First Nations, indigenous communities, and Global South cultures. *Buen vivir* is a movement with strong roots in Latin America and with connections to global indigenous networks that place value on relationships between communities and the places they live, people's knowledge of healing and food biodiversity, and culturally different ways of conceiving and living within environments. These perspectives recognize that the notion of sustainable development, including most of its critics, is inherently informed by American and European worldviews (referred to as a "Euro-American-centric" worldview). In contrast, a *buen vivir* ("living well") perspective is predicated upon culturally specific understandings of how human societies are embedded within their environments. For example, in some Latin American indigenous traditions, the earth is conceptualized as alive, a mother that provides for her children which include humans and nonhumans. This kind of conceptualization offers a profoundly different foundation for sustainability. The concept of "resources" make no sense in this worldview because all things on earth are assumed to have their place and function independent of human use. These narratives generally reject the Enlightenment separation of nature from society, and rather seek to reflect how humans are part of nature. In this perspective, living well with other species is not a goal to strive toward, but rather the fundamental starting point for all interactions. Thus while not always explicitly a critique of global capitalism, these narratives are rooted in radically different conceptions of economy and society based on respect for and sharing kinship with the nonhuman world, rather than control over it. It is therefore not surprising that the *buen vivir*, and many environmental justice, movements sit in opposition to sustainability narratives that prioritize technical interventions and expertise. Proponents also advocate for political systems that place kinship with the nonhuman world as a fundamental building block of governance mechanisms to tackle today's environmental challenges.

At this point, a core message of this book should be emphasized: all these concepts, categories, and narratives—such as society, economy, environment, capital, development pathways, and *buen vivir*—are analytical ideas. They are important categories through which we seek to understand the world, but they should not be mistaken for the world itself. The world itself is far more complex. As already stated, these perspectives lie on a spectrum. Some of them overlap in different ways and individuals or organizations may in fact operate from more than one of these narratives. We present them as analytical reference points, tools that can help us to wade through the complex and often overly idealistic propositions put forward under the flag of sustainability. When we understand that our narratives of sustainability and the analytical categories we use to talk about them are metaphors and abstractions, our imaginations can open up to other possibilities. In many respects, we advocate for moving towards socionatures and *buen vivir* framings, but that is because we see these alternatives to mainstream narratives as providing us with promising analytical tools to reimagine sustainable futures.

## Measuring Sustainability

As discussed above, the idea of sustainability is characterized by vague definitions, lending themselves to a multitude of conflicting agendas. This lack of clarity has not deterred attempts to measure sustainability, however. On the contrary, a wide number of measures have emerged, in part with an explicit purpose to bring more precision to the idea. Another important driver behind the urge to measure is the need to operationalize sustainability in policy. Policymakers want to assess progress and trajectories, and funding for sustainability projects is often tied to showing improvement, requiring an indicator or reference point. While extensive coverage is given to measuring sustainability in Chapter 4, in the following we review basic problems relating to measuring un/sustainability, and explore a core idea supporting many protocols for doing so—that of "limits."

The idea of measuring sustainability is itself highly contested. Some suggest that without quantitative measures, we cannot know if we are moving towards sustainability, nor can we understand which domains require most attention. In contrast, others argue that the desire to measure is part of an Enlightenment thinking paradigm that believes we can separate different dimensions of environment from society and understand their operations separately from each other. They therefore see the quest to measure as regressive for sustainability goals. Proponents of "living well/*buen vivir*" narratives, for example, see measurement as failing to take seriously the need for new metaphors and images of socionatures that can transcend the current "nature as resource," anthropocentric approach to society–environment dynamics which underpin why sustainability problems arise in the first place.

As a result of these contrasting positions, any attempt at measurement must be considered with a critical eye. The questions that need to be asked are: what is being measured, how do these quantifications reflect certain assumptions about environment–society dynamics, and what functions do they serve for whom? Of most importance for our purposes in this book, measures of sustainability give us great insights into what is regarded as sustainable and the domains of life that are considered most suitable for intervention. Indeed, in many respects, all indicators used to measure sustainability are aimed at sustainable development. There are very few, if any, that are simply used as measures of sustainability divorced from an effort to produce a more sustainable outcome.

To take one example, the idea that a growing population positively correlates to overuse and misappropriation of natural resources dominated sustainability practice and theory until the 1990s. Limiting population growth was a central theme at both Stockholm (1972) and the Rio Earth Summit (1992), and in 1987, the Brundtland report recommended population control by nation-states as essential for sustainability. Today, many sustainability efforts continue to be closely associated with population control, or the idea that the earth has a limited carrying capacity (see accompanying text box). Below we probe some of the attempts to define and measure limits and discuss what they mean for sustainability practice. While the idea of carrying capacity and ecological limits underpins many attempts at measurement, other kinds of limits are also discussed including economic limits and the need for social justice.

**Carrying Capacity and the Green Revolution**

"Carrying capacity" is normally understood as the maximum population of a particular plant or animal species that an ecosystem can support on a long-term basis. If the carrying capacity is exceeded, a precipitous decline in the population of the species in question will result. Carrying capacities of human societies, also part of this rubric, is defined as strongly linked to agricultural productivity. The so-called Green Revolution, which began about fifty years ago to increase productivity in agriculture, constitutes one attempt to increase human carrying capacity. Through modern plant breeding and ideal growing conditions (e.g., enough water), optimal fertilizer applications, and use of chemical pesticides, higher crop yields became possible. Much of the increase in food production includes irrigation from underground water supplies that are at risk of depletion, however, raising serious questions about whether productivity gains can be maintained. And intensive inputs of chemical fertilizers and pesticides pollute the surrounding lakes, rivers, and groundwater. As a result, some are questioning if the carrying capacity of humans has already been reached.

Neo-Malthusianism refers to a narrative that advocates for population control as the primary means to ensure resource availability for current and future societies. It gets its name from the work of Thomas Malthus, introduced in Chapter 2. Neo-Malthusianism is based on the notion that people will inherently seek to overexploit their resources unless the right kind of limits are in place. In 1968, the ecologist Garrett Hardin published the article, "The Tragedy of the Commons," in which he argued that people are unable to put collective interests before individual ones and, as a result, communally owned land (the commons) will necessarily be degraded. His ideas were rooted in a Malthusian understanding of population and resources, which led to his conclusion that the number of people who have access to land needs to be limited. In his view, this can only be accomplished through the breaking up of commons land into private property.

The tragedy of the commons has been thoroughly debunked by the Nobel Prize-winning economist, Elinor Ostrom, but it is remarkable how Hardin's ideas continue to hold sway in the popular imagination. Ostrom and her colleagues (1990) showed that when resources are managed by defined groups of people, with clear rules over resource extraction and abilities to punish violators, self-interested overexploitation of resources does not occur. Rather, these "common property regimes" have been able to maintain resources over many years of regular use. Despite these insights from social science, Hardin's ideas underpin many conservation strategies and are foundational to neoliberal economic approaches to natural resource governance and development that are predicated upon private property rights. In 1972, neo-Malthusianism was popularized in a sensational form when the Club of Rome published the highly influential book *Limits to Growth* by Donella H. Meadows (together with Dennis L. Meadows, Jørgen Randers, and William W. Behrens, III). The book explores the correlation between economic and population growth and depletion of natural resources, predicting the collapse of human civilizations if trends of the last several centuries continue. In 1974, Lester Brown set up the Worldwatch Institute, publishing annual assessments of the global use of natural resources, in an attempt to predict environmental collapses. Limiting population growth was a central theme at both Stockholm (1972) and the Rio Earth Summit (1992), and in 1987 the Brundtland report recommended that population control by nation-states is essential for sustainability.

Neo-Malthusian ideas have been criticized for, first, failing to take account of technological, institutional, and organizational improvements that can enhance the resource base and ease pressure on resources. There is not a direct correlation between population and resource use. Rather, resource availability is mediated through a variety of means, which include social relations, institutions, and technology. Second, populations have not increased at the predicted rate given the use of contraceptives and, perhaps more importantly, rising incomes across the globe have proven to be positively associated with decreased fertility rates. Third, and most profoundly, a narrow emphasis on population fails to account for how resource distribution is central to overall resource use. The United States, for example, consumes the most resources and energy per capita, while recently China has surpassed the United States as consuming the largest total amount of energy and resources for any one country. (Note however that China's per capita resource consumption rate is far below that of the United States.) Since the Brundtland report was issued, the focus has shifted from population control to consumption control, in part because recent UN reports suggest that the world's population will stabilize at 9 to 12 billion by 2100. Therefore, what is discussed today relates to increases in consumption, not population. These three critiques together show how neo-Malthusian ideas do not match what is actually happening in the world.

Finally, neo-Malthusian interventions have been increasingly considered disempowering for the world's poor. Countries that have been targeted for population control, mainly those in the Global South, use the least amount of resources per capita globally and, in most cases, use fewer overall resources than rich, industrialized countries. A focus on population control has led to sterilization campaigns, many of which fail to obtain proper informed consent from women before operations are performed, or provide incentives of a dubious ethical nature to people who volunteer for sterilization. These campaigns have deep racist and sexist undertones and have been almost exclusively targeted at poor people in the Global South. People's lack of education is used as an excuse for sterilization without consent. One can only imagine the lawsuits that would ensue if a white woman in the United States or Europe woke up after delivering a baby by caesarean section to discover the doctors had decided to sterilize her without consent. And yet this has happened repeatedly to women in India and Africa. As a result of these critiques, population control has now shifted into questions of agency, autonomy, reproductive rights, and health for women, and away from birth rates. From a sustainability perspective, if social equality is considered foundational to sustainability outcomes, then it requires us to reject neo-Malthusian ideas because of their unacceptable assumptions about unequal social relations.

Critiques of neo-Malthusianism have helped move sustainability discussions towards use and distribution of resources, which in turn has shifted concerns from population to ecological limits. The idea of limits is beginning to colonize a lot of sustainability narratives, although as we probe here and in Chapter 4, not all these limits are closely related to ecology or to measurement. The mainstream sustainability narrative continues to be dominated by a desire to define and measure sustainability with more precision.

## Conclusion

In this chapter we have shown how sustainability is not one knowable, achievable state. Rather, we have emphasized the importance of conceptualizations of society–environment dynamics for shaping a wide range of sustainability and sustainable development narratives. People using anthropocentric framings have promoted versions of sustainability that are quite different from the versions pursued by people championing more transformative, ecocentric worldviews. Many of these narratives have originated in the Global North, but they have been taken up and owned by different actors across the world, including in the Global

South. As sustainable development efforts have poured millions of dollars into the Global South, these projects and initiatives have been transformed by the ambitions of local people, graft and corruption at all scales, and by the different worldviews people working together towards sustainability bring to these projects.

At the same time, a number of social movements in the Global South have put forward their own worldviews and aspirations for sustainability. Sustainability is thus a coherent practice in the sense that it travels remarkably well. It is a global narrative that helps us to understand that globalization itself is not a coherent project. Like sustainability, globalization of the economy, of environmental goals, and of conceptualizations of society–environment relations all take different forms in different places. As the case studies in Part Two of this book help illustrate, sustainability is used to justify practices at both local and global scales, practices that demand radical transformations of current ways of life, and conversely, practices that justify the continuing industrial exploitation of environments and peoples.

Many narratives of sustainability are underpinned by the idea of limits, particularly of resources, thresholds, and populations. Here we have introduced carrying capacity, which in many senses underpins all other conceptualizations of limits. In the next chapter we look at several attempts at measuring *unsustainability* and how they have operationalized the idea of limits.

## References and Further Research

Anonymous. 1999. "September 11. Our Durable Planet." *The Economist* 352: 29–31.

Assadourian, Erik. 2012. "The Path to Degrowth in Overdeveloped Countries." In *State of the World 2012: Moving Toward Sustainable Prosperity,* edited by L. Starke, E. Assadourian, and M. Renner, 22–37. Washington, DC: Island Press.

Castree, Noel. 2005. *Nature*. London: Routledge.

Cronon, William. 1983. *Changes in the Land: Indians, Colonists, and the Ecology of New England.* New York: Hill and Wang.

Escobar, Arturo. 2011. "Sustainability: Design for the Pluriverse." *Development* 542: 137–140.

Haraway, Donna. 1991. *Simians,Cyborgs, and Women: The Reinvention of Nature*. New York: Routledge.

Harcourt, Wendy. 1997. *Power, Reproduction and Gender: The Inter-Generational Transfer of Knowledge.* London: Zed Books.

Hardin, Garrett. 1968. "The Tragedy of the Commons." *Science* 162, no. 3859: 1243–1248.

Hecht, Susanna B. 2014. "Forests Lost and Found in Tropical Latin America: The Woodland 'Green Revolution'." *Journal of Peasant Studies* 415: 877–909. doi:10.1080/03066150.2014.917371.

Hinchliffe, Steve. 2007. *Geographies of Nature: Societies, Environments, Ecologies*. London: Sage Publications.

IPCC. 2014. Summary for Policymakers. In *Climate Change 2014: Impacts, Adaptation, and Vulnerability. Part A: Global and Sectoral Aspects. Contribution of Working Group II to the Fifth Assessment Report of the Intergovernmental Panel on Climate Change*, edited by C. B. Field, et al., 1–32. Cambridge: Cambridge University Press.

Kosoy, Nicolas, et al. 2012. "Pillars for a Flourishing Earth: Planetary Boundaries, Economic Growth Delusion and Green Economy." *Current Opinion in Environmental Sustainability* 41: 74–79. doi:http://dx.doi.org/10.1016/j.cosust.2012.02.002.

Malthus, Thomas R. 1798. *An Essay on the Principle of Population, as it Affects the Future Improvement of Society*. London: Printed for Joseph Johnson.

Marland, Pippa. 2013. "Ecocriticism." *Literature Compass* 1011: 846–868. doi:10.1111/lic3.12105.

Marten, Gerald G. 2001. *Human Ecology: Basic Concepts for Sustainable Development*. London: Earthscan Publications.

McGinnis, Michael D., and Elinor Ostrom. 2014. "Social-Ecological System Framework: Initial Changes and Continuing Challenges." *Ecology and Society* 192. doi:10.5751/ES-06387–190230.

McShane, Katie. 2007. "Anthropocentrism vs. Nonanthropocentrism: Why Should We Care?" *Environmental Values* 162: 169–186. doi:10.3197/096327107780474555.

Meadows, Donella H., Dennis L. Meadows, Jørgen Randers, and William W. Behrens, III. 1972. *The Limits to Growth: A Report for The Club of Rome's Project on the Predicament of Mankind.* London: Earth Island.

Milton, Kay. 1996. *Environmentalism and Cultural Theory: Exploring the Role of Anthropology in Environmental Discourse.* London: Routledge.

Naess, Arne. 1990. *Ecology, Community and Lifestyle: Outline of an Ecosophy.* Cambridge: Cambridge University Press.

Nightingale, Andrea J. 2005. "'The Experts Taught Us All We Know': Professionalisation and Knowledge in Nepalese Community Forestry." *Antipode* 343: 581–604.

——— 2015. "Adaptive Scholarship and Situated Knowledges? Hybrid Methodologies and Plural Epistemologies in Climate Change Adaptation Research." *Area* 481: 41–47. doi:10.1111/area.12195.

Norton, Bryan G. 2005. *Sustainability: A Philosophy of Adaptive Ecosystem Management.* Chicago: University of Chicago Press.

Ostrom, Elinor. 1990. *Governing the Commons: The Evolution of Institutions for Collective Action.* Cambridge: Cambridge University Press.

Parris, Thomas M., and Robert W. Kates. 2003. "Characterizing a Sustainability Transition: Goals, Targets, Trends, and Driving Forces." *Proceedings of the National Academy of Sciences of the United States of America* 100, no. 14: 8068–8073. doi:10.1073/pnas.1231336100.

Peet, Richard, and Michael Watts. 2004. *Liberation Ecologies: Environment, Development, Social Movements.* London: Routledge.

Peluso, Nancy Lee. 2009. "Rubber Erasures, Rubber Producing Rights: Making Racialized Territories in West Kalimantan, Indonesia." *Development & Change* 401: 47–80. doi:10.1111/j.1467–7660.2009.01505.x.

Richardson-Ngwenya, Pamela. 2012. "A Vitalist Approach to Sugar-Cane Breeding in Barbados: In the Context of the European Union Sugar Reform." *Geoforum* 436: 1131–1139. doi:http://dx.doi.org/10.1016/j.geoforum.2012.07.005.

Sen, Amartya. 1999. *Development as Freedom.* Oxford: Oxford University Press.

Sluyter, Andrew. 2003. "Material-Conceptual Landscape Transformation and the Emergence of the Pristine Myth in Early Colonial Mexico." In *Political Ecology: An Integrative Approach to Geography and Environment-Development Studies*, edited by K. S. Zimmerer and T. J. Bassett, 221–239. New York: The Guilford Press.

Stern, David I., Michael S. Common, and Edward B. Barbier. 1996. "Economic Growth and Environmental Degradation: The Environmental Kuznets Curve and Sustainable Development." *World Development* 247: 1151–1160. doi:http://dx.doi.org/10.1016/0305-750X(96)00032-0.

UN World Commission on Environment and Development. 1987. *Our Common Future. (The Brundtland Report).* New York: Oxford University Press. Available electronically from the UN: www.un-documents.net/our-common-future.pdf.

Whatmore, Sarah. 2002. *Hybrid Geographies.* London: Sage.

# 4

# Enacting Sustainability

Andrea J. Nightingale, Tom Böhler,
Linus Karlsson, and Ben Campbell

In the previous chapter we introduced some of the narratives of sustainability and their conceptual underpinnings. In this chapter we take a selective look at the ways that these narratives have been put into practice. We begin with the idea of measuring sustainability, introduced in Chapter 3, and examine how approaches based in the biophysical sciences may differ from those in the social sciences. As a consequence of the ambiguity surrounding the concept of sustainability, and therefore what to measure, many attempts at measuring sustainability have emerged. The very large number of measures and indicators raises normative questions about what to measure, using what standards, and how to compare across contexts and scales. In the last three sections, we take an explicit look at sustainability from justice, governance, and globalization perspectives.

## Measures of Sustainability

Measuring sustainability and sustainable development poses a number of problems, not least of which is availability of data. In 2003, Thomas M. Parris and Robert W. Kates characterized different measures of sustainable development and counted over five hundred different efforts at measurement. Many of these draw on established indicators or sought to use available data sets. A few aim to establish new indicators, but even these are reliant on available sources of data like census figures, measures of socioeconomic status or change, and ecological indicators. Because sustainability questions seek to link together domains of life that are often held separately in science, policy, and practice, developing one "sustainability indicator" that can be applied to a wide range of contexts has proven elusive. Here we review some of the most common indices used.

Many indicators put ecology at the core of sustainability, assuming that ecological sustainability is a prerequisite for social and economic sustainability (see more on the three pillars of sustainability in Chapter 3). These indicators sometimes include social development measures as well. Most that have gained popularity are underpinned by an ontological position (discussed in Chapter 3) that humans have an impact on the earth's ecosystems (rather than being part of ecosystems), and therefore measuring the damaging effects of human activities is of central concern. Such measures include, among others, ecological footprints and planetary boundaries that we discuss in more detail in the following sections. These indicators begin from the idea of limits, revealing their inherently neo-Malthusian orientation, strongly informed by the idea of carrying capacity, introduced in Chapter 3.

## Ecological Footprints

University of British Columbia scholars Mathis Wackernagel and William Rees (1998) developed the idea of ecological footprints to capture how individuals living in one geographical location are reliant upon resources that span a much larger area. An ecological footprint represents the amount of biologically productive land and sea area necessary to supply the resources and energy a given human population consumes and needs to dispose of its waste. Using this assessment, it is possible to estimate how much of the earth (or, as the measure has shown, more than one planet earth) is needed to support humanity if everyone followed a given lifestyle. The footprint includes roads and other infrastructure, land required to produce food and energy, and the surface area necessary to bind carbon dioxide produced from emissions as well as other waste. A growing footprint means that a larger land mass is essential in order to meet a given level of material consumption, assumed to be increasing. Today, based on the ecological footprint measure, the earth has a biological productivity of 2.0 hectares (4.94 acres) for each person in the world, but the average person's ecological footprint is 2.7 hectares (6.67 acres), a sign that the human population uses more resources than is reproduced each year. If this process continues we would need two planet earths by the year 2050 to maintain present levels of consumption.

As the idea of ecological footprints has grown in popularity, protocols have been developed to measure footprints at different scales. Whereas the figures we present above represent the ecological footprint of humanity at large, individuals can consult online resources to try to estimate their own ecological footprint. It is important to recognize that as with all indices, ecological footprints are only useful if the data going into them are accurate. And of course, for individuals, accurately estimating how much of what resources you consume can be difficult if not impossible.

For more information on ecological footprints see:

*Greencred.me*. Greencred. www.greencred.me/footprintcalculator/.

To calculate your footprint try these sources, among others:

"Footprint Calculator." Global Footprint Network. www.footprintnetwork.org/en/index.php/GFN/page/calculators/.
"How Big is Your Environmental Footprint?" WWF–UK. http://footprint.wwf.org.uk/.

Aggregate measures, however, cannot account for distribution. So while it is perhaps useful to know that at present humans use more than one planet's worth of resources for present ways of life, this is not true for everyone. Rather, some populations are using far more than the 2.7 hectares average, while others are using substantially less. In this way, while ecological footprints can be used to make arguments about equitable use of resources, in itself an ecological footprint does not directly address redistribution of resources. From a social justice point of view, distribution of use, as well as availability, of resources is a key question, one which ecological footprints leave out.

## Socio-Ecological Systems and Resilience

In Chapter 3 we briefly touched on socio-ecological systems (SES) as one attempt to link society and environment by conceptualizing complex feedback loops among different processes that are identified as driving environmental problems. SES emerged from work on resilience. Resilience was first elaborated by C. S. Holling and Lance H. Gunderson in the late 1970s and early 1980s as a counter theory to *equilibrium* in ecological systems. They questioned whether ecosystems really do exist in states of dynamic equilibrium. *Dynamic equilibrium* refers to a model that represents healthy ecosystems as having abilities to absorb or respond to shocks or disturbances while maintaining their essential functions and structure (for more, see accompanying text box). They hypothesized that rather than stability, disturbance and change characterize ecosystems. While both ideas include an element of dynamic change, the difference between them is that resilience assumes *change* as an important part of maintaining ecosystems, whereas dynamic equilibrium assumes that *stability* ensures maintenance of an ecosystem.

Resilience thinking includes the idea of *panarchy*—a model of the world described by Gunderon and Holling in 2002, in which systems pass through various phases of change and disturbance until severe shocks or an accumulation of disturbance irrevocably forces the ecosystem beyond a threshold (known as a *tipping point*), causing a phase shift. *Phase shift* means the ecosystem has passed into a new state, such as a forest becoming a grassland, or vice versa. The concept of panarchy also describes how processes operating at different scales shape ecosystems. Smaller-scale, rapid-rate processes can accumulate to cause larger-scale, slower-rate changes. The way the decomposition of bacteria on a forest floor (a smaller-scale, faster-rate process) promotes better tree growth over time (a larger-scale, slow-rate process) is a good example of these kinds of cross-scale dynamics.

**Equilibrium and Climax States**

Ecological science has been primarily underpinned by the idea of stability at least since Frederic Clements' seminal text on climax vegetation in 1936. Clements proposed a succession model of ecosystems, whereby vegetation passes through a series of intermediate ecological states before achieving a climax vegetation type assumed to be the stable state for that ecosystem. An example of this is the succession of a grassland to shrubs and fast-growing tree species, culminating in a mature forest dominated by slower-growing, and longer-living tree species. The mature forest is considered the climax ecosystem. Disturbance can knock an ecosystem out of balance, and smaller disturbances can cause the succession process to start over again—for example wind knocking over trees in part of a forest that would then begin the process of succession from fast-growing species to the climax forest again. Disturbances that change the character of the climax vegetation over a wide area are considered destructive in this conceptualization.

These ideas from ecology inform many approaches to sustainability and conservation. They effectively frame disturbance as a negative force, and therefore the goal is to maintain life processes in what is known as *dynamic equilibrium*, such that while some change is expected, smaller disturbances do not affect the overall character (climax) of an ecosystem.

Resilience has helped open up ecological theories by reframing disturbance, although it should be noted that resilience is still a debated concept. Empirically, it has been impossible to demonstrate that any particular ecosystem behaves in the manner predicted by the panarchy model. In part this has to do with the complexities of the interrelated processes involved, and the difficulties of both modeling their dynamics and of collecting adequate data to represent them in models. But others suggest it is because the world works differently than the model proposes. Resilience thinking has been most successfully applied in climate change arenas and underpins the idea of planetary boundaries discussed in the next section. It has a limited influence on environmental management, but resilience has helped to redirect attention towards the positive role disturbances might play in maintaining core ecosystem properties. Managers now recognize that fire, for example, is crucial to some ecosystems, rather than a completely destructive event to be avoided at all costs (as the example of the oak savanna discussed in Chapter 3 illustrates).

Since the 1990s, there have been attempts to theorize socio-ecological systems. SES theory works with the same ecological principles as ecological resilience, but adds human processes as dynamic parts of the system. Social systems are assumed to work in more or less the same manner as ecological systems, so that learning, adaptation, and governance are modeled as part of system processes along with decomposition, plant growth, and the like. Some of these models are quantified while others are used more as qualitative sketches of how processes come together.

Critics of SES have pointed out that social systems behave very differently from ecological systems and therefore cannot be imported into resilience models easily. For example, in social systems, small-scale governance processes at the household level do not accumulate into large-scale governance systems at the national level in the same manner as microbiological processes shape macrobiological processes. In addition, critics argue that the patterns of stability and change in social systems are not comparable to ecological systems and do not follow predictable patterns. Nevertheless, resilience has taken hold in the popular imagination and is now a dominant narrative in discussions of sustainable development and in climate change debates.

## Planetary Boundaries

Recently, the idea of planetary boundaries has captured the imagination of sustainability communities and has helped to increase awareness of sustainability issues. Originally proposed by a group of scientists, including Nicolas Kosoy and Will Steffen, the planetary boundaries model merges concerns over carrying capacity with the concept of resilience (see Kosoy et al. 2012; Steffen et al. 2015). In particular, the concept of planetary boundaries focuses on the notion of "tipping point" (see accompanying text box), or the point beyond which ecosystems undergo fundamental change as a result of disturbance. The planetary boundaries model attempts to identify and quantify the interdependent biophysical boundaries of the current global ecosystem. If these boundaries are exceeded, there is the risk that tipping points will be crossed and we will experience a major phase shift in the earth's ecological systems. The theory postulates that there are limits within which a "safe operating system for humanity" is possible (see Leach, Raworth, and Rockström 2013). This safe operating system encompasses nine domains: climate change, biodiversity loss, ocean acidification, ozone depletion, the nitrogen and phosphorus cycle, freshwater use, deforestation and other land-use changes, particle pollution of the atmosphere, and chemical pollution. Many of these nine boundaries are deeply implicated in global warming and are believed to operate in positive feedback loops so that, for example, if the ocean acidifies too much, it will contribute more to carbon dioxide emissions.

The idea of planetary boundaries is thus far more complex than the notion of carrying capacity, but nevertheless is underpinned by a fundamental assumption that earth system processes are finite, according to Steffen and colleagues in 2015. It is also explicitly anthropocentric. While certainly driven by a concern for the earth's present ecological processes, the idea is defined in relation to safe systems for human beings. Concerns over climate

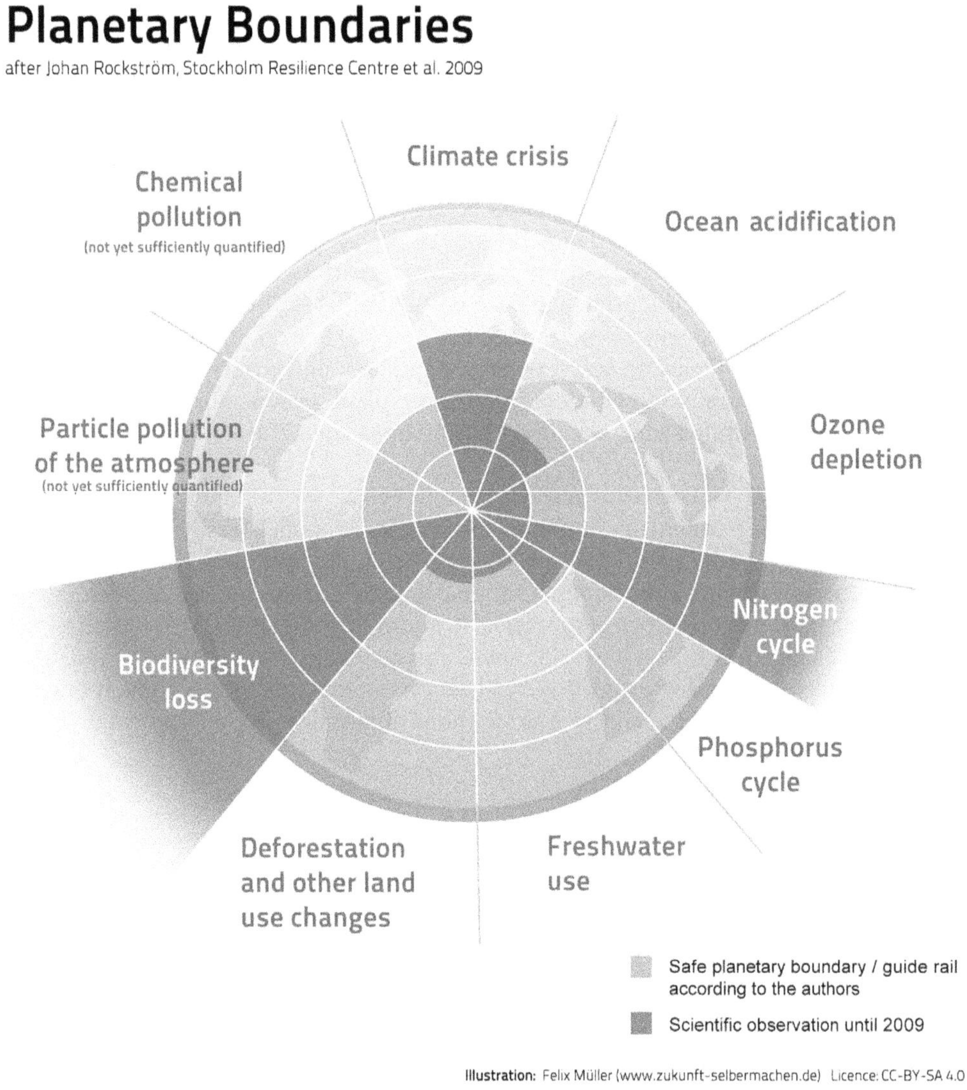


A model of planetary boundaries, based on Rockström et al. in *Nature* in 2009. The inner circle represents estimated boundaries, while the darker radiating bands refer to the estimated current boundaries in various domains.

*Source:* Felix Meuller/Wikimedia; https://commons.wikimedia.org/wiki/File:Planetary_Boundaries.png.

change have been central to the emergence of this idea, and indeed, the boundaries themselves correspond strongly to the different working groups within the Intergovernmental Panel on Climate Change (IPCC), which provides a review of climate science at regular intervals (see also the discussion of climate change in Chapter 6). There are real concerns that the atmosphere will be pushed over a tipping point and that this will cause major changes throughout the earth's ecosystems. And of most concern, these phase changes could mean that reducing the effects of global warming will become even more difficult or even impossible. In this sense, advocates of planetary boundaries argue that the resilience of the earth's ecosystems is not indefinite, but risks collapsing if tipping points are reached, with catastrophic consequences for humanity as well as many other species.

**Tipping Points**

The term "tipping point" originated in physics where it referred to the point when adding a small amount of weight caused an object in balance to tip over. In the social sciences, the American political scientist Morton Grodzins applied the term in 1957 to the social dynamics of gentrification. The "tip point" was when the share of non-white residents in a largely white community would increase such that white residents would relocate. Since then the term has been applied to many processes that in some way or another are reinforced by positive feedback loops.

In ecology and resilience thinking, a tipping point refers to a disturbance or accumulation of changes that can "tip" an ecosystem from one set of mutually reinforcing processes (referred to as stable states) into another. Positive feedback loops can reinforce changes and force the system into another state. For example, when enough trees in a forest fire are killed, any remaining living trees are more vulnerable to being destroyed by a big wind storm or disease, or are unable to regenerate themselves. The tipping point is the point at which there are no longer enough healthy trees to maintain the forest. When a tipping point is passed, a transition to a new state occurs—in the forest example, a grass or shrub land replaces a forest. From an anthropocentric perspective, such transitions often refer to change from a desirable state to a non-desirable state. The process that "tips" the ecosystem can either be part of the system itself, or it may be produced from outside.

The term "tipping points" has recently come into widespread popular and scientific use to talk about some of the major social and environmental changes happening today. In climatology, it is used to refer to the point when the volume of greenhouse gases in the atmosphere causes irreversible change, such as increased average temperatures and melting glaciers. These changes then cause other changes, such as sea level rise and ocean acidification, which then contribute to more greenhouse gases in the atmosphere and so on, resulting in an exponential rate of change. The idea of tipping points has also been used in the analysis of social and political movements—for instance, to explain the Arab Spring, the political revolts in the Middle East and North Africa beginning in 2010. In this context, it refers to the point at which social discontent erupts into widespread popular protest, leading to the overthrow of governments. It is very difficult to predict or anticipate when tipping points are reached, although some have tried to identify and quantify them. People working with the planetary boundaries model are engaged in that kind of research now. Today, there are profound fears that the climate system is being, or has already been, pushed over a tipping point and we are facing major transformations in earth-system processes as a result.

## Other Biophysically Oriented Measures

In addition to planetary boundaries, socio-ecological systems, and ecological footprints, dozens of other biophysical measures of sustainability have been developed. Some examples, with the environment at their core, include the following: the environmental performance index, measuring the effectiveness of a country or region's policies aimed at environmental targets; the environmental sustainability index, measuring overall progress toward sustainability; and the environmental vulnerability index, measuring the severity of environmental dangers and issues facing countries and regions of the world.

In addition, some measures focus on peak usage or consumption of resources that are considered necessary for present ways of life including oil, gas, coal, water, metals, and so forth. For example, the Hubbert Peak Theory is used to assess when a key resource (especially oil) reaches its extraction or production peak as a way to understand when sustainability thresholds will be crossed. Other measures are similar to planetary boundaries in the sense that they draw upon the limits to environmental systems, and the idea of life-cycle assessments. Life-cycle assessment is a technique used to think about how water, carbon, phosphorus, nitrogen, sulfur, and oxygen, among other resources, cycle through social–environmental systems. Here we are deliberately not calling such systems "socionatures" because life-cycle assessments are predicated upon defining ecosystem boundaries as closed systems. This means that resources cannot flow into or out of the system for the purposes of the assessment and humans are not part of that system.

For details of what various sustainability indexes include see:

- "Environmental Performance Index (EPI)." Socioeconomic Data and Applications Center (SEDAC). NASA.
  http://sedac.ciesin.columbia.edu/data/collection/epi.
- "Environmental Sustainability Index (ESI)." Socioeconomic Data and Applications Center (SEDAC). NASA.
  http://sedac.ciesin.columbia.edu/data/collection/esi/.
- "Building Resilience in SIDs: The EVI." Environmental Vulnerability Index.
  www.vulnerabilityindex.net/.

All of these measures have an anthropocentric orientation, despite putting environmental systems first. The idea of limits is not aimed at the limits to life, but rather limits to *human* life, at least as we know it now. There is deep concern that, without major changes in how we extract, use, and dispose of resources, we will reach, or extend beyond, ecological limits. Indeed, some scientists have begun to worry that we have already crossed the threshold for biodiversity loss. If too many species are lost, it could lead to ecological collapse in a variety of different ecosystems, which has implications not only for ecology, but also for human societies. Many are worried that without concerted efforts now to save threatened species, societies and economies will be radically reshaped by the environmental conditions that ensue.

## Measures of Sustainable Development

In contrast to indices that place most emphasis on environmental domains, some focus primarily on social and economic processes. For instance, Gross Domestic Product (GDP) per capita, a common measure of economic development, features in many attempts at measuring sustainability. GDP per capita measures average economic output in a nation-state as a proxy for economic performance, and divides that by the total population of the nation. This means that if the population increases but production remains the same, the GDP per capita will decrease, or if population remains the same, and production increases, GDP per capita will rise. There are important assumptions behind using this measure for sustainable development. First, it is neo-Malthusian inasmuch as a larger population is assumed to have a negative effect on economic growth if GDP cannot keep up with population growth.

Second, it is assumed that economic productivity will have a positive influence on resource use. Rarely does a rising GDP per capita result in a negative measure of sustainable development. Rather, it is used to measure whether the population and economy are in balance with each other over time, drawing largely from "development as growth" perspectives.

Yet, as discussed in Chapter 3, many approaches to sustainable development are quite critical of the "development as growth" narrative. The Human Development Index (HDI) is therefore often preferred over GDP per capita as a foundation for measuring sustainability. In addition to income, the HDI includes life expectancy, literacy rates, and average years of schooling. As a result, it tries to capture more dimensions of well-being than simply income. This measure is often adopted by "development as freedom" advocates, although the HDI is not without its critics. It is seen as over-privileging certain aspects of human development that reflect Global North priorities for living well. Others seek to nuance the way they measure sustainability by considering other indices that also put social concerns at their core, including:

Democracy index
Sustainable governance indicators
Child development index
Corruption perception index
Education index
Gini coefficient (used to measure income inequality)
Life expectancy index
Gender parity index
Gross national happiness
Happy planet index
Gender empowerment index

The target of most of these indices is fairly evident from their titles, which reveal certain themes about what sustainability means to those who use these indices. These themes include:

- social equity (gender empowerment, gender parity, child development, life expectancy, education)
- governance (sustainable governance, corruption perception, democracy)
- well-being (happy planet, gross national happiness, child development, gender empowerment).

These measures have been incorporated into sustainability indices in a variety of ways and their scope helps to illustrate what some consider required for sustainability. The underlying assumption is that environmental sustainability cannot be achieved if human basic needs are not met first. Understanding precisely how different measures are chosen is complicated because each one gives more weight to one dimension or another. There are at least as many indicators as there are understandings of what sustainability ought to be.

A number of global indices are quite ambitious, aiming to compare all countries in relation to one another; some compile over sixty different measures to arrive at one index, according to the 2003 analysis by Thomas M. Parris and Robert W. Kates. In addition to globally oriented indices, Parris and Kates count over four hundred efforts to measure sustainability that have been developed at the municipal, state, and country scale. The Human Wellbeing Index, for example, developed by the International Union for Conservation of Nature (IUCN), is a composite of 88 indicators for 180 countries. These include GDP, HDI, and other indicators that are commonly collected through census data or compiled by the UN and other development agencies, in addition to others that are less well known. One of the most important efforts in this regard is the UN Commission on Sustainable Development's effort to formulate indices that could be used to compare one country to another. In an effort to be both comprehensive and to simplify, the indices were developed by first identifying 134 existing measures and then cutting them down to 58. Initially, the goal was to develop a set of indicators that could be applied in all countries. However, the UN Commission concluded that each country was unique enough that the specific set of indicators used needed to be tailored for each context. As a result, each country can choose how they measure sustainable development, making it perhaps most relevant internally, whereas comparisons across countries are impossible.

Another international effort relevant to the measurement of sustainability is the UN's Sustainable Development Goals (SDGs) program, adopted by the UN Commission in September 2015, to build on the Millennium Development Goals (see Chapter 2). The SDGs are intended to set the development agenda until 2030, through 17 goals and 169 targets. (Note that targets are not the same as indices. The indices help measure whether the targets are met.) The SDGs are presented as universally applicable and significant, informing the world's population about what challenges lie ahead and hence what areas to focus on. While the SDGs are framed to recognize differences among and within countries, they are formulated on the national scale. The nation-state is the level at which activities will be aimed and targets evaluated, making comparisons across nation-states more or less irrelevant and potentially masking a lack of progress made in one part of a nation if big gains can be demonstrated overall. The three pillars of sustainable development are easily recognizable within the goals, and conceptually, they continue to hold social, economic, and environmental domains separate. Inclusive language, such as "for all," "everywhere," "in all its dimensions," and "at all levels," attempts to illustrate how social justice is central to the goals themselves, but what this means and how to achieve it are not elaborated. Rather, the terms "sustainable development," "sustainably increase," and "sustainably use" are persistently used. The SDGs lean heavily on the ability of "sustainability" to resonate widely for diverse audiences. As we have seen in this book, however, "sustainable development" can invoke a remarkably wide set of practices. A month after their publication, the goals were already being criticized as elusive, prohibiting any meaningful evaluation and easy implementation. As a result, by 2016, there were already many controversies over how the SDGs should be put into effect.

### Sustainable Development Goals

The seventeen Sustainable Development Goals were announced following the United Nations Sustainable Development Summit 2015, held in New York City, September 25–27, 2015. Each goal has specific targets to be achieved by 2030. The goals are stated on the UN web page "Sustainable Development Goals" as:

- Goal 1. End poverty in all its forms everywhere
- Goal 2. End hunger, achieve food security, improve nutrition, and promote sustainable agriculture
- Goal 3. Ensure healthy lives and promote well-being for all at all ages
- Goal 4. Ensure inclusive and equitable quality education and promote lifelong learning opportunities for all

- Goal 5. Achieve gender equality and empower all women and girls
- Goal 6. Ensure availability and sustainable management of water and sanitation for all
- Goal 7. Ensure access to affordable, reliable, sustainable and modern energy for all
- Goal 8. Promote sustained, inclusive economic growth, full and productive employment and decent work for all
- Goal 9. Build resilient infrastructure, promote inclusive and sustainable industrialization, and foster innovation
- Goal 10. Reduce inequality within and among countries
- Goal 11. Make cities and human settlements inclusive, safe, resilient, and sustainable
- Goal 12. Ensure sustainable consumption and production patterns
- Goal 13. Take urgent action to combat climate change and its impacts
- Goal 14. Conserve and sustainably use the oceans, seas and marine resources for sustainable development
- Goal 15. Protect, restore and promote sustainable use of terrestrial ecosystems, sustainably manage forests, combat desertification, and halt and reverse land degradation and biodiversity loss
- Goal 16. Promote peaceful and inclusive societies for sustainable development, provide access to justice for all, and build effective, accountable, and inclusive institutions at all levels
- Goal 17. Strengthen the means of implementation and revitalize the global partnership for sustainable development.

*Source:* United Nations. "Sustainable Development Goals." www.un.org/sustainabledevelopment/sustainable-development-goals/.

Of most interest to us in this book are the consequences of such ideas. When whole communities of practice are galvanized around the idea that planetary boundaries are being crossed, or that Sustainable Development Goals are capable of bringing together vastly different definitions and ambitions for sustainability, it has far-flung political consequences. But those consequences may be hard to predict, and they are not necessarily linked to the specific planetary boundaries that are crossed or the specific SDGs that are achieved, nor when or how those outcomes might occur. Rather, a whole range of responses is possible—all of them constrained by how we think about sustainability problems.

Furthermore, the planetary boundaries approach brings us back to neo-Malthusian ideas of carrying capacity, even if at core the notion takes us well beyond such a simple relationship between population and resources. It holds environmental domains separate from social domains. Similarly SDGs promote a notion that a supranational body (the UN) is capable of defining what is universally good, and that economic, social, and environmental sustainability goals sit side by side, rather than being embedded within each other. These conceptual assumptions mean that the planetary boundaries model refocuses our attention back on resources and resource use, rather than on more holistic notions of living well, or on connections between humans and other species. SDGs similarly uphold conceptual and pragmatic separations between the three pillars of sustainable development.

In contrast, a socionature narrative leads us to think more carefully about how human alienation from other species can limit possibilities for living well. Socionatures perspectives reject the clear divide between social and environmental domains that underpin both planetary boundaries and the sustainable development goals. As a result, a socionatures approach argues for an entirely different set of planetary boundaries, ones that do not separate out biophysical from social processes. It is only by thinking about how humans are part of the world in a profoundly different manner that we can imagine new possibilities for living well. A partial application of this approach has been taken by Melissa Leach and colleagues (2013) when they tried to add social–political mediating processes between the causes of biophysical change (i.e. greenhouse gas emissions), and the effects that might be observed within the domains defined as planetary boundaries (i.e., ocean acidification). Their intervention has been welcomed by those who lament the return to neo-Malthusian thinking and the dominance of the natural sciences in defining the sustainability agenda, even if it does not fundamentally (that is, ontologically) challenge the notion of planetary boundaries.

The compelling resonance of ideas like planetary boundaries and SDGs also contributes to a revival of Malthusianism in popular thought. Many people today continue to accept a straightforward "overpopulation means environmental degradation" narrative. As we have argued throughout this book, challenging such simplistic understandings is fundamental to moving towards a sustainable future. It is crucial that we probe the bases of these narratives about how the world works and replace them with others if they prove to be rooted in racism, sexism, and other dimensions of social injustice. The goal is not about getting the right narratives, but rather fostering critical thinking around the power of narratives to influence what happens on the ground and for whom. Few early advocates of population control as a tool to solve environmental problems imagined it would result in coercion, lack of consent, and disempowerment for women in developing countries, for example. Yet their inability to foresee such consequences lay in the emphasis on the single issue of population control. Thinking holistically and critically about how we frame every dimension of the problem, therefore, is necessary. Such an approach probes how the narratives of sustainability themselves, including the three pillars, create particular kinds of possible futures and close down others.

## Sustainability and Justice

As has been shown above, the problems of measuring sustainability raise normative issues that bring us squarely to questions of justice. Like sustainability itself, different assessments of justice and how to achieve it exist. Most definitions of justice, however, are underpinned by a commitment to equality, peace, fairness, and respect for people. Note how these normative commitments are often different from those linked to measures of sustainability that emphasize resource use and other economic dimensions. In relation to sustainability, four different narratives that include considerations of justice are explored below: ecofeminism, ecosocialism, environmental justice, and climate justice.

### Ecofeminism

Ecofeminism emerged in the 1970s when scholars and activists brought a feminist perspective to the environmental movement. Ecofeminism and subsequent narratives of environmental feminisms and feminist political ecology explore how gender, race, class, ecology, and nationhood are all linked. Early versions of ecofeminism were particularly concerned with Enlightenment dualistic thinking that set women apart from men, and nature from culture and society, leading to a variety of exploitative relations for women and environments. Carolyn Merchant's book *The Death of Nature* (1980) explored how the rise of a mechanistic understanding of the earth was directly tied to both the oppression of women and overexploitation of the earth. This idea was taken up Vandana Shiva in her book *Staying Alive* (1988), which linked development, patriarchy, oppression of women, and overexploitation of resources. She argued that women had an inherent desire to protect the environment because of their role as mothers and caretakers of their households. This early version of ecofeminism closely parallels deep ecology (see Chapter 3). Both narratives argue for the need to reimagine our kinship with environments in a manner that emphasizes connections between humans and nonhumans, rather than separation, or hierarchical relations.

Other environmental feminists, spearheaded by the work of Bina Agrawal, rejected this kind of essentialist reading of women; rather they showed how women's productive roles (farming and forest-harvesting, for example) and roles within the household (providing food and caring for children, for example) were at the root of why women were more likely than men to understand the importance of protecting the environment. This shift in focus from inherent qualities to everyday practices and structural social relations helped to reorient feminist thought on the environment. The focus shifted to the ways that economic and household relations, culture, and people's everyday experiences shaped their relationships with the environment.

Environmental feminisms gained momentum after the Brundtland report (1987), when opposition to the dominant sustainability narrative was formed around feminist thinking. The Bruntland report focused on development and economic growth, whereas feminists stressed the interdependency between environmental degradation and gender inequalities. At the same time, the work of Donna Haraway (1991) was articulating early formulations of socionatures and related ideas that ontologically link the environment and society. Together, feminists' critiques of the Bruntland report along with Haraway's work helped to show the relevance of feminist thinking for understanding how economy and culture produce practices that serve to dominate nonhumans as well as women.

In the 1990s, feminist political ecologists took up ecofeminist ideas and linked them to critiques of development. Feminist political ecologists have shown how many projects implemented in the name of sustainability fail to take account of their impact on women. Indeed, in many cases, sustainability projects seek to capitalize on the presumed interest of women to protect their environments by enrolling them in voluntary activities aimed at environmental sustainability. The result is a greater burden for women without a transformation in gender relations. Feminist political ecology and ecofeminism continue to have very important influences both within research and social movements. Most in this wide field place a critical focus on the relationship between environmental degradation and the subordination of women, arguing that capitalism and an ontological divide between environment and society detaches production and social life from nature through gender and class divisions.

## Ecosocialism

Many ecofeminists consider themselves ecosocialists, who argue, like ecofeminists, for the need to change material conditions and the social structure of society to overcome both environmental crises and injustice. Ecosocialism draws on the writing of Karl Marx and Friedrich Engels on society's relation to the environment. Engels wrote in 1883 that, "we by no means rule over nature like a conqueror over a foreign people, like someone standing outside nature—but that we, with flesh, blood and brain, belong to nature, and exist in its midst" (Engels 1940) . While Marx is strongly associated with the forms of communism that emerged in the twentieth century, most of his writings were devoted to analyzing capitalism as an economic system. It is this analysis that ecosocialists draw upon to understand how the modern global economy creates social and ecological problems.

For ecosocialists, social relations typical of capitalist societies are at the heart of unsustainability. Without addressing basic discrimination based upon class and global capitalism (i.e., disparities between the Global North and the Global South, as well as inequalities within nations), society–environment dynamics will remain exploitative. Here note some of the parallels with the concerns about economy in sustainability narratives that have less emphasis on social justice. One could argue that inequality can be overcome through economic means. The difference is that ecosocialists are concerned about the economic system and the relationships between people that produce inequality in the first place. Simply trying to give people a bigger share of economic wealth will not overcome the roots of inequality. This commitment to social relations leads them to see a common linkage between struggles for justice and environmental protection.

## Environmental Justice

Environmental justice movements originated in the United States within poor urban areas largely populated by people of color, on Native American reservations in the United States and Canada, and amongst poor women of all races. It represents a fundamental resistance to mainstream sustainability by insisting that "environment" needs to include respect for health, cultural diversity, spiritual values, and educational rights. Environmental justice activism came from populations who mainstream observers least expected to protest the negative consequences of waste generated by capitalist economies. These communities began organizing into grassroots movements in the 1980s to protest specific projects that they believed were a threat to their health and well-being. A growing number of Hispanic, African American, and Native American peoples stressed that indigenous groups and low-income minorities were isolated from, or explicitly ignored by, mainstream sustainability because their neighborhoods and livelihoods were not considered worthy of consideration. For example, none of the big environmental organizations would help a group of women from southeast Los Angeles fight the establishment of a garbage incinerator in their neighborhood. Eventually, Greenpeace took up their cause, according to research by Giovanna Di Chiro (1996), but all the other so-called "big ten" environmental groups ignored air quality in poor, urban neighborhoods. Their focus was rather on the so-called pristine environments of American Wilderness areas and the preservation of threatened species.

Environmental justice activists have made linkages with each other since the 1980s and now embrace a wide variety of local and global issues. The environmental justice movement's struggle materialized in September 1991 when more than a thousand activists gathered for the First National People of Color Environmental Leadership Summit in Washington, DC. The seventeen principles adopted at the Summit came to define the future environmental justice movement (see accompanying text box). The movement is founded on the idea that economic expansion and social inequalities contribute to an unequal distribution of wealth, power, environmental degradation, and ecological conflicts. *Buen vivir* (covered in detail in Chapter 3) is part of this movement and is one attempt to include non-Western understandings of environment–society interconnections into an environmental agenda. Indigenous rights, spiritual and cultural values, and the inviolability of nature are invoked, shifting the focus from the nature–economy nexus to the socio-spatial distribution of environmental benefits and burdens, as well as the ontological foundations upon which our narratives of sustainability are based.

**The Seventeen Principles Adopted at the First National People of Color Environment Leadership Summit, 1991**

1) Environmental Justice affirms the sacredness of Mother Earth, ecological unity and the interdependence of all species, and the right to be free from ecological destruction.
2) Environmental Justice demands that public policy be based on mutual respect and justice for all peoples, free from any form of discrimination or bias.
3) Environmental Justice mandates the right to ethical, balanced and responsible uses of land and renewable resources in the interest of a sustainable planet for humans and other living things.
4) Environmental Justice calls for universal protection from nuclear testing, extraction, production and disposal of toxic/hazardous wastes and poisons and nuclear testing that threaten the fundamental right to clean air, land, water, and food.
5) Environmental Justice affirms the fundamental right to political, economic, cultural and environmental self-determination of all peoples.
6) Environmental Justice demands the cessation of the production of all toxins, hazardous wastes, and radioactive materials, and that all past and current producers be held strictly accountable to the people for detoxification and the containment at the point of production.
7) Environmental Justice demands the right to participate as equal partners at every level of decision making, including needs assessment, planning, implementation, enforcement and evaluation.
8) Environmental Justice affirms the right of all workers to a safe and healthy work environment without being forced to choose between an unsafe livelihood and unemployment. It also affirms the right of those who work at home to be free from environmental hazards.
9) Environmental Justice protects the right of victims of environmental injustice to receive full compensation and reparations for damages as well as quality health care.
10) Environmental Justice considers governmental acts of environmental injustice a violation of international law, the Universal Declaration On Human Rights, and the United Nations Convention on Genocide.
11) Environmental Justice must recognize a special legal and natural relationship of Native Peoples to the U.S. government through treaties, agreements, compacts, and covenants affirming sovereignty and self-determination.
12) Environmental Justice affirms the need for urban and rural ecological policies to clean up and rebuild our cities and rural areas in balance with nature, honoring the cultural integrity of all our communities, and provided fair access for all to the full range of resources.
13) Environmental Justice calls for the strict enforcement of principles of informed consent, and a halt to the testing of experimental reproductive and medical procedures and vaccinations on people of color.
14) Environmental Justice opposes the destructive operations of multi-national corporations.
15) Environmental Justice opposes military occupation, repression and exploitation of lands, peoples and cultures, and other life forms.

16) Environmental Justice calls for the education of present and future generations which emphasizes social and environmental issues, based on our experience and an appreciation of our diverse cultural perspectives.
17) Environmental Justice requires that we, as individuals, make personal and consumer choices to consume as little of Mother Earth's resources and to produce as little waste as possible; and make the conscious decision to challenge and reprioritize our lifestyles to ensure the health of the natural world for present and future generations.

*Source:* "Principles of Environmental Justice." EJNet. Environmental Justice/Environmental Racism. www.ejnet.org/ej/principles.html.

## Climate Justice

Earlier we discussed how climate change has contributed to a revitalization of neo-Malthusian thinking in sustainability narratives, and in Chapter 6 we will elaborate more on climate change as an environmental concern. Here, however, the spotlight is on how it is framed as a justice issue. Climate justice is fundamentally concerned with how the populations primarily responsible for global climate change are not the ones who will be affected most by projected changes in the water cycle and corresponding impacts on agriculture and livelihoods. In this sense, climate justice narratives and activist movements are closely connected to ecosocialist and ecofeminist perspectives as well as global environmental justice movements, although they cannot all be subsumed into each other.

The first Climate Justice Summit was held alongside the sixth Conference of the Parties (COP6) to the United Nations Framework Convention on Climate Change (UNFCCC) in The Hague in 2002. The statement issued jointly by the Summit and the UNFCCC reads, "climate change is a rights issue. It affects our livelihoods, our health, our children and our natural resources" (EJNet; Agostino and Lizarde 2012). The climate justice movement stresses the economic and social causes of climate change and brings attention to questions of global solidarity and justice. Climate change, climate justice movement activists argue, is fundamentally a product of a global capitalist system that simultaneously overexploits environments and oppresses marginalized people. Of most concern, many parts of the world projected to be hardest hit by climate change are not the places that are responsible for producing emissions that are fueling global warming. The movement has not only criticized world trade and corporate-driven development models, but also exposed how the effects of climate change unfold in local struggles over land dispossession, water rights, and urbanization. In other words, the climate justice movement tries to connect different scales of impact, responsibility, and inequality in how climate change plays out, rather than viewing it simply as a global environmental issue.

For climate justice and related initiatives see the Mary Robertson Foundation–Climate Justice, Global Justice Ecology Project, and the What's Next Forum at these URLs:

- "Principles of Climate Justice." Mary Robertson Foundation–Climate Justice. www.mrfcj.org/principles-of-climate-justice/.

- "Climate Justice." Global Justice Ecology Project. http://globaljusticeecology.org/climate-justice/.
- United Nations. "The Convention on Biological Diversity: A Visual Introduction." www.whatnext.org/climatewatch/index.php?published-max=.

## Sustainability and Governance

Justice concerns raise another important issue related to sustainability: who is responsible for ensuring that sustainability goals are set and, perhaps more importantly, implemented? In this book we have discussed how such questions are political and normative in nature. In terms of practice, they bring us squarely into issues of governance. For the purposes of this discussion, governance—the types of organizations, laws, and practices through which resource use is managed—must be understood separately from institutions—the rules and norms that serve to guide and shape how groups of people manage and use resources. Institutions are thus subsumed under governance, but understanding governance itself encompasses more than just the rules and norms, or institutions; it also includes the scale at which organizations are conceived and operate (see also Chapter 5), the actors who are considered appropriate for different kinds of actions, and the mechanisms for enforcement. Agenda 21, the action plan that followed the UN's 1992 Rio Earth Summit, made reference to the need for governance mechanisms, and within sustainable development practice a strong emphasis is placed on setting up the right institutions to govern resources. Today, nearly all global environmental governance mechanisms are predicated upon control over environmental laws at the national level. In Chapter 6 some of the global governance mechanisms for climate change are covered, but these nevertheless are based on individual nation-states signing on to global agreements and enforcing them at the national scale.

Governance mechanisms reflect the logic that underpins them; many are predicated on isolating an environmental problem from its wider context, whereas others take a more holistic view and seek to transform society–environment dynamics and the ontologies through which we understand ourselves in the world. In the following sections we take a look at four different dimensions of governing sustainability: the role of knowledge in managing environment; calls for good governance; privatization; and common property and the commons.

### Knowledge in Managing Environmental Change

Many narratives of sustainability draw upon scientific knowledge, which by most definitions is considered to be objective facts. Indeed, as we explore below, the endeavor to measure sustainability seeks to provide consistent parameters and quantifiable indicators of environment–society dynamics. However, a growing number of scholars and activists have shown that the production of scientific knowledge struggles with cultural contradictions and boundaries, drawing into question how to evaluate objectivity in science. Knowledge is not detached, but situated, contingent upon historical trajectories, and bound to its social, ecological, and economic context. As a result, some facts are only true within the contexts they refer to, yet they are often applied as universal truths, outside of context. Retaining knowledge within its cultural, historical, and political context helps to increase objectivity, rather than dilute it, because the limitations of what can be known is clearer. Advocates of *buen vivir* and ecofeminism argue that in order to live more sustainably, it is crucial to shift how we seek to know the environment. It is not a case of denying climate change, for example, but rather to think through a profoundly different lens about the relationships between human activities and wider processes of change. Such worldviews preclude many of the industrial practices that maintain our current economies and lifestyles on the basis that they violate alternative principles underpinning society–environment conceptualizations and corresponding ethics. As discussed previously, these points of view push us to create whole new categories through which we try to understand change in the world.

But the critique of science has come from many other quarters as well. Most of these criticisms do not challenge the importance or utility of scientific knowledge. Rather, they emphasize the social nature of science, arguing that the knowledge science produces is as embedded in culture, history, and structural power relations as any other form of knowledge; see Donna Haraway's *Simians, Cyborgs, and Women: The Reinvention of Nature* (1991) and Helen E. Longino's *Science as Social Knowledge* (1990) for this line of reasoning. If we understand science as a product of the social world, the dominant truth reflects structural power relations, which control people's ability to access resources as well as decision-making authority. If we accept that science is socially and culturally embedded, it opens up critical space to recognize how modern science tends to subsume other ways of understanding, defining, and addressing what is considered to be the reality of environmental problems. While science or the quest for facts should not be abandoned, it is critical to think more analytically about how, why, and when we mobilize different kinds of knowledge in sustainability debates. These insights bring us back to the core questions of this textbook: sustainability of what, and for whom? To these questions another can be added that is equally important in this context: who decides?

And in fact, today, sustainability is most often governed by reference to scientific knowledge. Planetary boundaries, discussed above, is simply the latest narrative that privileges knowledge and measurement of specific biophysical processes over other concerns in defining how sustainability should be governed. In this book we are not challenging the usefulness of planetary boundaries as a scientific concept—although within the scientific community it is still being challenged and some argue it inherently cannot be quantified—but rather we are insisting on probing the work that the narrative of planetary boundaries does in the name of sustainability. At the moment, the planetary boundaries narrative is being accepted as common sense by a remarkably wide range of actors. Yet it is not common sense; rather it is a particular construct used to make sense of complex, intersecting processes. Critics challenge whether as currently framed the planetary boundaries approach foregrounds the "right" processes about whether boundaries, and their normative underpinnings, are knowable. These critiques do not make the idea of planetary boundaries useless. On the contrary, when we approach the concept with a critical stance, asking who developed it, what conceptual foundations it assumes, and what it includes and excludes, we can use the concept in a more nuanced way, keeping open to the consequences for justice, governance, and society–environment dynamics as we do so.

The power of narratives to shape society–environment dynamics highlights the ambiguity in the so-called "common vision" of sustainable development. It raises issues about why certain ontologies are foregrounded while others become marginal. The link between power and narratives is key: powerful narratives are powerful because of the knowledge they employ, and the knowledge is considered legitimate because of who promotes it. When devising governance strategies for sustainability, it is crucial to hold these issues in view.

## Good Governance and Managing the Environment

If science is the dominant knowledge regime underlying many narratives of sustainability, good governance is the dominant narrative employed in relation to environmental management. Good governance was identified at the World Summit on Sustainable Development in 2002 as a key factor for ensuring successful sustainable development. "Good governance" thus is often used as an umbrella term for effective stewardship of the environment.

While "good governance" sounds like it matches well with the social aspects of sustainable development and to be a perfectly reasonable goal, as a buzzword in the international development industry, the term, like "sustainable development," has become nearly meaningless since it can mean many different things to different groups. It has come to represent political participation, accountability, representativeness, efficiency, effectiveness, transparency, responsiveness, inclusiveness, adherence to the rule of law, consensus-oriented decision making, legitimacy, individual freedom, and so forth. Without taking issue with any of these goals, it's easy to see that the term is open to criticism for being vague and easy to bend to any purpose. Who could defend bad governance, for example? And also, which goals are considered more important in which contexts? Since these decisions are made within the political realm, they are a product of power relations embedded in societies.

Furthermore, good governance points to forms of governing with philosophical and historical roots in Europe and North America. While these traditions are useful in some contexts, deeper moral questions need to be asked about imposing ideas of good governance on the Global South, where people have very different histories and traditions. Even when good governance is framed in a way that makes it impossible to oppose, it can close down discussions about why certain organizational forms should or should not be exported to the Global South. As a result, difficult questions about who has the right to define and address environmental problems need to be placed at the center of good governance debates. At the moment, good governance is seen as a tool for resolving such difficult issues, and yet as movements like *buen vivir* and other indigenous rights movements highlight, other ways to conceptualize living well are possible. Once again, we are back to a core message of this book: the need to critically evaluate what is being done by whom in the name of governance for sustainability.

## Privatization

One very popular environmental governance mechanism today is privatization. As discussed in Chapter 2, the idea that private property will necessarily be better managed has deep historical roots. During the 1990s, it became one of the cornerstones of the neoliberal Washington Consensus, which promoted free markets and deregulation as the most effective and efficient way to grow economies and ensure good environmental management. The logic underlying private property is that the property holder can use his or her assets as collateral to access credit, and hence invest in land improvements and environmental protection. Private ownership is believed to incentivize the owner to use the resource in question responsibly and effectively because they have control over it.

A number of resources have been privatized in recent decades that traditionally had been managed collectively. Globally, governments have sold off state-owned resources—forests, water resources, oil and mineral deposits, and even lands set aside for conservation—in the belief that they will be used more efficiently and managed more effectively by private owners. These moves have prompted many criticisms, particularly by social movements, and the record is inconsistent in terms of whether resources are being more effectively managed or not. In contrast, many social movements promote the idea of common management instead.

Even for resources used collectively, the privatization of resources is increasingly given a central role in environmental governance. One of the most important is the idea of payment for ecosystem services. The ecosystem services model is discussed in more detail in Chapter 6, but here we want to emphasize how market mechanisms are being used to allocate and govern access to resources. Payment for ecosystem services (PES) works on the principle that keeping environmental resources intact (rivers, forests, and the like) has financial benefit for people downstream and therefore they should pay people living in close proximity to those resources for conserving them. There have been a number of PES water schemes implemented in South America through which more urbanized water users pay communities upstream to keep the water source clean and not to overexploit it. Also, a large research program in the United Kingdom explores the potential to use PES for poverty alleviation but so far the results are quite mixed, resulting from the complexity of relationships between resources and users with conflicting needs. Another well-known example of PES today are carbon payments under reducing emissions from deforestation and degradation (REDD+) schemes. Again, it is unclear if these schemes will result in the desired ecological outcomes, how they are reshaping the scale of environmental governance (see also Chapter 5), and even more unclear if communities receiving payments benefit overall.

## The Commons and Collectives

Many of the resources and landscapes that are of concern for sustainability advocates present problems for privatization. Such resources are known as common-pool resources, meaning those that are difficult to fence or bound. Forests are often considered common-pool resources, as are mobile species like fish. As discussed in Chapter 3, some have argued that common-pool resources will inevitably be overexploited and therefore should be privatized. However, Elinor Ostrom and others have argued that collective management of such resources does not necessarily lead to overexploitation. Rather, they distinguish between common-pool resources and common property. Common property, often known as "the commons," are resources or landscapes that have governance regimes in place with institutions that manage harvesting, limit the total number of users, and sanction those who break the rules. Common-pool resources can be overexploited when common-property institutions are not present, a situation known as open access rather than a commons.

Commons were an important part of pre–Industrial Revolution land-use regimes in Europe and continue to persist in many parts of the world today. Some examples include pasture areas, forests, fisheries, and water sources. Commons were used for grazing animals, collecting wild nuts and berries, firewood, and small game animals. In Europe very few of these commons remain, although some places have tried to reinstitute them. Historically, the advent of modern, large-scale industrial agriculture led to the enclosure or privatization of the commons across Europe. Land that had been open for the extraction of some resources was enclosed by landlords, resulting in the decline of peasant livelihoods. (A notable exception was the hunting of deer, which had largely been forbidden.) One explicit motivation for enclosing the commons was to push poor people off the land and into paid labor in the growing factories of industrializing Europe. The "enclosures," as it is known, was one of the driving forces behind large-scale out-migration from Europe to North America, Australia, and New Zealand since without access to the commons, people could no longer support themselves on the small pieces of tenant farmland that were left to them (some owned this land, most rented it). In Scotland, which continues to have the most unequal landownership in Europe, social movements against the enclosures led to the passing of laws to protect the right of tenant farmers to the commons. Today, these lands are managed by grazing committees consisting of tenant farmers (called "crofters") as common lands on large, private estates. Private landowners cannot deny crofters access to the common grazing resources even if they legally own the land. Access to common grazing land is a crucial part of Scottish Highland landownership laws, without which only very large, private estates and small plots of private land would exist.

After the enclosures in Europe, many Europeans settled in the United States triggering a reorganization of property rights. As a result, no tradition of commons exists among

Highland cow grazing on Isle of Lewis in Scotland.

*Source:* Mlm42/Wikimedia; https://commons.wikimedia.org/wiki/File:Highland_cow.jpg.

immigrant populations in the United States. Most Native American groups had no concept of private property (or even common property), at least not in relation to land and water resources, so the privatization of land in the United States had significant impacts on their lives. Many government lands are effectively managed as commons in the United States today, although in terms of their governance, they are not considered commons. For example, large parts of the western United States (Nevada, Utah, Wyoming, Montana, and so forth) are owned by the government and managed by the Bureau of Land Management (BLM). The BLM leases out grazing rights to private ranchers who are then given access to the land for their cattle and sheep. In Utah and Nevada, in particular, some of these lands are commons in the sense that often, but not always, many individuals are given access to the same piece of land for extracting resources (that is, grazing). They are not governed as commons, meaning resource users do not collectively decide how to manage the land. Users are unable to make management decisions such as how many individual ranchers can gain access, whether to close particular areas for grazing, and so forth; the BLM makes all those decisions independently. However, political pressure means that the BLM rarely departs from historical practices, whereas nongovernmental groups like the Nature Conservancy are able to engage in more experimental grazing practices that try to implement the latest scientific findings (see the accompanying resource box for more information). Other government lands that are managed in a similar manner include U.S. Forest Service timber contracts and wilderness areas (see Glen O. Robinson 2013, for more information). Wilderness areas limit access of users based on permit systems (in some places) and type of transport (nonmechanized only), but the government retains all rights over setting such regulations.

Bureau of Land Management (BLM) lands are managed under several different governance mechanisms. The commons-type arrangement described in the text is one model. In most places, individuals are given exclusive allotments to grazing areas which effectively privatizes the grass, while the land is still owned by the government. For more information on grazing in the western United States, see the following sources from the Bureau of Land Management (BLM), American Prairie Reserve, and the Nature Conservancy:

- "About Livestock Grazing on Public Lands." U.S. Department of the Interior Bureau of Land Management. www.blm.gov/programs/natural-resources/rangelands-and-grazing/livestock-grazing/about.
- American Prairie Reserve. www.americanprairie.org/.
- "The Matador Ranch." The Nature Conservancy. www.nature.org/en-us/get-involved/how-to-help/places-we-protect/matador-ranch/.

Scholarship on the commons since the late 1980s has provided an alternative to the neoliberal drive to privatize all resources in the name of ensuring good environmental management. Indeed, after Agenda 21, many sustainable development projects have been predicated upon some form of collective ownership or governance of resources and landscapes. Nevertheless, the idea of common property does not fundamentally overturn the logic of private property. Elinor Ostrom's concept of new institutional economics is based in rational choice theory and individualistic conceptualizations of society. Rational choice refers to the idea that society is composed of rational individuals who make choices based on self-interest so that institutions and laws are necessary to ensure commitment to collective interests. No room for altruism or a sense of collective commitment exists within rational choice theory. Therefore, while Ostrom's work on the commons has been important in providing sound evidence that collective management of resources and landscapes works, it fails to confront mainstream narratives of sustainability that are rooted in ideas of property, rational choice, and economic self-maximization. In contrast, *buen vivir* and similar narratives aim to base environmental governance on principles of interspecies connections and emotional attachments to land.

**Design Principles for Common-Pool Resource (CPR) Institutions**

Opposing Hardin's (1968) tragedy of the commons thesis, Ostrom has identified eight design principles that make local common-pool resource management stable. These can be found in several of her writings including her 1990 book, *Governing the Commons: The Evolution of Institutions for Collective Action*. Here we present them as they appear on the International Association for the Study of the Commons website (www.iasc-commons.org/).

1. Clearly defined boundaries (effective exclusion of external un-entitled parties).
2. Rules regarding the appropriation and provision of common resources that are adapted to local conditions.

3. Collective-choice arrangements that allow most resource appropriators to participate in the decision-making process.
4. Effective monitoring by monitors who are part of or accountable to the appropriators.
5. A scale of graduated sanctions for resource appropriators who violate community rules.
6. Mechanisms of conflict resolution that are cheap and of easy access.
7. Self-determination of the community recognized by higher-level authorities.
8. In the case of larger common-pool resources, organization in the form of multiple layers of nested enterprises, with small local CPRs at the base level.

From work on the commons, scholarly consensus is emerging regarding the conditions for successful governance of local *and* regional commons. However, global commons, such as air and water, are more difficult to govern, as evidenced by the failure to find common ground to address climate change. (The 1987 Montreal Protocol against ozone depletion is a notable exception.) This is in part because a larger number of participants with diverging agendas makes it harder to organize, agree on, and enforce rules, leading some to argue for privatization of such resources. Others have suggested that common property is being enrolled in neoliberal attempts to privatize resources such as fisheries, which previously had been open commons.

Under the banner of the commons, however, a new movement is emerging that explicitly seeks to resist capitalist models and practices predicated upon private property and centralized control over resources. Commoning, as it is generally known, is emerging in many different parts of the world, but in particular in the United States, Europe, and Australia. It is an attempt to recapture the collective nature of both environmental resources and society. Notions of justice that seek to redefine society–environment dynamics to support commoning are linked to or similar to ecofeminism, ecosocialism, and environmental justice. Many of these initiatives are focused around open access to electronic and technological resources, but others include efforts to make agricultural research and on-farm systems collectively available, such as Green Start, a group based in New Hampshire. Commoning movements are often concerned with public spaces that have been neglected in order to engage in activities within them to build community and a collective sense of attachment to place.

For some examples of commoning, see these pages from the websites Commons Transition and Green Start:

- "The Catalan Integral Cooperative (or CIC) Is a Political Project Seeking to Unite Consumer and Labor Initiatives." Commons Transition. http://commonstransition.org/catalan-integral-cooperative/.
- "Sharing Technology for Healthy Soil, Food and Energy." GreenStart. www.greenstartnh.org.

Although a relatively new movement, commoning seems to be gaining ground quickly in the Global North, reflecting a growing dissatisfaction with current society–environment dynamics shaped by capitalism and private property. It can be seen as one of the attempts to create a "deep sustainability" that begins by trying to address the relationships among people and between societies and their environments, relationships that are believed to be at the root of unsustainability today. These movements also need to be approached with a critical stance, however. Research on common property regimes has shown how collective action can be riddled with coercion and exclusions that are often structured by unequal social relations. Commoning is an attempt to foster collective action and more just economic, social, and ecological relations, but such attempts need to hold in view how preexisting relationships are not automatically erased. Rather, structural inequalities require concerted efforts to promote new, more equitable relations if deep sustainability is the goal.

## Global Governance

Much of the discussion on governance so far has centered on national and subnational arrangements, but global governance is gaining importance, especially in the environmental realm. The United Nations is one of the most important institutions and in Chapter 2 we outlined some of the key international treaties and conferences wherein agreements over environmental governance have been made since the 1970s. The UN does not have legal authority to enforce compliance with these arrangements, however. Some treaties have legally binding agreements written into them, but most rely on voluntary compliance. In fact, the lack of a clear supranational, global authority for governing the environment is leading to calls for the development of such a body.

The European Union is an example of a supranational body that has legal authority to enforce environmental laws across its member states. It has implemented a number of important laws that require changes at the nation-state level in water quality, fishing regulations, and conservation areas, for example. These laws are not without their critics, however, and indeed, environmental regulations along with lack of democratic accountability have been some of the reasons that critics have called for individual countries to leave the European Union over the past several years—a trend that played out in Brexit, the British vote to leave the EU in June 2016. The EU therefore struggles to retain legitimacy in the eyes of many of its citizens and it remains to be seen if, over the long term, this experiment in supranational governance will succeed. The EU experience suggests that, from a justice and governance perspective, thorny questions remain about how a supranational environmental governance body should be constituted, who should be in charge, and what the relationship between it and nation-states should be. In Chapters 5 and 6 we return to specific examples of global governance in terms of various UN treaties and climate change.

## Conclusion

In this chapter we have taken a selective look at some of the ways that sustainability has been enacted around the world. Measuring sustainability is used primarily by nation-states and managers at a variety of levels to help guide their efforts and to show that some progress has been made. The assumption of progress and attempts to develop achievable targets is firmly rooted in conceptualizations of sustainability that hold the three pillars of sustainability (environment, society/culture, and economy) separately, and that understand humans as somehow affecting their environments. In contrast, many of the calls for environmental justice are predicated upon very different understandings of the society–environment dynamic. Environmental justice proponents want more attention paid to everyday environments and to our relations with other species as a foundation for sustainability. These different narratives and approaches to enacting sustainability have unsurprisingly been accompanied by divergent approaches to governing sustainability. Here we have reviewed some common narratives including the role of scientific knowledge, good governance, privatization, the commons, and global governance.

Enacting sustainability draws our attention to the scales at which environmental problems are conceived, the levels at which governance mechanisms are proposed, and the actors who are involved in governing and achieving sustainability. Increasingly, our narratives of sustainability are shifting to the global scale. In the next chapter we look at how environmental problems are scaled and the connections between the local and the global. Here we want to highlight that, while sustainability is an agenda that diverse groups seem able to accept politically, when viewed from a globalization perspective, more diversity exists. What is being done in the name of sustainability can take radically different forms, with radically different goals, and proposed by groups that are sometimes in direct opposition to each other. It is crucial to probe how sustainability questions seek to make comparisons across very diverse contexts and bring everyone into similar relationships with their environments, like the Sustainable Development Goals do. And yet, sustainability questions in fact represent radically different commitments to ways of life and socionatures. It is in this sense that environmental issues are being increasingly globalized, and yet localized at the same time. It is to these questions of scale that we now turn in Chapter 5.

## References and Further Research

Agarwal, Bina. 1992. "The Gender and Environment Debate: Lessons from India." *Feminist Studies* 181: 119–158.

Agostino, Ana, and Rosa Lizarde. 2012. "Gender and Climate Justice." *Development* 551: 90–95. https://doi.org/10.1057/dev.2011.99.

Bardi, Ugo. 2009. "Peak Oil: The Four Stages of a New Idea." *Energy* 343: 323–326. https://doi.org/10.1016/j.energy.2008.08.015.

Bollier, David, and Silke Helfrich, eds. *Patterns of Commoning*. Amherst, MA: The Commons Strategies Group.

Clements, Frederic. 1936. "Nature and Structure of the Climax." *Journal of Ecology* 24: 252–284.

Cooke, B., and Uma Kothari. 2001. *Participation: The New Tyranny?* London: Zed Books.

Cote, Muriel, and Andrea J. Nightingale. 2012. "Resilience Thinking Meets Social Theory: Situating Social Change in Socio-Ecological Systems (SES) Research." *Progress in Human Geography* 364: 475–489. https://doi.org/10.1177/0309132511425708.

Cronon, William, ed. 1996. *Uncommon Ground: Rethinking the Human Place in Nature*. New York: W. W. Norton.

DeSombre, Elizabeth R. 2000. "The Experience of the Montreal Protocol: Particularly Remarkable, and Remarkably Particular." *UCLA Journal of Environmental Law & Policy* 19: 49.

Di Chiro, Giovanna. 1996. "Nature as Community: The Convergence of Environment and Social Justice." In *Uncommon Ground: Rethinking the Human Place in Nature*, edited by W. Cronon, 298–320. New York: W. W. Norton.

EJNet. "Bali Principles of Climate Justice." From "Environmental Justice Principles" developed at the 1991 People of Color Environmental Justice Leadership Summit, Washington, DC. www.ejnet.org/ej/bali.pdf.

Engels, Friedrich. 1940 (written 1883). *Dialectics of Nature*. New York: International Publishers.

Färe, R., S. Grosskopf, and F. Hernandez-Sancho. 2004. "Environmental Performance: An Index Number Approach." *Resource and Energy Economics* 264: 343–352. https://doi.org/10.1016/j.reseneeco.2003.10.003.

Furubotn, Eirik Grundtvig, and Rudolf Richter. 2005. *Institutions and Economic Theory: The Contribution of the New Institutional Economics*. Ann Arbor: University of Michigan Press.

Grodzins, Morton. 1957. "Metropolitan Segregation." *Scientific American* 1974: 33–41. https://doi.org/10.1038/scientificamerican1057–33.

Gunderson, Lance H., and C. S. Holling, eds. 2002. *Panarchy: Understanding Transformations in Human and Natural Systems*. Washington, DC: Island Press.
Haraway, Donna. 1991. *Simians, Cyborgs, and Women: The Reinvention of Nature*. New York: Routledge.
Hardin, Garrett. 1968. "The Tragedy of the Commons." *Science* 162, no. 3859: 1243–1248.
Holling, C. S. 1973. "Resilience and Stability of Ecological Systems." *Annual Review of Ecology and Systematics* 41: 1–24.
Hunter, James. 1999. *Last of the Free: A Millennial History of the Highlands and Islands of Scotland*. Edinburgh: Mainstream Publishing.
IPCC. 2014. "Summary for Policymakers." In *Climate Change 2014: Impacts, Adaptation, and Vulnerability. Part A: Global and Sectoral Aspects. Contribution of Working Group II to the Fifth Assessment Report of the Intergovernmental Panel on Climate Change*, edited by C. B. Field, et al., 1–32. Cambridge: Cambridge University Press.
Kosoy, Nicolas, et al. 2012. "Pillars for a Flourishing Earth: Planetary Boundaries, Economic Growth Delusion and Green Economy." *Current Opinion in Environmental Sustainability* 41: 74–79. https://doi.org/10.1016/j.cosust.2012.02.002.
Krotscheck, Christian. 1997. "Measuring Eco-Sustainability: Comparison of Mass and/or Energy Flow Based Highly Aggregated Indicators." *Environmetrics* 86: 661–681. https://doi.org//:10.1002/(SICI)1099-095X(199711/12)8:6<661::AID-ENV278>3.0.CO;2-N.
Leach, Melissa, Kate Raworth, and Johan Rockström. 2013. "Between Social and Planetary Boundaries: Navigating Pathways in the Safe and Just Space for Humanity." *World Social Science Report* 2013: 84–89.
Lemos, Maria Carmen, and Arun Agrawal. 2006. "Environmental Governance." *Annual Review of Environment and Resources* 311: 297–325. https://doi.org/10.1146/annurev.energy.31.042605.135621.
Lenton, Timothy M., et al. 2008. "Tipping Elements in the Earth's Climate System." *Proceedings of the National Academy of Sciences* 1056: 1786–1793. https://doi.org/10.1073/pnas.0705414105.
Longino, Helen E. 1990. *Science as Social Knowledge: Values and Objectivity in Scientific Inquiry*. Princeton, NJ: Princeton University Press.
Mansfield, Becky. 2004. "Neoliberalism in the Oceans: 'Rationalization,' Property Rights, and the Commons Question." *Geoforum* 35: 313–326.
Merchant, Carolyn. 1980. *The Death of Nature: Women, Ecology, and the Scientific Revolution*. San Francisco: Harper & Row.
Moran, Daniel D., Mathis Wackernagel, Justin A. Kitzes, Steven H. Goldfinger, and Aurélien Boutaud. 2008. "Measuring Sustainable Development—Nation by Nation." *Ecological Economics* 643: 470–474. https://doi.org/10.1016/j.ecolecon.2007.08.017.
Muradian, R., et al. 2013. "Payments for Ecosystem Services and the Fatal Attraction of Win–Win Solutions." *Conservation Letters* 64: 274–279. https://doi.org/10.1111/j.1755-263X.2012.00309.x.
Ostrom, Elinor. 1990. *Governing the Commons: The Evolution of Institutions for Collective Action*. Cambridge: Cambridge University Press.
———. 1992. "The Rudiments of a Theory of the Origins, Survival, and Performance of Common-Property Institutions." In *Making the Commons Work: Theory, Practice, and Policy*, edited by D. Bromley, 293–318. San Francisco: ICS Press.
Ostrom, Elinor, Joanna Burger, Christopher B. Field, Richard B. Norgaard, and David Policansky. 1999. "Revisiting the Commons: Local Lessons, Global Challenges." *Science* 284, no. 5412: 278–282.
Parris, Thomas M., and Robert W. Kates. 2003. "Characterizing a Sustainability Transition: Goals, Targets, Trends, and Driving Forces." *Proceedings of the National Academy of Sciences of the United States of America* 100, no. 14: 8068–8073. https://doi.org/10.1073/pnas.1231336100.
Pestre, Dominique. 2008. "Challenges for the Democratic Management of Technoscience: Governance, Participation and the Political Today." *Science as Culture* 172: 101–119. https://doi.org/10.1080/09505430802062869.
Plumwood, Val. 1991. "Nature, Self, and Gender: Feminism, Environmental Philosophy, and the Critique of Rationalism." *Hypatia* 61: 3–27.
Robinson, Glen O. 2013. *The Forest Service: A Study in Public Land Management*. New York: Routledge.

Rockström, Johan, et al. 2009. "A Safe Operating Space for Humanity." *Nature* 461: 472–475.
Roth, Robin J., and Wolfram Dressler. 2012. "Market-Oriented Conservation Governance: The Particularities of Place." *Geoforum* 433: 363–366. https://doi.org/10.1016/j.geoforum.2012.01.006.
Schroeder, Richard A., and Roderick P Neumann. 1995. "Manifest Ecological Destinies: Local Rights and Global Environmental Agendas." *Antipode* 274: 321–324.
Shiva, Vandana. 1988. *Staying Alive: Women, Ecology, and Survival in India*. New Delhi: Kali for Women.
Steffen, Will, et al. 2015. "Planetary Boundaries: Guiding Human Development on a Changing Planet." *Science* 347, no. 6223. https://doi.org/10.1126/science.1259855.
Swyngedouw, Erik. 2005. "Dispossessing $H_2O$: The Contested Terrain of Water Privatization." *Capitalism Nature Socialism* 161: 81–98.
United Nations Environment Program. 1992. *Agenda 21*. New York: United Nations.
United Nations World Commission on Environment and Development. 1987. *Our Common Future (The Brundtland Report)*. New York: Oxford University Press. Available electronically from the UN: www.un-documents.net/our-common-future.pdf.
Wackernagel, Mathis, and William Rees. 1998. *Our Ecological Footprint: Reducing Human Impact on the Earth*. Vancouver, CA: New Society Publishers.

# 5

# Issues of Scale in Producing Sustainability

Andrea J. Nightingale and Linus Karlsson

As demonstrated in previous chapters, implementing sustainability goals is complicated by the lack of a clear definition of sustainability, the wide range of indicators and measures that could be used as a baseline, and the huge diversity of interests in what sustainability ought to be. Cutting across all of these complications is the problem of scale.

"Scale," in the context of sustainability, refers to the dimensions of processes or entities that refer to a quantifiable measure, for example, territorial extent or temporal duration. Scale is crucial in sustainability debates because some solutions that are considered sustainable by a group of actors for relatively localized problems, for example, are unable to be satisfactorily scaled out to encompass national issues. Similarly, some practices that might be sustainable at a small scale, like use of fossil fuels, are totally unsustainable when undertaken at a global scale. Therefore, observers and analysts must probe the scale at which sustainability projects are conceived and how they depend upon, or are related to, processes happening at other scales. Scale questions arise in all sustainability contexts but in order to think carefully about why scale is so important, we focus the following discussion on three current approaches to sustainable development: conservation, participation, and market-oriented public–private partnerships. But first we explore the concept more generally.

## Making Sense of Scale

Explicit discussions of how scale is important to a sustainability debate or program are rarely undertaken. Rather, the scope of a project is often defined by the resources available (for instance, how many places can be included given the money or personnel available) or the type of problem addressed (for instance, what area or population uses a given resource) rather than the scale at which the relevant ecosystem and social processes operate. Such decisions also rarely consider the way that ecosystem processes—including human activities—operate on multiple scales simultaneously, with very small-scale processes influencing large-scale processes over time and space. Describing a phenomenon is very different from understanding its causes. To quote the philosopher Paul Thompson: "The way that we conceptualize a system is deeply value laden, and reflects judgments about what is thought to be problematic, as well as likely guesses about where solutions might lie" (Thompson 2010, 239). As this quote suggests, sustainable development is inherently about conceptualizing social–ecological systems and situating possible solutions within achievable scales.

While its meaning varies across many disciplines, scale generally refers to the extent or dimensions of an entity, practice, or problem. It is often represented in quantitative terms, but it does not have to be. In the United States, for instance, the local scale can be radically different in terms of territorial area from one region to another; defining the local scale is not quantitative or fixed therefore, but rather refers to a particular level at which some political decisions are taken. Also, scale should not be confused with "level," which can be understood as points along a scale, for example local municipalities, county governments, and national government are levels within the national scale. Because scale is used to represent complex processes, our conceptualizations of scale can have important implications for how we approach those processes. Global climate change is a good example. Climate change has been conceptualized as a global problem, and as a result, most models of climate change have been developed for the global level, and require data at the global scale to work. This has meant that much research has been concentrated at that level—for example, predicting how much the global average temperature will increase—with the result that our understanding of regional and local climate change is significantly more limited. Researchers are therefore turning their attention to developing better models and data to understand regional climate change, not least because that is the level at which responses are being planned and implemented. While climate change certainly is global in its level of impact, by conceptualizing the problem as occurring at the global scale, rather than a global problem developing out of many regional-scale problems added together, research has been focused in a particular manner.

Similarly, the overuse or contamination of a resource at one level can reduce availability for users at other, spatially and temporally separate, levels. Climate change is often brought up as an example of a problem that will extend far into the future, but all human (and non-human) activities have imprints that travel through time and space; think of nuclear waste and its long-lasting temporal impacts. There is a need to scale up our responses to environmental challenges. Yet, efforts to protect environmental resources by changing the rules or conditions of use tend to focus on a single level, and more often than not the temporally present and spatially local. For example, many pollutants cross political and institutional boundaries in ways that affect the people who did not produce them. This is true for climate change but also nuclear waste, toxic chemicals, and other forms of air pollution. Dealing with these pollutants is challenging because they cross from one institution to another (e.g., from one nation-state to another). As we have seen, there is tremendous disagreement over how best to tackle particular environmental challenges across levels, institutions, interest groups, and so forth. And yet all these groups must coordinate together to address pollutants that cross jurisdictional boundaries. As we discussed briefly at the end of Chapter 4, calls for large-scale (supranational) environmental governance bodies are further complicated by major questions over democratic accountability, representation of often highly divergent interests, and a lack of good mechanisms for deciding how to force compliance with supranational environmental laws and agreements. Another source of complexity is how differing views on nature, society, sustainability, and sustainable development influence which sorts of responses are favored by whom.

Many attempts aim to overcome these problems of scale. In the following sections we highlight three: conservation, which is an attempt to stabilize *on a temporal scale* the ecosystem properties of a particular territory; participation, a strategy to overcome collaboration problems *across levels on a spatial scale*; and market-oriented public–private partnerships, which try to draw differently scaled actors together. First, however, we take a look at different approaches to mitigate the issues of scale on a more general level.

## Ecosystems, Natural Resources, and Scale

Ecosystems are a good example of a concept in science that has inherent assumptions about scale which are important for how we interact and manage them. Ecosystems are understood to exist at multiple levels on a spatial scale ranging from the very local to the global, and interconnect through flows of resources, information, commodities, capital, norms, values, and legislation. Simultaneously, ecosystems are conceptualized on a temporal scale so that change is expected within certain time frames and influenced by a wide range of factors—from diseases and pollutants to financial crises—but whether that change is considered normal is dependent upon the type of ecosystem and the cause of change in question. These changes can also spread across scales to influence other ecosystems. Acid rain is a good example of pollutants crossing scales to influence other ecosystems. Acid rain is caused by industrial pollutants entering the atmosphere from smoke stacks that are then transported by the earth's atmospheric circulation to other parts of the world (from in the mid-regions of North America, for instance, to the eastern United States and Canada). Those pollutants change the chemical composition of rain: it becomes more acidic than normal, causing damage to forest and lake ecosystems that exist at another scale and in a different location than the industries causing acid rain. Spatial connectivity also allows positive forces to spread throughout a system. Today it is easier than ever before to disseminate knowledge and share responsibility among global, regional, national, and local levels, for example. In fact, this is what sustainable development interventions aim for, even if they encounter a number of conceptual, scalar, governance, and justice problems that we have highlighted throughout this book.

Perhaps most significantly, resources and ecosystems are often understood very differently depending on the scale of observation and the values attached to those observations. For

Smoke stacks, such as these, from Sudbury in Ontario, Canada, produce pollutants that drift east and contaminate lakes and forests.

*Source:* Tony Webster/Wikimedia; https://commons.wikimedia.org/wiki/File:Vale_Copper_Cliff_Nickel_Refinery_Smoke_Stacks_-_Sudbury,_Ontario_(30886962846).jpg.

some, a resource might be worth going to war over. Oil and diamonds in particular have been blamed for generating a lot of violent conflict in relatively localized areas, particularly as they gained value on global markets. The resources themselves are not the source of conflict, however. Rather, it is the *value* different societies attach to them that can generate conflict. Some societies may not even consider oil or diamonds (for example) to be a resource with value and function.

Another important example of how conceptualizations of ecosystems and scale come together is found in forest carbon. In recent years, the concept of "forest carbon," or carbon stored in soils and forest plants, has been invented in relation to climate change mitigation efforts. Previously, very few specialists were concerned with carbon in soils and forests, and indeed, forest carbon was thought of very differently: the concepts used were forest "biomass," or "organic matter" in soils, rather than "carbon." Current understanding and conceptualization of climate change—with a large focus on the effects of carbon dioxide (a molecule where a carbon atom is bonded to two oxygen atoms, or $CO_2$) and methane (composed of one carbon atom and four hydrogen atoms, or $CH_4$) in the upper atmosphere—has created carbon as a tradable resource. Carbon dioxide and methane are two potent greenhouse gases that trap the sun's heat, preventing it from escaping into outer space. While some $CO_2$ in the atmosphere is necessary for the kind of climate we have today, industrial emissions have drastically increased the concentration of greenhouse gases, and the earth is getting warmer as a result. Concerns over carbon, therefore, are derived from a simplification of $CO_2$ and $CH_4$ into "carbon," and entire industries are now emerging to try to trap or bond carbon on earth so it doesn't circulate in the atmosphere. A new financial market has been designed to help limit emissions where carbon credits can be bought and sold (see "Carbon, Forests and the REDD Paradox" by Sandbrook et al. 2010 for more information and analysis on this trend). The idea is to limit overall emissions of carbon but still allow some industries access to emissions through carbon trading. As carbon has become a tradable good, farmers and foresters in economically marginal areas are now linked to the global carbon market as their soils and trees become an object of the global commodity carbon—they can "store" carbon in their soils and forests, as the thinking goes, to "trade" with industrial emitters. Such linkages force people to rethink how they value different resources and, in effect, how they choose to use them. (See the section below titled "The Commodification of Conservation" for more on this trend.)

## Scale, Connectivity, and Cross-Level Governance

Today, two principal strategies address the challenges associated with the increasing spatial connectivity of our world. One approach is to transfer responsibility to a higher level of public authority, or centralization. While it is true that such authorities may possess the institutional capacity to set norms and standards across scales and the financial means to implement change, local resource users are often considered to have a more comprehensive and context-sensitive understanding of the potential and limits of local socionatures. Thus, many have pointed to how top-down interventions can be detrimental to local users, become illegitimate, and contribute to resistance at lower levels. The other approach therefore seeks to transfer responsibility to a lower level of authority, or decentralization. However, local users are often unaware of or unable to adequately coordinate with users in other places or at other levels, requiring a higher authority to ensure that interests are negotiated across communities.

Decentralization programs have been promoted all over the world following the publication of Agenda 21. For more information see "Democratic Decentralization of Natural Resources: Institutionalizing Popular Inclusion" by Jesse C. Ribot, available at this website:

- World Resources Institute. www.wri.org/publication/democratic-decentralization-natural-resources.

Another useful source is "Decentralisation," an article in *The Economist* available here: www.economist.com/node/14298890.

Both centralization and decentralization strategies therefore have merits and faults. As a result, development practitioners have sought to create a hybrid governance system that builds on local knowledge but at the same time is sensitive to spatial connectivity. For instance, Elinor Ostrom's eight design principles try to explain the long-term sustainability of resource governance systems (see the text box in Chapter 4 on Design Principles for Common-Pool Resource (CPR) Institutions). While these design principles have proven useful to explain how *local* socio-ecological systems can be sustainably governed, lifting these principles to a higher level has remained a major challenge. The problem lies in the eighth principle, "nested enterprises," which indicates that institutions at different spatial levels need to cooperate. If "nested enterprises" is juxtaposed with the first principle, "clearly defined boundaries and users," the enormity of cross-level and scale challenges becomes visible. It is difficult to have clearly defined users at the same time as having nested enterprises. This has remained a central challenge for not only Elinor Ostrom, but for many sustainable development practitioners. Ostrom attempted to address such issues towards the end of her career, but empirical studies have struggled to implement her framework on the ground. The question is: how to maintain clearly defined boundaries of spatially and temporally connected ecosystems while allying responsible institutions at multiple levels? This question has given rise to a number of different strategies, all with different approaches to and conceptualizations of sustainable development.

One strategy that has been used to address this problem is to pragmatically fix, or localize, an ecological system or environmental problem to make it legible and governable. Since the most urgent problems tend to manifest locally, a given problem is often considered to have its roots in local activities. Projects intended to tackle these problems therefore emphasize the local scale even if the ecosystem in question may have a wider territorial extent or if local land users may be constrained by wider scale political economic processes (see also the coverage of political ecology in Chapter 3). The use of agrochemicals, such as pesticides and chemical fertilizers in agriculture, are good examples. Soil depletion, erosion, and chemical pollution have immediate local effects so most interventions are focused on educating farmers on appropriate production techniques. However, overuse of agrochemicals also has long-term effects, such as decreases in yield and fouling of waterways through overfertilization, that have regional and global ramifications. Agricultural chemicals can pollute water sources and ecosystems well beyond agricultural areas. Yet surprisingly little attention is paid to responses that might address these scales. For example, efforts to restrict corporations from manufacturing and selling environmentally destructive pesticides, or to incentivize governments to subsidize biological fertilizers, are limited. For the local farmer, pollution still pays, and these pollutants do not stay in the level at which they are applied.

For more information on agrochemicals, view these sites from the U.S. Environmental Protection Agency and the Toxics Action Centre:

- "Herbicides." EPA. www.epa.gov/caddis-vol2/caddis-volume-2-sources-stressors-responses-herbicides.
- "Insecticides." EPA. www.epa.gov/caddis-vol2/insecticides.
- "Pesticides." Toxics Action Center. www.toxicsaction.org/problems-and-solutions/pesticides.

For a more industry-oriented interpretation of agrochemicals, see this page from the International Union of Pure and Applied Chemistry: http://agrochemicals.iupac.org/.

While spatially defined problems might be more governable, they offer no safeguards if the problem transcends spatial or institutional boundaries, as in the agricultural chemical example. Cross-scale and cross-level flows take us back to the initial puzzle of this chapter: territorial traps for which we have to find solutions. By portraying an environmental problem as situated at a certain level, be it local or global, who should be responsible is also defined for that problem, as is who should have input into which measures are adopted. It is even possible that, for some issues, the blame for sustainability failures is displaced from the global level and rescaled on to local populations by situating a problem in a certain place and time. Questions of scale thus embody questions of justice, bringing us back again to the core questions of this book: sustainability of what, and for whom? These questions force us to think about how sustainable development distributes responsibility across scales, to the benefit or disadvantage of different actors. The remainder of this chapter will take a closer look at attempts to address problems of scale and responsibility through, as already mentioned, three specific strategies of sustainable development: conservation, participation, and public–private partnerships.

## Conservation

Often a core goal in sustainability projects, "conservation" has been a key word in the environmental vocabulary for more than one hundred years and, as discussed in Chapter 3, it is one of the foundations to modern sustainability thought. In this section we explore the logic behind conservation and how it is implemented in different parts of the world. Conservation can be seen as a strategy to enclose a fragile environment—or spatially frame it—to ensure that a certain resource or ecosystem is sustained in a particular state into the future. As such, it is also a strategy of non-development, as conservationists often consider that human activity upsets ecosystem processes. While resilience thinking (discussed in Chapter 4) has established that ecosystems are characterized by dynamic change, many in the natural sciences are strongly committed to managing environments in such a manner that disturbance is minimized. Human activities in particular are generally considered unnatural and to cause problematic disturbances. Conservation projects therefore most often emphasize the preservation or restoration of particular ecosystems or species, beginning from defined baseline conditions.

Conservation strategies are usually place-based, even for mobile species like large game and fish—hence the establishment of national parks and marine conservation areas—and this has significant implications in terms of scale. Many ecosystem processes cross the jurisdictional scale of institutions intended to manage those processes. In addition, many conservation efforts are predicated on a well-defined separation of society from environment. Even in conservation areas that retain human populations, management strategies are underpinned by an inherent understanding that many human activities are detrimental to ecosystem processes.

The very act of conservation entails the establishment of jurisdictions and boundaries that define access and use rights to natural resources. In its essence, conservation is therefore an act of environmental control. Generally, an external managerial body implements conservation strategies to which local resource users, tourists, and scientists are all beholden. In this sense, conservation is neither "natural" nor "neutral," but a social and political construction of territory and landscape, transferring rights to and control over resources from local users to an external managerial body. Conservation thus contributes to both a redistribution of authority and responsibility across scales, and a transformation of local socio-ecological relations.

Orthodox forms of conservation focus on excluding humans from certain territories in order to reduce destructive behaviors in fragile environments. Such "fortress conservation" often results in the eviction of local inhabitants and establishment of border controls to hinder intruders from entering the conservation area. Environmental fortresses are often established in geographically peripheral areas, which are inhabited by economically marginalized, and often vulnerable, populations. Such measures are usually justified on the grounds that they guarantee the protection of a public good, including virgin forests for climate change mitigation purposes; wetlands for flood prevention, water purification, or erosion control; coral reefs for the protection of biological diversity; and savannas for game and safari tourism. However, the establishment of conservation areas often constrains local populations from assuming traditional sustenance strategies, such as seasonal migration, collection of wild foods and timber, and hunting for animals, which raises the question of who has the right to define what environments are subjects of conservation, and, fundamentally, what is a common good. Even though fortress conservation has been criticized as a colonial practice, it is still implemented today.

However, as a result of local resistance and critics pointing to the environmental injustices inherent in fortress conservation strategies, we have seen a shift to a conservation-as-development paradigm, wherein local users and communities co-manage resource use in collaboration with external authorities to ensure the welfare of local inhabitants. (Participatory development practices are discussed in detail below.)

## The History of Conservation

Conservation has long been linked to both the preservation and protection of ecosystems, and to normative ideas of the nation and morality. Gifford Pinchot, a pioneering American conservationist, argued in his 1910 book *The Fight for Conservation* that conservation is imbued with moral, political, and economic virtues that ensure the health and welfare of nations. Pinchot witnessed firsthand the geographical expansion of the United States and expressed concern over how land was converted for human use. Most forests were clear-cut and the timber burned. For Pinchot, conservation came to represent order and harmony with nature. It was not environmental destruction itself he opposed, however. Rather, the problem he identified was how to use "the earth for the permanent good for man" (Pinchot 1937, 262–263). From his perspective, conservation was not a strategy to sustain a certain natural state on a temporal scale, but a form of resource management that would guarantee durable human development. Consider, however, that Pinchot's view of environmental destruction was based on a flawed view of pristine nature. Native American land-use practices were crucial in how the ecosystems of America evolved, and need to be read as human landscapes, as William Cronon explores in *Changes in the Land: Indians, Colonists, and the Ecology of New England* (1983), and yet in Pinchot's narrative, the people who were driven off the land are not mentioned alongside clear-cutting and timber burning. This legacy—of ignoring the importance of local populations to landscapes targeted for conservation as well as their historical and current relationships to the land—is largely felt today in current thinking and practices.

Pinchot's ideas were foundational to the establishment of national parks and other kinds of conservation areas in both developed and developing countries. As conservation areas have expanded and experiences from around the world have accumulated, the logic underpinning conservation has changed profoundly, while in other respects, it has remained the same. Today, conservation practices continue to be focused on landscapes, ecosystems, and species that are of high value to humans. What is considered "high value" has expanded in recent years through greater recognition of the importance of biodiversity and non-charismatic species like microorganisms or insects. In addition, conservation areas, increasingly, rarely exclude humans. Rather, new parks and conservation designations generally assume that humans will continue to inhabit and use those places, although conservation goals will curtail some activities. This has come to represent the form of conservation enforced by national and international government bodies. In the 1980s, the Brundtland Commission noted that "The challenge facing nations today is no longer deciding whether conservation is a good idea, but rather how it can be implemented in the national interest and within the means available in each country," (UN World Commission on Environment and Development 1987). Despite these important changes in the logic of conservation, most conservation efforts are still oriented toward minimizing the amount of human disturbance and protecting environments that are appealing to humans. (Chapter 9 explores this in more detail in relation to community forestry in Nepal.)

The idea that the goal of environmental conservation is primarily to ensure use of those environments by humans over the long term has also been challenged. Aldo Leopold (1887–1948), considered by many to be one of the most important American conservationists, posed an early challenge to such assumptions. He argued, in contrast to Pinchot, that humans have moral obligations to nature and therefore must extend rights to the nonhuman world. Drawing on ecology, Leopold explored, in *A Sand County Almanac* and *Sketches Here and There* (1949), how the land can be considered a living organism, or a source of life in a biotic community, and human beings biotic citizens of the land. This perspective forces humans to include plants, soils, animals, and waterways in their moral consciousness, rather than treating nature as an object to be manipulated for utilitarian purposes. His idea that land is a living organism contributed to a resource conservation movement that emphasized the extent to which the manipulation of one component in an ecosystem will affect the ecosystem at large. The goal of conservation therefore is to maintain an ecosystem in equilibrium so that it is not pushed beyond the point of self-repair and self-renewal (see coverage of dynamic equilibrium in Chapter 4). Leopold summed up his work in a land ethic: "A thing is right when it tends to preserve the integrity, stability, and beauty of the biotic community. It is wrong when it tends otherwise" (Leopold 1949, 224–225). Leopold's legacy has been inspirational for rethinking how society–environment boundaries are conceived and has brought ethical discussions about the intrinsic value of nonhumans to sustainability debates. Deep ecology, introduced in Chapter 3, builds directly on Leopold's challenge to anthropocentrism in most conservation thought.

In practical terms, conservation strategies following Leopold's land ethic often focus on preserving wilderness or pristine nature, even if they are not about conserving the natural by excluding humans from interactions with nature. Yet most conservation practices that include humans emphasize that human development is only acceptable as long as ecosystem integrity is maintained, or as long as all species of an ecosystem can thrive free from negative effects. From this perspective, conservation is concerned with restricting or regulating those (human) activities that produce distress, such as the generation of waste products, pollutants, and greenhouse gas emissions, or the overuse or misappropriation of natural resources. Many have therefore argued that conservation is in the interest of *local resource users* as it hinders global actors from such overuse and misappropriation. For example, in Scotland, communities on the outside of designated national parks want the park boundaries expanded since they regard being inside the park as valuable for them economically. The goal of this kind of conservation is not to preserve the natural, but to support the natural values and processes that allow humans to coexist and coevolve with nature.

Despite these efforts to think of humans as part of ecosystems, environmental conservation is often criticized by ecofeminists and activist movements around the world for being too anthropocentric and based upon Global North ideas of environment–society dynamics. They argue that conservation conceived in this way reinforces the idea that humans are somehow separate from the rest of the environment. Rather than overcoming the modernist alienation of humans from nature, conservation serves to cement such alienation by fostering the idea that conservation is inherently about protecting other species from human activities. Many activists and alternative sustainability advocates challenge these notions, as discussed in Chapter 3. They believe that conservation cannot create the fundamental transformations in environment–society dynamics that are required conceptually, politically, culturally, and economically. Worse, local inhabitants and their livelihoods are often portrayed as a threat to the objects of conservation, leading to a variety of coercive and violent relationships between conservation authorities and local populations. As such, conservation unfairly limits access to resources for local people, and participation in conservation policies and projects is coerced and at times enforced by authorities literally wielding military grade weapons. There are reports from game parks in Africa, for example, of park authorities killing poachers on sight with AK-47 rifles when poaching is not a capital crime in those places (see Büscher 2016; Peluso 1993; and Neumann 2004 for detailed examples).

For more analyses of how national parks are sites of violence against local populations see:

- "Conservation Meets Militarisation in Kruger National Park." Conservation and Society. www.conservationandsociety.org/article.asp?issn=0972-4923;year=2015;volume=13;issue=4;spage=356;epage=369;aulast=Lunstrum.
- "The Forgotten History of 'Violent Displacement' that Helped Create the National Parks." The Huffington Post. www.huffingtonpost.com/entry/national-park-service-anniversary-indigenous-people_us_55dcdd7ce4b0a40aa3ac9998.
- "The Human Victims in the Fight Over Rhino Poaching in Africa." Al Jazeera America. http://america.aljazeera.com/multimedia/2015/1/the-human-cost-ofrhino poaching.html.

Considering the development of conservation from Pinchot to modern national parks, which are aimed at economic as well as environmental sustainability (like those in Scotland), one can see that conservation debates are underpinned by questions of whether humans are considered to be internal or external to the natural environment, and what rights we as humans have to utilize and manipulate the natural world to our benefit. While some would say that the object of conservation is human development, others would say it is the natural state that is to be conserved. Yet, if human beings are indeed biological citizens of the land, at what point does the natural become unnatural and hence in need of repair? The answer is as political as it is philosophical. In the case study chapters in Part Two we take a closer look at how humans live with their environments in different parts of the world and see the wide range of not only environment–society dynamics but also the implications of different efforts to shape or improve such dynamics. In Chapter 15, for example, Matthew D. Turner explores transhumance in the Sahel, where pastoralists follow the resource (livestock), rather than containing and manipulating it for human use. Chapter 12 shows how permaculture, an agricultural practice sensitive to local ecological circumstances, is based on a narrative of mutually dependent and integrated socio-ecological relations whereby humans are positioned as participants in a system in which other parts of that system (animals, plants, waterways, etc.) also have significant rights and roles to play.

Conservation is therefore not an apolitical form of environmental management. There is no question that many conservation projects have been vital for protecting some ecosystems and species, but these gains have been accomplished by bringing ecosystem components (including humans) into particular relationships with each other. It is these relationships that can never be politically neutral, whether we deem them beneficial or not. Conservation strategies often transfer control over resources from the grassroots to higher authorities and set limits on the type and intensity of human activities within conservation area boundaries. In this sense, conservation is as much about spatial distribution of environmental control as it is about temporal durability of ecosystem processes. Conservation thus embodies struggles over what to conserve, in what state, for what purpose, and for whose benefit.

## The Commodification of Conservation

Paradoxically, conservation sites often become a matter of commoditization, drawing local areas into the global economy. Not only can they be a source of revenue through tourism, but they also attract foreign aid and development assistance, and become strategic targets of international agreements. Consequently, they are also subjects of trade and, increasingly, market integration that crosses scales. For example, the climate change mitigation program Reducing Emissions from Deforestation and Degradation (REDD+) is designed to trade conservationist behavior in the Global South in return for funding from the Global North. In effect, REDD+ has created a new commodity, carbon stored in forests and soils (discussed above) that is rescaling the global problem of carbon mitigation to more localized scales. While this may seem like an efficient way to promote mitigation activities, fast-growing, young forests sequester more carbon than slow-growing, old forests, and programs often come into conflict with other local-level livelihood strategies. As a result, goals around the enhancement of forest carbon stocks can result in conversion of old forests into industrial plantations and displacement of people's livelihoods. This example suggests that caution is needed: the commodification of environmental goods and services sometimes results in unintended outcomes. From many points of view, the conversion of mature forests to industrial plantations of fast-growing trees is not the most desirable outcome even if the numbers indicate that more carbon has been stored.

REDD+ projects also illustrate how conservation is becoming a process whereby economic growth and ecological decay are reconciled through technological innovation and market integration. Conservation becomes, in other words, a practice of ecological modernization, allowing governments to retract from environmental regulation and rely on market forces to regulate instead. As the case study of oil exploration in the Amazon shows (Chapter 13), market forces can act to create sustainability outcomes within an ecological modernization framing that contrasts sharply with the original goals of sustainability-related activist movements. Some analysts have suggested that the involvement of local people will overcome such contradictions, but we remain less convinced.

## Participation

Participatory development emerged because top-down and expert-oriented development practices met with resistance from local resource users. To be successful, experts argued, development interventions need to be inclusive and locally anchored, ensuring that resource users have influence and control over planning and implementation. Participation came to represent a socially sensitive way to achieve this. Participatory development, according to Anne M. Larson and Fernanda Soto, can be divided into three types: (1) decentralization of government authority, (2) devolution of responsibility to local communities, and (3) local participation to mobilize indigenous knowledge for sustainable self-governance. The participatory process is often described in terms of co-management and co-production, or the sharing of knowledge and responsibilities for natural resource management among actors at multiple levels to ensure equity and accountability. In many respects, participation marks a new form of politicization of sustainable development and a reversal of hierarchies of power. Under ideal circumstances, it shifts power hierarchies so that local and marginalized people have greater control over and access to environmental governance. At the same time, it also serves to authorize various higher-level authorities charged with overseeing large numbers of local user groups (see Chapter 9 on community forestry in Nepal), and to authorize locals to lead such groups. As we explore below, both these relationships have introduced new political and sustainability challenges into participatory processes.

Agenda 21 set the stage for participatory sustainable development in 1992. At the time, decentralization policies were being rolled out in many countries of the Global South, and development practice was shifting from top-down, external interventions, to projects that explicitly sought the inclusion of local people. Participation was recognized not only as a means for sustainable development, but also as an end in itself. Participatory environmental management typically reallocates responsibility for resource management to local users by devolving governance mechanisms and financial resources in order to make local users respond appropriately to environmental problems as they manifest locally. The lowest level institution is often a group of geographically bound resource users with common characteristics, which forms a community or a user group. Local participation has been celebrated as a democratic approach to development, and as such, sustainable. Participation was intended to help address the scale problems that arise when cross-level interventions can result in competition, domination, or collaboration. By more clearly authorizing local-level groups and having clear rules on how higher-level authorities oversee those groups, participatory development resolves problems of who has authority for what, and at what level.

Despite gains made from the decentralization of environmental governance through participation, a number of criticisms require careful consideration. In participatory development schemes, communities by default are seen as uniform, spatially bound, and fixed in a local socio-political matrix, which can conceal power relations and mask patterns of oppression and domination based on gender, ethnicity, caste, and class. The case of community forestry in Nepal detailed in Chapter 9 shows this clearly. While community forestry gives management rights to local communities, the government forest office also retains oversight of these groups. In addition, the committees that emerge at the local level are usually dominated by educated, local elites, most often men. In community forums, many marginalized community members may in fact speak up, but are unable to influence decisions, in effect only sitting in, rather than participating. Participatory approaches can, in this sense, contribute to a cooptation of power by elites at different levels.

Another problem with participatory projects is their tendency to pin down who uses what resources in time and space. This often leads to a mismatch between the scale of resource use and the scale of resource governance. Nomadic communities such as Sahelian pastoralists (discussed in Chapter 15) building their livelihoods around mobile resources are a good example of these mismatch problems. Pastoralists have long been seen as obstructive to sustainability efforts. Neither the community nor the resources they utilize are easily bound, making it difficult to define property and use rights and to determine responsibility for environmental governance as a result. In fact, many (particularly African) governments are trying to make nomadic populations sedentary in order to make them governable. In many cases, such efforts have imposed demands that push nomadic populations to adopt a fixed habitat or operate within a given geographical territory in the name of sustainable development. Forced sedentarization then requires vulnerable communities to adapt to a changing climate, and alters the way people use, manage, and relate to nature. What is considered an inclusive and locally embedded development practice can, in other words, contribute to increased vulnerability at the local level.

Participation can also shift power relations from vertical (global–local) to horizontal (local–local). Latent power relations often materialize and contribute to increased inequalities and the construction and deconstruction of communities around participatory development projects. Two new categories of cross-level interactions have been introduced: incorporation (market integration) and cooptation by local elites. Many empirical examples show how local strongmen can become aligned with the interests of national and international actors in a way that the rhetoric of participation conceals. This is referred to as the centralization of power in the name of decentralization.

Participatory projects also draw local and often vulnerable populations into the development processes and, in turn, the global market economy. REDD+ projects are one example, and community forestry, covered in Chapter 9 in Nepal, is another. Both projects have focused on the selling of forest products (carbon in the case of REDD+ and various non-timber forest products in the case of community forestry) as tools to achieve both conservation and development. Yet marginalized people rarely have much power to set their own terms when they enter into such market relations. Furthermore, their resources are brought into global markets that are characterized by demands for growth and economic activity. This has led critics to suggest that participation shuts down alternative paths of development. Perhaps most deviously, it sidesteps the real challenges of how to empower people to participate in the articulation of their own future, and replaces that with participation in economic development only. These pitfalls of participation have led Bill Cooke and Uma Kothari to ask whether participation had become a tyranny in their book, *Participation: The New Tyranny?* (2001).

These drawbacks don't invalidate participation as a strategy; they just illustrate that no simple ways exist through which democratic inclusion and sustainable development can be wed. Chapter 12 discusses a permaculture project in Zimbabwe wherein the local community has tried to address head-on these problems of market integration and local–local as well as local–global relations. While the community is trying to address them, relations remain constant challenges.

Another critical problem concerning participation is the way that local people's ways of life are enrolled into global processes. It is a common belief that indigenous peoples have a closer connection with nature or cultural attachments to place, and these attachments are used to help motivate and foster sustainable development projects. While on the surface this seems like a way to respect and take seriously local knowledge, in practice it often ends up being another form of tyranny. Local knowledge, which is embedded in practice, history, and culture, becomes compartmentalized into knowledge of local plants that might have medicinal uses, for example. Mobilizing local knowledges and ways of life does not necessarily lead to long-term sustainability, nor does it necessarily respect the foundations of those knowledge systems and ultimately the lifestyles they give rise to (see also Chapter 13).

## The Role of Markets and Public–Private Partnerships

Development practice shifted somewhat towards the end of the millennium. Economists found positive correlations between economic growth and societal welfare indicators, such as levels of democracy and equality. In addition, the economies in southeast Asia (so called "Asian Tigers") and Brazil, Russia, India, and China (so-called "emerging powers") were showing such strong rates of growth in the 1990s that globalization emerged as the new hope of sustainability. If Agenda 21 helped to cement a commitment to participatory, bottom-up, decentralized development, the success of these fast-growing economies helped to cement a belief in neoliberal, deregulated markets as the mechanism of choice for fostering development. This section probes this turn towards economics-led sustainability through fostering of markets, the push for public–private partnerships, and corporate sustainability.

### Economic Growth as a New Sustainable Development Regime

In the early 1990s economists found a positive relationship between income and environmental quality, known as the inverted-U Environmental Kuznets Curve (EKC). According to the EKC, environmental degradation and pollution increase at an early stage of economic growth, but decrease again when a certain level of income (measured by GDP per capita) is reached. The logic underpinning this correlation is that poor countries fast-track industrialization processes and infrastructure development based on the use of cheap energy sources. In other words, energy-intensive and environmentally destructive activities are assumed to be necessary for developing countries to take off economically. Once a certain level of wealth is reached, however, nations can afford to invest in green technologies and environmental protection. As the World Bank put it in 1992, "As incomes rise, the demand for improvements in environmental quality will increase, as will the resources available for investment" (39). This paved the way for private interests promoting business for development and justified economic growth as a means to environmental improvements. Importantly, environmental concerns, in this model, are secondary or at least assumed to be impossible to achieve without first obtaining high levels of industrialization.

These findings have led some to argue that poor nations should be allowed to emit more carbon dioxide and pollutants than rich nations. Note that this argument is based not on climate or distributive justice, but on economic rationality. From this perspective, the best sustainability practice is to initiate market reforms to privatize and deregulate public assets and operations. This practice then sets up the conditions to ensure the safe and free operations of economic processes, which will evolve into long-term environmental sustainability.

However, several factors complicate the relationship between economic growth and ecological benefits, which can be described in temporal and spatial terms. First, the EKC does not incorporate feedback from environmental degradation and resource depletion into economic production. A more holistic view insists that the costs to the environment be factored into economic growth models. Second, environmental degradation can be irreversible and can thus contribute to intergenerational inequity and undermine the resource base in the future. In other words, satisfying more immediate material conditions may not be related to long-term well-being.

Furthermore, suggesting that economic growth leads to long-term environmental protection also disregards the spatial connectivity of production chains. Today, a product is rarely produced and consumed in an isolated system. Externalities of production, such as waste, pollutants, and greenhouse gases, often leave the site of production only to end up in other areas. Another factor to consider is that consumption levels increase with income, implying that, as the income level increases, so will production (albeit often in another place) and its accompanying environmental imprints. While many Western countries are proud of their light "ecological footprint," most production occurs in other parts of the world, and, increasingly, waste is exported as well. Environmental problems, especially those related to production, are thus not containable in space and time, making assessments of the costs of wealth highly inaccurate if they do not account for these spatially and temporally dispersed impacts. As discussed in Chapters 3 and 4, this makes it very difficult to measure environmental sustainability, and therefore to establish relationships between production and environmental protection. Advocates of a more holistic approach to sustainability insist that these externalities must be included in any assessment of environmental costs for highly industrialized countries.

Another critique of assuming economic growth will produce sustainability is the risk of undermining the idea of "public goods." Public goods can be understood as resources that are held in common, not privatized or put into common property regimes because they are recognized as important for everyone. The "public" in "public goods" can be both local users and the global community as a whole. Since what are considered public goods varies along temporal and spatial scales, it is difficult to ensure their protection when market forces are promoted as the best mechanisms for sustainability. As conservation practices like carbon payments show, even the most eco-sensitive strategies can be commoditized, turning resources that had been available for everyone, or even just considered to have intrinsic value in themselves, into products that are incorporated into global markets.

Many have pointed out how economic globalization and spatial connectivity facilitate international corporations' access to natural resources and low-cost labor in peripheral areas where environmental regulations and social standards are minimal. Simultaneously, however, spatial connectivity also provides for the articulation and implementation of environmental and social standards across scales. International agreements and trading regulations can help to raise the environmental standards of poor countries that often have few regulations or difficulties enforcing the regulations they do have. Furthermore, it is believed that the private sector has the means and capacity to implement change through market mechanisms and thereby reach corners of the globe to which governments do not have easy access. At an event organized by Business Action for Sustainable Development in 2002, during the World Summit on Sustainable Development, Kofi Annan noted that "it is only by mobilizing the corporate sector that we can make significant progress. . . . The corporate sector has the finances, the technology and the management to make all this happen. The corporate sector need not wait for governments to take decisions for them to take initiatives" (UN Department of Public Information). What he suggests is that the public sector is now dependent on the resources possessed by the corporate sector for the implementation of sustainable development interventions.

## The Push for Public–Private Partnerships

The shift to economic growth within development practice has not been without some contradictions. Nation-states still carry the responsibility for the health of their environments and populations and these duties can only be partially handed over to private entities without governments losing significant legitimacy in the eyes of their citizens. This contradiction has led to the idea of public–private partnerships wherein public bodies join with private-sector actors to make investments into everything from hospitals, schools, and bridges to businesses that seek to promote green industries. Within the aid community, this has become known as Business for Development (B4D). B4D aims at stimulating economic development by creating business-friendly environments, facilitating investment, and setting up public–private partnerships within and across nations. On an increasing scale, businesses and corporations are now the agents by which sustainable development is enacted.

For more information on Business for Development (B4D), see this website: http://businessfordevelopment.org/.

Public–private partnerships and the promotion of economic growth to achieve sustainability are, like all the issues explored in this book, complex and do not present simple or nonpolitical outcomes. In many cases, it is the state or the local government that is eventually held accountable for repaying the costs of infrastructure investment. As a result, many

The Skye Bridge in Scotland, purportedly the most expensive road bridge in Europe, was constructed under the auspices of a public–private partnership, which passed on the costs to users in the form of high bridge tolls. The Scottish government abolished the fees in December 2004, nearly ten years after its construction.

*Source:* Mike Peel/Wikimedia; https://commons.wikimedia.org/wiki/File:Skye_bridge,_2010.jpg.

public–private enterprises have led to radically increased costs for bridge tolls or hospitals that users are expected to pay. There have also been criticisms over the loss of standards and controls that can come through such partnerships. On the other hand, getting local entrepreneurs to invest in community enterprises can help to keep capital local and provide for much-needed enterprise development.

## Corporate Sustainability

Corporate sustainability can also be understood as one component of a new system of governance, wherein it is no longer states alone that are responsible for the welfare of the planet and its inhabitants, but also the corporate world. Many governments have therefore seen it as a duty to assist the private sector in providing social services. In countries lacking governments willing or capable of providing social well-being, some argue, corporations might have the capacity to do that in their place. This raises a number of important questions: To whom are they accountable? Why would they care about the environment? Can we make them care?

Corporate sustainability revolves around aligning profit-oriented activities with social and environmental responsibilities. This logic marks a radical departure from Agenda 21–inspired participatory, inclusive user groups expected to provide equitable and accountable environmental governance. Businesses are believed to be efficient wealth producers, not despite their profit-oriented strategy, but because of it; their abilities to promote social and environmental welfare are assumed to follow from their economic contributions. This privileging of economic growth produces clear divisions among the three pillars of sustainable development—environment, economy, and society—not by putting them in a triangle or as overlapping domains, but in a hierarchical order where one is supposed to lead to another.

There are several forms of corporate sustainability. We have already explained public–private partnerships, but in addition, corporate sustainability can refer to market and private-sector driven development as well as efforts within corporations to make their operations more socially and environmentally sustainable, often known as corporate social responsibility or CSR. While terms such as "corporate social responsibility" (CSR) or "corporate citizenship" are still used, corporate sustainability is gaining traction as a concept signifying the combination of social and environmental considerations within the logic of corporate profit. Fundamentally, corporate sustainability is a form of self-regulation to contribute to social welfare, environmental protection, or fair economic distribution.

For more information on Corporate Social Responsibility see:

- "What is Corporate Social Responsibility?." *Business News Daily.* www.businessnewsdaily.com/4679-corporate-social-responsibility.html.
- "CSRlive: Corporate Social Responsibility." CSR Wire. www.csrwire.com/corporate_social_responsibility.

For a critical view see:

- Aneel Karnani. 2010. "The Case Against Corporate Social Responsibility." *Wall Street Journal.* www.wsj.com/articles/SB10001424052748703338004575230112664504890.
- Deborah Doane. 2005. "The Myth of CSR." *Stanford Social Innovation Review.* http://ssir.org/articles/entry/the_myth_of_csr.

Corporate sustainability can imply community involvement, socially or environmentally sensitive production, or employee relations and labor standards. Corporate sustainability has, in this sense, transformed from philanthropy to a component of business and management strategy itself, evidenced by the fact that more and more corporations communicate their CSR activities to stakeholders and consumers. Corporate sustainability is thus in part a strategy for seeking legitimacy on the market and gaining competitive advantage by taking responsibility for externalities such as waste products or pollutants. In part, it is also a strategic necessity for ensuring that the resource base can keep providing inputs to future production.

Corporate sustainability thus needs to be understood as more than abiding by laws and regulations; it is also good for business. Advocates further argue that shareholders are increasingly demanding that products such as coffee, chocolate, and fruits are produced in an environmentally and socially sensitive manner. These demands are pushing corporations to invest in fair production and good employee environments to satisfy consumers. Of course, these products tend to command a premium price as well, ensuring that such investments stay within the profit logic of corporations. It can therefore be useful to distinguish between proactive and reactive responsibility. While proactive responsibility ensures that social–ecological systems are maintained, reactive responsibility is more about disaster or image management.

Critiques of corporate sustainability abound, and while the meaning of corporate sustainability is contingent upon its application, it is deeply contested. Some critics argue that corporate sustainability is inherently contradictory, as profit is often incompatible with environmental and social protection. Many point out how business managers' responsibilities are to shareholders, and shareholders only, which makes corporations incapable of true social responsibility. The economist Milton Friedman argued in a 1970 *New York Times* article, with the telling title, "The Social Responsibility of Business Is to Increase Its Profits," that corporations should leave the provision of social services to governments. The power of consumer demand in relation to the ability of corporations to dictate market conditions has been seriously challenged. In addition, by referring to consumer demand, the responsibility for corporate activities is displaced from boardrooms to the general public. Perhaps most vociferously, critics have pointed to the danger of corporations using the rhetoric of sustainability for image-making or greenwashing purposes. Others claim that corporate sustainability is a strategic necessity. This debate leaves us with a fundamentally normative question: If corporations bring socio-ecological benefits, does the intent matter?

## Conclusion

In this chapter we have probed how sustainability is scaled. Scale has consequences not only in terms of whether sustainability projects can be considered successful (i.e., a local solution may depend on producing waste, unequal social relations, and other unsustainable relations elsewhere), but also in terms of which communities of practice are invested in addressing them. Enacting sustainability necessarily involves enrolling differently scaled actors into projects that help to bring everyone towards a common set of goals. Early sustainability projects had a strong emphasis on conservation of ecosystems and participation by local people. More recently, attention has been paid to the private sector and the need to enlist corporations in a sustainability agenda. Such attempts often shift the emphasis away from ecological sustainability and more towards economic sustainability.

From a globalization perspective, the issue of scale becomes paramount. First, a number of justice questions emerge when people around the world become invested in and attempt to govern the uses of local environments. We see this occurring with REDD+ schemes and other efforts at addressing climate change. Second, concerns are being raised over whether the scale at which problems are defined and the scale at which we are attempting to address them match. Global problems are also often local problems and require local action, yet local actions do not necessarily add up to a global response. The next chapter investigates the range of environmental concerns and explores a number of challenges to this localized framing of sustainable development, both in terms of the scale of the problem and in terms of governance.

## References and Further Research

Ahlborg, Helene, and Andrea J. Nightingale. 2012. "Mismatch Between Scales of Knowledge in Nepalese Forestry: Epistemology, Power, and Policy Implications." *Ecology and Society* 17, no. 4: 16.

Brockington, Dan. 2002. *Fortress Conservation: The Preservation of the Mkomazi Game Reserve, Tanzania*. Oxford: James Currey.

Büscher, Bram. 2016. "'Rhino Poaching Is Out of Control!' Violence, Race and the Politics of Hysteria in Online Conservation." *Environment and Planning A* 8, no. 5: 979–998. https://doi.org/10.1177/0308518X16630988.

Cooke, Bill, and Uma Kothari. 2001. *Participation: The New Tyranny?* London: Zed Books.

Cronon, William. 1983. *Changes in the Land: Indians, Colonists, and the Ecology of New England*. New York: Hill and Wang.

———. ed. 1996. *Uncommon Ground: Rethinking the Human Place in Nature*. New York: W. W. Norton.

Friedman, Milton. 1970. "The Social Responsibility of Business Is to Increase Its Profits." *The New York Times Magazine*, September 13, 2017.

Goldman, Mara. 2009. "Constructing Connectivity: Conservation Corridors and Conservation Politics in East African Rangelands." *Annals of the Association of American Geographers* 992: 335–359. https://doi.org/10.1080/00045600802708325.

Larson, Anne M., and Fernanda Soto. 2008. "Decentralization of Natural Resource Governance Regimes." *Annual Review of Environment and Resources* 331: 213–239. https://doi.org/10.1146/annurev.environ.33.020607.095522.

Leopold, Aldo. 1949. *A Sand County Almanac and Sketches Here and There*. New York: Oxford University Press.

Neumann, Roderick P. 2004. "Moral and Discursive Geographies in the War for Biodiversity in Africa." *Political Geography* 23: 813–837.

Newell, Peter. 2012. *Globalization and the Environment: Capitalism, Ecology & Power*. Cambridge: Polity Press.

O'Neill, R. V., and A. W. King. 1998. "Homage to St. Michael; or, Why are There so Many Books on Scale?" In *Ecological Scale. Theory and Applications*, edited by D. L. Peterson and V. T. Parker, 3–16. New York: Columbia University Press.

Ostrom, Elinor. 1992. "The Rudiments of a Theory of the Origins, Survival, and Performance of Common-Property Institutions." In *Making the Commons Work: Theory, Practice, and Policy*, edited by D. Bromley, 293–318. San Francisco: ICS Press.

———. 2009. "A General Framework for Analyzing Sustainability of Social–Ecological Systems." *Science* 325, no. 5939: 419–422. https://doi.org/10.1126/science.1172133.

Peluso, Nancy. 1993. "Coercing Conservation?" *Global Environmental Change* June: 199–217.

Pinchot, Gifford. 1910. *The Fight for Conservation*. New York: Doubleday, Page.

———. 1937. "How Conservation Began in the United States." *Agricultural History* 114: 255–265.

Roth, Robin J. 2008. "'Fixing' the Forest: The Spatiality of Conservation Conflict in Thailand." *Annals of the Association of American Geographers* 982: 373–391. https://doi.org/10.1080/00045600801925557.

Sandbrook, Chris, et al. 2010. "Carbon, Forests and the REDD Paradox." *Oryx* 44, no. 3: 330–334. https://doi.org/10.1017/S0030605310000475.

Stern, David I. 2004. "The Rise and Fall of the Environmental Kuznets Curve." *World Development* 328: 1419–1439. https://doi.org/10.1016/j.worlddev.2004.03.004.

Stern, David I., Michael S. Common, and Edward B. Barbier. 1996. "Economic Growth and Environmental Degradation: The Environmental Kuznets Curve and Sustainable Development." *World Development* 247: 1151–1160. https://doi.org/10.1016/0305-750X(96)00032-0.

Thompson, Paul B. 2010. *The Agrarian Vision: Sustainability and Environmental Ethics*. Lexington: University Press of Kentucky.

United Nations Department of Public Information. 2002. "Summit Historic Opportunity to Further Business Role in Sustainable Development Says Secretary-General in Remarks to 'Business Day' Event." World Summit on Sustainable Development. New York. www.un.org/events/wssd/summaries/envdevj15.htm.

United Nations Environment Program. 1992. *Agenda 21*. New York: United Nations.

United Nations World Commission on Environment and Development. 1987. *Our Common Future. (The Brundtland Report)*. New York: Oxford University Press. Available electronically from the UN: www.un-documents.net/our-common-future.pdf.

West, Paige, James Igoe, and Dan Brockington. 2006. "Parks and Peoples: The Social Impact of Protected Areas." *Annual Review of Anthropology* 351: 251–277. https://doi.org/10.1146/annurev.anthro.35.081705.123308.

World Bank. 1992. *World Development Report 1992: Development and the Environment*. New York: Oxford University Press.

# 6

# The Range of Environmental Concerns

Andrea J. Nightingale, Linus Karlsson, Laura Saxer, and Ben Campbell

In this chapter, we map out the spatial and conceptual range of environmental concerns that relate to sustainability and sustainable development in a globalizing world. We have organized the discussion into three themes: fragile earth, supporting ecosystems, and energy and technology. Our purpose is to be illustrative rather than comprehensive and to show how the themes around scale introduced in Chapter 5, and the four core themes of the book (see "The Conceptual Framework of the Book" in Chapter 1) highlight the complexity of sustainability questions. The analysis here should be understood as a conceptual guide that can be applied to other kinds of sustainability issues. Learning how to analyze sustainability of what, for whom, and at what cost is particularly important given the rapidly changing landscape of environmental concerns as the Sustainable Development Goals and the Paris Agreement on climate change are implemented over the next several years.

We start with the *fragile earth*—the planet as a whole—to explore how different perceptions of the natural world have produced and informed responses to environmental concerns. The role of charismatic species in sustainability debates is introduced and connected back to conservation areas introduced in Chapter 5. We then look at the idea of *ecosystems supporting life*. Within this section, we focus particularly on ecosystem services and counter-narratives in *buen vivir* ("living well") and similar conceptualizations, food production, and climate change. Finally, we delve into *energy and technology* and probe several different dimensions of technology that are at the forefront of sustainability debates. In that section, we begin by examining energy and its importance to our present ways of life and concerns over energy technologies within sustainability debates. We then introduce the sustainability issues around genetic engineering and nanotechnologies. These discussions reveal tensions and contradictions within and among approaches to sustainability. As such we return to the core themes of this book introduced in Chapter 1: narratives of environment and sustainability shape what is considered a problem that demands attention; these narratives bring with them a range of justice and governance questions that play out in uneven ways across the globe.

## Fragile Earth

The notion of the earth as a fragile ecosystem has a very long history, as we outlined in Chapter 2. This notion was thrown into relief in 1972 by the famous photograph of earth, taken by U.S. astronauts aboard *Apollo 17*, showing a blue and green planet floating in empty space, an image known as the "blue marble." On the one hand, the photograph became a symbol of the fragility of the planet in that the photo clearly shows its outer boundaries, making it possible to see firsthand that the carrying capacity of planet earth was indeed finite. On the other hand, the photograph seemed to show a manageable system. Since the boundaries of the planet were clearly visible, many people believed that humans also could control and manipulate it to their advantage. The narrative of a global community nurturing a fragile home planet has remained a powerful image ever since, contributing to views that humans are the stewards of the planet, responsible for maintaining its health and supporting its life-supporting processes.

For a color image of the earth known as the "blue marble," refer to "Apollo 17: Blue Marble" web page from NASA:

www.nasa.gov/image-feature/apollo-17-blue-marble.

The "blue marble," an image of earth taken from space by *Apollo 17* astronauts in 1972, has come to symbolize the fragility of the planet.

*Source:* Yvette Smith/NASA; www.nasa.gov/imYage-feature/apollo-17-blue-marble.

Throughout the twentieth century, geoscientific knowledge taught the world that the physical, chemical, and geological processes that sustain and balance life on earth are destabilized by human activity. Today these lessons appear to be even more true as the effects of human practices increasingly take place on a global scale. Such human behaviors as habitat destruction (e.g., in tropical forests, fresh waters, or coastlines), introduction of invasive species, biodiversity loss, overexploitation of natural resources, industrial pollution, and recent manifestations of global climate change, are all seen as major threats to both biological and cultural diversity. The root causes of such problems are believed to stem from humans' failure to stay within planetary boundaries and uphold a balance between life-sustaining ecological processes and human development. These narratives around limits introduced in Chapters 3 and 4 are powerful metaphors through which the idea of the fragile planet is perpetuated. While we do not question the importance of these very real biophysical transformations, in this book we are most interested in thinking about the consequences of how these problems are framed, by whom, and what sorts of actions are therefore proposed to deal with them.

## Earth as a Closed System

Many of our assumptions about fragile ecosystems and the earth as a closed system come from studies of islands. In 1976, the social anthropologist Margaret Mead noted that an isolated island constitutes a social–ecological model of the planet as a whole, but before her, early naturalists were fascinated by the unique ecosystems and species that had developed on the very large islands of Australia and New Zealand. Since then, many have drawn parallels between island communities, surviving in closed social–ecological systems surrounded by water, and the earth as an equally isolated system in space. Similar to islanders, it is argued, the global community has to manage the environment wisely in order for it to sustain life, habitats, and ecosystems. That populations have survived on secluded islands, many believe, is a testimony to the ability of humans to make informed decisions on how best to govern limited resources and sustain them over time in fragile ecosystems.

However, the island metaphor has also given rise to various collapse theories. Rapa Nui (Easter Island) in the southeastern Pacific Ocean is often brought up as an example of how an island population can outgrow the carrying capacity of an ecological system and self-destruct under population growth and overexploitation of natural resources in a Malthusian manner. The proposed explanations of why the ways of life on Rapa Nui collapsed are many. Some believe that deforestation and habitat destruction culminated in an ecocide, where exploitation by the inhabitants produced an uninhabitable ecosystem. Others believe that the demise of the community was caused by disease and pests brought to the island by foreign merchants. These contrasting narratives suggest informative possibilities. In the first one, Rapa Nui is assumed to be a closed system and if the ecosystem there could no longer support the population, no other means of feeding themselves would exist. While part of this could certainly be the case, a lot of evidence also exists that islanders in the Pacific, in addition to being skilled fishers, were engaged in trade networks among quite far-flung islands. The second narrative presents Rapa Nui as an open system, exposed to outside influences, not all of which were beneficial. Given the evidence for inter-island exchange in the Pacific, it seems more accurate to view it as an open system, even if that does not prove that disease was the prime cause of the population's collapse. Even in an open system, ecological destruction could have precipitated the decline in Rapa Nui society, and of course, a combination of these two factors is also quite likely—namely, that disease combined with a declining resource base led to Rapa Nui becoming uninhabited.

Many have argued that the earth is comparable to Rapa Nui. "We only have one planet" has become a mantra in modern sustainability thinking. Starting as a catchphrase used by environmental movements, this concept is now gaining traction in both political and corporate circles. The use of measurements such as ecological footprints, discussed in Chapter 4, is a clear example of this. Yet, even if it is true that we only have one planet, those words actually conceal a debate over how to understand the earth, how it functions, and what impact human activities have on the so-called "earth system." While some believe that the planet is fragile and at risk of collapse, others believe the island metaphor is misguided, and that the planet has capacity for self-renewal and self-repair. However, other narratives point to the influx of energy from outer space. Light energy from the sun, which drives the entire planetary system (discussed below), is the most plentiful incoming resource, but smaller amounts of space dust as well as asteroids and meteoroids enter the earth's system through the atmosphere, bringing with them new material. Conceptualizing earth as a closed system from a material point of view is therefore not entirely accurate.

Evolutionary biology also presents a narrative that all species strive to sustain and expand their habitats. Some believe that the human species has succeeded too well in expanding its habitat into nearly every realm of the planet, so well that it threatens not only many other species, but also its own existence. It is from such concerns that sustainable development has become a defining concept of our time, trying to reduce the conflicts between human development and ecosystem degradation.

Economic globalization is said to further increase the fragility of the planet as it facilitates redistribution of environmental problems globally. Capital, information, commodities, and norms all transform the way humans perceive and use the environment and natural resources. Via human interactions, destructive behaviors as well as cultural values can travel around the world, transforming society–environment dynamics. This is part of the reason we insist that society and environments cannot be understood as separate entities merely exchanging information and resources. Neither can the biophysical planet be understood as isolated from social networks. Rather, they intertwine to form global socionatures tied together by biophysical, physico-chemical, politico-legal, and socioeconomic currents that can both provide stability and create new forms of fragility. In this view, the idea of fragility begins to seem a bit too simplistic. It is possible for the earth to be both fragile *and* self-renewing at the same time. Our perspective in this book also draws attention to social–political and cultural processes when evaluating fragility and the impacts of our actions on earth.

## Earth as a Resilient System

In contrast to narratives of fragility and collapse, a parallel narrative presents the earth as a resilient system with abilities to withstand significant shocks and transformations. The historical record indicates, for example, that the extinction of one species does not necessarily lead to the collapse of a whole ecosystem, but rather a re-composition of it. The American chestnut and the dodo bird, to take two quite different examples, disappeared without any measurable effects on ecosystems, although cultural nostalgia over their demise has been profound. The decline of a predator can create spaces for other species to expand their habitats, such as the expansion of white-tailed deer with the rapid decline in wolf populations in northern states of the United States in the second half of the twentieth century. The destruction of a natural habitat can also bring new life. For instance, now many wild animals are found in urban areas, and military firing ranges in waters immediately off the coast of Scotland appear to be teeming with marine life. Similarly, wildlife has flourished by the old reactor facilities after the Chernobyl nuclear meltdown in 1986 despite the continuing high radiation levels as described in William Cronon's book, *Uncommon Ground* (1996).

These examples draw into question just how fragile life is on earth, but at the same time, many of them lend support to the idea that humans are a major transformative force. We create new kinds of habitats for other species, we are one of the most significant predators, and we have altered not only the face of the earth, but increasingly the subsurface environment through mineral resource extraction (i.e., oil, gas, coal, precious and semi-precious stones). New hills have been formed from waste rock left over from coal strip mining, while entire mountains have been leveled from the same type of mining.

## Biodiversity, Charismatic Species, and Emblematic Landscapes

Within the fragile earth narrative, much discussion surrounds smaller scales as well, particularly around biodiversity and charismatic species. Here, sustainability concerns work across, and in some sense merge, scales. So while many species live within fairly confined habitats at smaller scales—obvious exceptions to this include, among others, many species of birds and fish—when discussing biodiversity, global narratives about protecting species are strong. Indeed, biodiversity as an idea is far less about individual species, and more about how many different kinds of species exist in a given spatial unit, whether that be a particular place, or the earth itself.

Since the 1980s, biodiversity conservation has become an increasingly popular strategy to ensure that human activity does not contribute to the extinction of any species. The strategy was firmly rooted in the global agenda during the Rio Earth Summit in 1992 (see Chapter 2) when the Convention on Biological Diversity (CBD), or simply the Biodiversity Convention, was opened for signature. Since then, it has been ratified by all UN member states (with the exception of the United States). The definition of biodiversity that was elaborated during the Earth Summit, and included in the CBD, is probably the most widely accepted one. It understands biodiversity as the "variability among living organisms from all sources including, inter alia, terrestrial, marine and other aquatic ecosystems and the ecological complexes of which they are part; this includes diversity within species, between species and of ecosystems." Since the biodiversity conservation was popularized in Rio, scientists, politicians, and the public have all turned their eyes to endangered species.

Despite the flourishing of species in rather surprising contexts like Chernobyl and efforts under the CBD, the current rate at which species are disappearing is unprecedented, and most evidence points to human development as the root cause. The biological diversity of life on earth is the outcome of 3.5 billion years of evolution, a process which by definition implies change. More than 99 percent of all species that ever have lived are now extinct. Thus, species extinction is not unnatural. Nevertheless, as a result of industrialization and large-scale landscape transformations, concerns over a mass extinction of species have intensified. Many scientists now predict that a large number of organisms will be extinct by the end of this century if we follow the current path of development, with potentially catastrophic consequences for ecosystems and life itself. One of the main concerns is the permanent removal of genetic resources from the earth's system. Species carry with them unique genetic codes, and while gene sequencing has revealed far more commonality between even very distantly related species, each species carries unique genetic combinations which, once lost, are not likely to reappear. We discuss in more detail some of the issues around genetics below in relation to genetic engineering.

Of concern to us in this section is the basis upon which species are considered worthy of conservation. Not surprisingly, conceptions of resilience and adaptability do not provide clear answers. On one end of the spectrum is the so-called Noah principle that takes an ecocentric view and argues that the very existence of a species is sufficient to warrant its preservation. On the other end are those who argue that only species useful to or valued by humans should be protected. History tells us that even if people increasingly believe biodiversity has intrinsic value, we are more selective in our conservation approaches. For a species to become included in a conservation scheme, it has to be selected through a process that includes a heterogeneous and historically situated array of people, practices, technologies, and data that inform species description, surveillance and research, evaluation, and prioritization. Once priorities are established, all those elements must be translated into an action plan for implementation. Conservation therefore can never be a neutral or value-free practice, it is always informed by history, competing values and worldviews, and a variety of normative judgments that bring some species into active management plans and exclude others.

As we discuss in more detail in the section "Supporting Ecosystems" below, because some species and ecosystems provide vital services for humans, rallying support for their preservation if threatened is easier. Rainforests work as natural climate regulators, wetlands work as natural water purification systems, and coral reefs protect coastlines from storms and erosion, to name a few. Also specific species are vital for human welfare. For example, in recent years the honeybee and other pollinators have received considerable attention by scientists and politicians alike. Honeybees are a vital part of any agricultural system. Even in large-scale commercial agriculture, honeybees pollinate 70 out of the top 100 food crops. In the last ten years, it was discovered that pesticides, pathogens, air pollution, and habitat destruction contributed to what is called Colony Collapse Disorder (CCD), threatening to lead to a dramatic decline in bee colonies. Given the crucial role honeybees play in agriculture, a rapid decline in colonies poses a serious threat to farmers and food producers, as well as to long-term food security. So while on the one hand many would argue for the preservation of species in their own right, in reality, species and ecosystems are valued differently by humans, and those we value more have a greater likelihood of being targeted for preservation.

For more information on honeybee declines see:

- "Bees in Decline: The Risks for Agriculture in Europe." Greenpeace. http://sos-bees.org/.
- "Colony Collapse Disorder." U.S. Environmental Protection Agency. www.epa.gov/pollinator-protection/colony-collapse-disorder.
- "Death and Extinction of the Bees." Centre for Research on Globalization. www.globalresearch.ca/death-and-extinction-of-the-bees/5375684.

The way that humans value different aspects of the environment unevenly is also starkly evident in the importance placed on what are known as charismatic species. Charisma refers to the characteristics of a certain species that condition human perception and evaluation; its detectability; how easy it is to encounter, identify, differentiate, and record by a surveyor; and so on. Plants or animals that communicate visually or orally with humans receive more attention than non-charismatic insects or worms, for example, that are difficult to survey or identify with. The perceived charisma of species is fundamentally anthropocentric since it is linked to how well we can relate to that species. At the same time charismatic nonhumans are also not simply names in a scientific taxonomy the way many microorganisms or other non-charismatic species are. Rather, through charisma, awareness is created about nonhuman differences as well as the importance of nonhuman organisms. For example, Orcas, or

This attractive image of a panda in the San Diego Zoo illustrates why the use of charismatic species helps in conservation efforts.

*Source:* Johann Balleis/Wikimedia; https://commons.wikimedia.org/wiki/File:Xiao_Liwu_im_San_Diego_Zoo_-_Foto_2.jpeg.

so-called "killer whales," became a popular charismatic species after the success of the movie *Free Willy* in 1993. Through the artificial construction of an animal's charisma, species that are feared or even hunted by local populations are reframed as species to love and protect, not because they provide services for humans, nor because they have higher intrinsic value, but because they are forced into our moral consciousness.

Through such harnessing of our moral consciousness, charismatic species are often used as icons in fundraising campaigns by environmental organizations (e.g., the panda used by World Wide Fund for Nature) since they trigger sympathy and awareness, which is beneficial in terms of financial support. For example, in the Unites States over 50 percent of funding for wildlife conservation is leveraged by fewer than 2 percent of the species listed as endangered; the top ten species receiving the most funding are at the same time the most popular charismatic species in the United States. These species include the Florida Panther, the California Condor, and the Northern Spotted Owl as they possess some appealing characteristics for the public. Having understood the success of applying charismatic species internationally in their campaigns, many initiatives are aiming to identify more species by applying systematic frameworks and determining what animal characteristics resonate with the public. The polar bear emerged in the 2000s as a charismatic species to represent and raise awareness about the threat of climate change, for example. Funds raised in the name of a charismatic species are not always used for preserving only that species, but rather go into all conservation efforts by environmental organizations that raise those funds.

As a result, the inclusion or exclusion of species into the conservation agenda largely relies on interconnected environmental, social, political, and strategic factors. So, while the desire to protect biodiversity seems to be a narrative encompassing all life on earth, in reality, a heavy bias exists towards species that are deemed either important or, in the popular vernacular, warm and cuddly, to humans. Loud public outcry against extinction of nematode species are rare, for example, as Jamie Lorimer points out in a 2006 article on biodiversity conservation policies in the United Kingdom.

By selecting what species and ecosystems to protect, conservation interventions contribute to the constitution of future ecosystems and environments. An intricate interplay exists between humans and environment in conservation management. For instance, in order to protect a territory from deforestation, conservationists are often forced to find a rare, charismatic, or service-providing species inhabiting the forest in order for it to become a subject of conservation. The spotted owl that lives in old growth rainforests in Oregon and Washington in the United States is a good example of a charismatic species used to protect the entire ecosystem from logging. This becomes a crucial strategy for conservationists when particular ecosystems are popularly believed to be worthy of conservation in their own right. This example also illustrates how those taking an ecocentric view can borrow from anthropocentric narratives in order to protect certain territories. Environmental management is thus largely a product of how humans understand and relate to nature, and our environmental concerns are fundamentally limited by our knowledge, senses, and moral consciousness.

## The Sustainability of Fragile Earth

Two important contrasts help define the fragile earth debate. On the one hand, the earth is believed to be sustainable in itself and capable of absorbing shocks. Some have gone so far as to argue that the planet can be seen as a single self-regulating organism. In the 1970s, James Lovelock first proposed the Gaia hypothesis, or Gaia theory, suggesting that living organisms coevolve with their abiotic environments to maintain the conditions for life on earth. If a natural cycle or state is disturbed, Gaia theory tells us, other forces will work to bring it back to balance. This view of the earth is much closer to medieval metaphysical understandings of the earth, discussed in Chapter 2, than it is to modern, mechanistic models.

Socionatures perspectives that downplay the idea of human impacts nevertheless can also embrace the idea of fragility. These narratives are similar to the Gaia hypothesis in that they focus on the integration of human societies with the rest of life to explain the co-creation of places that are more or less habitable, or more or less fragile. Of particular concern are the ways of thinking and ways of being that produce a sense of alienation from the rest of life. Such alienation opens the door for a variety of practices and relationships that are detrimental not only to human societies but to the planet as a whole.

On the other hand, within the fragile earth narrative, humans are cast as the agents by which sustainability or unsustainability is produced. How much human activity the earth can endure before it goes beyond the point of self-adjustment remains disputed. Even if some ecosystems have shown resilience to human activity, most scientists now agree that human impacts on nature are too severe for self-adjustment without significant losses of both human and nonhuman life. Climate change can be considered one manifestation of such severe shocks.

This contradiction within the fragile earth narrative creates a dilemma. To return to the debates on conservation in Chapter 5, when does biodiversity or environmental conservation become a form of manipulation of nature and evolutionary processes, and when are they acts of preservation? What ecosystems and species need protection? What ecosystems and species *deserve* protection? While some consider biodiversity to perform vital ecological services for human beings and therefore need protection, others consider biodiversity to be justified in its own right. For some sustainability advocates, the answer to these questions is crucial. In one vein is the "leave nature to take care of itself" school; officially designated wilderness areas in the United States where humans are "only visitors" are a noteworthy outcome of this kind of perspective. In contrast is the "humans need to clean up their mess" school. The U.S. Environmental Protection Agency's Superfund clean-up sites, where attempts are made to restore landscapes heavily polluted by industry, are a good example of how this school has been put into action.

For more information on U.S. wilderness areas and the ethos of leaving nature untouched see the following:

- "The Beginnings of the National Wilderness Preservation System." Wilderness Connect. The University of Montana. www.wilderness.net/NWPS/fastfacts.
- "Wilderness." U.S. Forest Service. www.fs.fed.us/recreation/programs/cda/wilderness.shtml.

For more information on U.S. EPA's Superfund clean-up sites see:

- "Super Fund." U.S. Environmental Protection Agency. www.epa.gov/superfund.
- "St. Louis River–U.S. Steel Superfund Site." Minnesota Pollution Control Agency. www.pca.state.mn.us/waste/st-louis-river-us-steel-superfund-site.
- Nelson, Bryan. "10 Superfund Sites: Where Are They Now?" *Mother Nature Network*. Narrative Content Group. www.mnn.com/health/healthy-spaces/photos/10-superfund-sites-where-are-they-now/toxic-turnaround.

Of course choosing between these two schools isn't necessary; but it is crucial, when proposing sustainability solutions, to think critically about the assumptions that underlie the solutions we support. If life on earth is indeed threatened, what will require our immediate attention? What can wait? Again, we find ourselves in a conceptual and moral web when we try to assess what is sustainable, and what is not. The answers are dependent upon how we perceive environment and the role of humans within it. As we have argued throughout this book, these unspoken perceptions drive which options are chosen and by whom; which knowledges are presumed necessary to make those choices; and who will enforce them through which governance mechanisms. Using global agreements related to sustainability is one approach. But as the discussion on conservation in Chapter 5 showed, these governance questions are far from resolved, and we are not trying to resolve them in this book. We believe they must be decided at multiple scales, by multiple actors, and perhaps with multiple answers. We advocate for continuing to ask these questions and continuing to probe the assumptions that underlie the stance on sustainability each one of us takes.

## Supporting Ecosystems

As we have introduced above (and earlier in Chapter 2), one of the most fundamental concepts within sustainability narratives is the idea that the earth supports human life. Like nearly all dimensions of sustainability debates, this concept represents a number of contrasting perspectives. As already mentioned, the spectrum is bounded on one end by those who advocate for an ecocentric viewpoint: that all life and biophysical systems are valued in themselves. On the other end of the spectrum are anthropocentric approaches that value only those nonhuman elements that support human life. In this section we take a closer look at the concept of supporting ecosystems through ecosystem services. We then contrast this approach briefly to *buen vivir* and similar perspectives that also consider the earth as supporting life, but in substantially different ways.

## Ecosystem Services

The narrative of ecosystem services emerged in the 1990s as represented in the 1997 book, *Nature's Services*, edited by Gretchen Daily. This approach, further cemented in the 2005 Millennium Ecosystem Assessment, has become a dominant concept for understanding the value of different dimensions of ecosystems. Despite its prominence, the narrative is not without question or controversy. Evaluating the importance of ecosystems in terms of their services is a deeply anthropocentric and instrumentalist vision of the world. Ecosystem services is almost the opposite of the Gaia hypothesis in terms of reasserting a mechanistic narrative of the earth rather than a holistic one.

Ecosystem services are organized around four main types as explained on the Millennium Ecosystem Assessment website:

1. **Provisioning services** or the products available from ecosystems such as food, fresh water, wood, fiber, genetic resources, and medicines.
2. **Regulating services** or the benefits attained from the regulation of ecosystem processes such as climate regulation, natural hazard regulation, water purification, and waste management, pollination, or pest control.
3. **Habitat services** or how ecosystems provide a diversity of spaces for species to live.
4. **Cultural services** or benefits for people from ecosystems such as spiritual enrichment, intellectual development, recreation, and aesthetic values.

For more resources and information on ecosystem assessments see:

- Millennium Ecosystem Assessment. www.millenniumassessment.org/en/index.html.
- "Global Biodiversity Information Facility." Govern d'Andorra. www.gbif.org.
- Biodiversity Information Service for Europe. http://biodiversity.europa.eu/.

As the list elaborates, the ecosystem services model divides up the world into sets of benefits that are provided to humans and other species by the biophysical world. Different types of ecosystems are assumed to provide somewhat different qualities and quantities within each of these four services, but without all four, an ecosystem could be considered unsustainable.

Ecosystem services have begun to replace other ways of evaluating the health of ecosystems, particularly within policy and environmental governance circles. Some ecologists, however, are concerned by this trend because understanding ecosystem function is neither straightforward nor agreed upon scientifically. Different analyses may provide completely opposite results, as the understanding of the ecosystem can be very different. What might look sustainable in one narrative becomes unsustainable in another. And perhaps most importantly, evaluations of sustainability are always embedded within certain assumptions about sustainability of *what* and for *whom*, where "whom" needs to be understood as including nonhumans.

For our purposes in this book, understanding how ecosystem services are framed is essential. There is a strong utilitarian and mechanistic underpinning to each of the four services. The Biodiversity Information System for Europe (BISE) website (see accompanying resource box) names cultural services as "non-material benefits that people obtain from ecosystems," betraying an explicitly anthropocentric stance. Ecosystem services is also inherently linked to economics. It is a mechanism for trying to ascribe value to different dimensions of ecosystems. Some critics have argued that the concept of ecosystem services has contributed to the commodification of things like carbon and water in payment for ecosystem services schemes (see Chapter 5). While in some sense the idea of ecosystem services is an attempt to show that ecosystems have intrinsic value, ironically, the way they are framed places so much emphasis on their benefits to human societies that ultimately they come across as deeply anthropocentric. It is interesting to compare ecosystem services as a sustainability narrative to conservation narratives based on charismatic species. "Provisioning services" are certainly not charismatic, but nevertheless, they are framed in a manner that makes their utility for humans obvious and as such, ecosystem services is another attempt to bring dimensions of ecosystems that are unfamiliar to most nonexperts into the popular moral consciousness.

## Socionatures and Other Concepts of Supporting Ecosystems

In contrast to ecosystem services, *buen vivir* and other conceptions of socionatures that reject Enlightenment framings of the world (introduced in Chapter 3) also understand ecosystems as supportive, but from radically different conceptual underpinnings. Indeed, the word *supporting* is somewhat inappropriate; socionatures ontologically shifts the relations between humans and environment. Socionatures joins them analytically and imaginatively, and therefore, a more accurate framing is that human societies have no meaning without their interrelations with the rest of the environment. Here, aesthetics and cultural values are not simply nonmaterial benefits but foundational to how people seek to be in the world. Emotional, spiritual, and aesthetic connections between the human self and the rest of the planet are the cornerstones from which people construct ways of life. Human actions therefore have values that cannot be separated from other interactions with environments or economies. In this sense, ecosystems are the foundation of all human life. Sustainability in this narrative emerges not from evaluating the utility of different dimensions of ecosystems, but rather from a radical reconceptualization of how humans understand themselves within the planetary system.

The socionatures challenge to the dominant sustainability narrative is therefore to make the concept less to do with human-centric instrumental purpose, and to recognize the diverse worlds that people make in distinct kinds of places. *Buen vivir* and similar social movements and voices from within the scientific community have been particularly important in insisting on this shift in focus. When Amazonian peoples speak of their cassava plants as their children, and wild game as their in-laws, the notion of ecosystem services becomes irrelevant. Service is replaced by kinship and with this shift, socionatures and similar narratives invoke an extra-human dimension of awareness and ethical relationships with entities that many natural sciences cast as mute or inert. Humans are not seen capitalizing on, or even retrieving, services from ecosystems, but rather come into being precisely *in the relationship with ecosystems*. People engage with their surroundings and interact consciously with species and places as socio-natural persons. As a result, community is no longer considered a human domain, but a socio-natural system of mutual exchange and becoming.

## Food Production

The importance of ecosystems for supporting life becomes clear when we look at food. Food itself is, of course, vital for life, but also all agricultural systems require sunlight, water, and nutrients. These three necessities help to show how different dimensions of ecosystems are connected to and in many respects dependent upon each other. While some greenhouse and hydroponic systems are able to replace natural sunlight and soils with alternatives such as artificial light and nutrient-rich growing mediums, plants nevertheless require these basic inputs to grow.

With climate change and other ecosystem transformations, such as the decline in honeybees discussed previously, our ability to produce food for the entire planet is of great concern. This issue revolves around Malthusian-inspired debates over carrying capacity versus technological advances to overcome so-called natural limits or to use resources more efficiently—discussed in other sections of this book.

Food production has inspired many sustainability attempts at different scales, a theme echoing throughout this book. Local communities such as those discussed in the case studies of Chapters 8 and 12 have experimented with organic agriculture, local food, and other initiatives to try to connect community and food production more strongly. On a much larger scale, agribusinesses are promoting the use of chemical inputs such as fertilizers and pesticides to try to increase production. Some argue that without chemical inputs, agricultural systems will never be able to produce enough food for the planet's population, and not surprisingly, others argue that chemical inputs are causing an overall decline in ecosystem health and therefore challenge long-term food production. Both ends of this spectrum claim to be working for sustainable food production.

In this book we are not trying to resolve such debates. Rather, our focus is how these different narratives lead to radically different solutions, all under the name of sustainability. If food security is directly linked to carrying capacity, which is assumed to be extended through the use of genetically engineered seeds that produce more usable food per hectare (see also Chapter 11), then research and human ingenuity is poured into new crop varieties and new chemicals. This has been the logic of the Green Revolution (see Chapter 2) and most forms of large-scale commercial farming. In the ocean, this logic has been extended to fish farms, which are dependent on wild-caught fish to produce food for commercially valuable species kept contained within cages in the ocean. There is no question that such techniques have allowed more food to be produced by fewer farmers, but increasingly, people are questioning whether such systems can be sustained into the future because of their heavy use of groundwater and the damaging effects on ecosystems of chemical inputs. In fish farming, concerns exist about overexploitation of wild fish in distant parts of the world to feed salmon that are raised closer to shore.

If carrying capacity is assumed to be integrally bound to the long-term health of ecosystems, then more emphasis is placed on soil development, efficient irrigation systems, and how to foster insects and microorganisms that are beneficial to plants or that help to combat destructive insects. Organic agriculture is an example of this kind of farming logic. Within today's economy, the definitions of organic are fluid. In some places, it means the limited use of some chemicals, particularly pesticides, whereas in other contexts, it excludes all chemical inputs. A number of different systems are emerging under this narrative including permaculture (see also Chapter 12) and agroforestry. At stake are understandings of how an agricultural system sits within an ecosystem and therefore what ways are best to foster more sustainable farming practices.

If increasing food security is assumed to fundamentally rely on our connection to food and where it comes from, then the focus falls on involving more people in food production, either directly or indirectly, thereby increasing humans' link to food production. Biodynamic farming (also a form of organic agriculture), for example, emphasizes the interconnections between crops, soils, animals, and human cultural and spiritual relationships with the non-human world to foster more sustainable agricultural systems. An example of indirect involvement is the rise of community-supported agriculture schemes wherein people agree to buy produce on a weekly basis from a local organic farm, knowing that what they receive is dependent on what the farm has been able to produce in that particular growing season.

All of these different approaches to food production are justified under the name of sustainability. In each, questions need to be asked about who is producing food, for which markets, under what conditions, and with what consequences. Today's global food system is deeply interconnected. People in the Global North have become accustomed to buying fruits like bananas, mangos, and citrus that are produced in tropical climates all year long. Few people question where this food comes from or how it is possible that fruits which normally only produce one crop a year can be available all year round. The answer lies within both food storage and travel. Food is produced all over the world, but fresh vegetables and fruit in particular are produced in many countries of the Global South and then shipped to the Global North. Depending on the type of crop, they are either stored for surprising amounts of time (using artificially controlled atmospheres to prevent rot), or crops are produced in different climatic zones so that they can be available all year. All of these practices use a tremendous amount of energy. This is partly why advocates of local food insist on attention to food miles—concerns revolving largely around climate change. And yet despite these critiques, many new narratives have emerged supporting large-scale commercial agriculture as the solution to potentially declining crop yields. All this continues apace, even as changes, resulting primarily from energy use, in temperature and precipitation in core agricultural regions of the world help lead to global climate change.

For more information on organic agriculture, permaculture, food miles, and other responses to the global food system see:

- The Organic Consumers Association. www.organicconsumers.org/.
- Food Miles. www.foodmiles.com/.
- The Biodynamic Association. www.biodynamics.com/what-is-biodynamics.

## Climate Change

Currently, one of the most publicized environmental concerns is global climate change. As briefly introduced in earlier chapters, scientists as early as the 1970s began to worry about the dramatic increase in carbon dioxide emissions into the atmosphere that the growing industrialized economies of the world produced. They developed the hypothesis that carbon dioxide, along with other gases that occur in much smaller concentrations, such as methane and nitrous oxide, act as greenhouse gases trapping heat within the earth's system, making life as we know it possible. Other planets in our sun's solar system do not have the same combination of ozone to help block solar radiation, and carbon dioxide, methane, and nitrous oxide to trap heat, moderating temperatures on earth so that we lack the extreme heat of Mercury and Venus and the freezing cold of Mars and the planets beyond it. The geological record suggests that the earth has gone through significantly warmer periods as well as significantly colder periods than the present climate. A major change in the earth's climate has been accompanied by the increase in some forms of life and a decrease in other forms. During the last 11,700 years (approximately) the earth has been in an interglacial period so there is evidence that the climate would be warming without anthropogenic emissions.

Despite this history of a dynamic climate and our present interglacial (warm) period, there is widespread scientific consensus that the scale and rate of emissions over the past 150 years are causing the climate to change more rapidly than it would otherwise. The increase in greenhouse gases means that the average temperature of the earth is rising, and with it changes in precipitation as well as profound changes in a number of the earth's processes. With these atmospheric changes, computer models predict that drought periods will be longer in some regions, some regions will be warmer, others cooler, the oceans will rise, the risk of flooding will increase. Also, certain species will disappear, some diseases will spread more easily, some crops will be more difficult to grow, while other plants might benefit from a warmer climate.

The concept of planetary boundaries introduced in Chapter 4 encompasses many of the interrelated biophysical–chemical processes that are believed to operate in feedback loops to produce global warming. For example, as temperatures increase, ocean acidification increases, causing a decline in carbon sequestering phytoplankton (tiny plants), increasing carbon dioxide levels in the atmosphere even more. Feedback loops that accelerate warming temperatures are only part of the concern, however. More importantly for human life, shifts in the climate are predicted to cause an increase in the frequency and strength of extreme events such as heat waves, drought, extreme rains, and storms. The relatively predictable patterns of rain, temperature, and dry periods that many parts of the world presently experience will become more random and punctuated with extreme events that can threaten agriculture. Already the monsoon rains in South Asia have become less predictable. Agricultural systems on the Indian Subcontinent are heavily dependent upon the timely arrival and cessation of the monsoon rains, so when the rains are late, crops don't mature on time or dry up in the fields, and when they persist too late into the season, the mature crop is ruined. So while there are real concerns over increasing or decreasing temperatures and total rainfall in core agricultural zones in the world, the bigger threat comes from extreme events.

The biggest threat of all, however, comes from the possibility of runaway climate change. As discussed in Chapter 4, a lot of climate science works with the concept of tipping points. It is believed that we are already, or will soon be, at the point where the concentration of greenhouse gases in the atmosphere will have reached levels such that transformations in other earth processes will feed global warming in an exponential manner. If this hypothetical tipping point is reached, levels of warming far beyond what has been projected could occur (see also Chapter 4 on planetary boundaries and tipping points; Lenton et al. 2008; Steffen et al. 2015).

Many leading scientists therefore use the idea of the carbon budget to think about limits and tipping points in relation to carbon concentrations in the atmosphere. The carbon budget refers to the amount of carbon ($CO_2$) emissions we can emit while still having a chance to avoid the global average temperature rising more than two degrees Celsius above preindustrial levels. However, when other greenhouse gases are taken into consideration, the budget decreases drastically. Methane ($CH_4$), for example, is about twenty-five times more potent as a greenhouse gas than carbon dioxide, although it doesn't linger as long in the atmosphere. It is also not emitted in as high quantities, but commercial animal farming and industrial processes do emit significant quantities of methane.

The carbon budget is framed as a proposed "safe operating space" for carbon emissions, and a strategy to quantify, measure, and regulate impacts of climate change. This of course raises questions of how and on what grounds to distribute this budget, what will happen when the budget dries up, and whether the budget represents a safe space in the first place. Fundamentally, the two-degree target and the carbon budget are political constructs rather than scientific thresholds.

Climate change thus has the potential to drastically transform ecological systems, affecting biodiversity, water systems, and of course agriculture. Due to the magnitude of these possible effects, many people believe it is the number one environmental sustainability issue today. Because of the large number of interconnected processes and variables, predicting precisely what will occur is impossible, so modelers work with variables instead. They input baseline conditions and a range of expected emissions volumes so the models can produce different scenarios. While we do not know precisely what will happen, it is quite clear that major changes in ecosystems around the world are likely. The uncertainty in the predictions is one of the challenges facing the governance of climate change, which we return to below. It is important to note here, however, that the most severe effects are predicted to be located in the Global South—in South America, Africa, and Southeast Asia. Changing environments will lead to new migration patterns, and indeed "climate refugees," most of whom are presumed to originate in the Global South, has become a key term in international climate change negotiations. At the same time, the Global North is historically responsible for the vast majority of greenhouse gas emissions, raising a number of crucial justice questions.

In many respects climate change shakes the very foundations of sustainable development and transfers the locus of intervention from the local to the global level; in contrast, sustainable development is presently oriented around local, participatory projects, as discussed in Chapter 5. It is no longer a question of contamination or degradation of particular ecosystems, or the inclusion of marginalized people into environmental conservation projects, but of the planetary system as a whole. Because climate change is not spatially bound, it calls for a holistic approach, one that can motivate large numbers of people to protect the global bio-ecological world that sustains all life forms, including humans. But this statement brings us back to the core ideas of this book: who will decide what actions are best and who will be given the authority to enforce agreed upon rules?

One attempt at producing consensus on climate change is the Intergovernmental Panel on Climate Change (IPCC). The IPCC sits under the United Nations Framework Convention on Climate Change (UNFCCC) as a network of scientists who produce reports at periodic intervals to evaluate the scientific consensus on the issue. The reports (published at roughly seven-year intervals) compile a huge range of literature from across the natural and social sciences and then publish a condensed summary for policymakers. There are many criticisms of the IPCC and its process—lack of representation of scientists from the Global South is one—but very few question the overall scientific consensus that emerges from the reports, namely that climate change impacts will have wide-reaching and significant implications for human societies around the world, albeit in an uneven manner.

The IPCC reports have recently tried to grapple with a core critique that has been leveled at the process, namely the lack of integration between the natural and social sciences in climate change science. Climate change and sustainable development are historically anchored in two different philosophies of science. While climate is essentially rooted in the natural sciences, sustainable development has historically been associated with social sciences, and to some extent economics. Up until the Rio+10 Summit in 2002, separate circles of researchers and politicians alike tended to participate in climate change and sustainable development sessions. Yet, the social sciences are gaining greater ground in the IPCC only recently. The 2014 report represents significant progress in this regard. One consequence, however, has been relatively large inconsistencies across different chapters over how key concepts related to social systems are framed, in part because they draw from different debates within both the natural and social sciences.

In addition to the challenges of coordinating the global research community, the recent history of climate change negotiations illustrates well the challenges of global governance of the climate. The most important international body governing both the coordination of climate science and global efforts to address climate change is the UNFCCC. The UNFCCC was constituted at the Rio Conference in 1992 as a treaty to bring together nations concerned about the potential impacts of climate change and to help coordinate the reduction of emissions. The UNFCCC furthermore has acted as a coordinating body for commitments at the global scale to address climate change including the Intergovernmental Panel on Climate Change, the 1997 Kyoto Protocol, and the 2015 Paris Agreement.

Resources on the UNFCC, the IPCC, the Kyoto Protocol, and the Paris Accord can be found at these websites:

- United Nations Framework Convention on Climate Change. https://unfccc.int/.
- Intergovernmental Panel on Climate Change. www.ipcc.ch/.
- "Kyoto Protocol: Targets for the First Commitment Period." United Nations Framework Convention on Climate Change. http://unfccc.int/kyoto_protocol/items/2830.php.
- "The Paris Agreement." United Nations Framework Convention on Climate Change. http://unfccc.int/paris_agreement/items/9485.php.

And updated information on the implementation of the Paris Agreement is posted here:

- "Paris Agreement: Status of Ratification." United Nations Framework Convention on Climate Change. https://unfccc.int/process/the-paris-agreement/status-of-ratification.

The 1997 Kyoto Protocol came into effect in 2005 until the 2015 Paris Agreement replaced it, but it nevertheless reflects well the kinds of global governance mechanisms put forward for climate change—and their challenges. This international treaty sought to operationalize and extend the UNFCCC by creating binding targets on emissions reductions. It also set guidelines to make it cheaper and easier for industrialized countries to reduce their emissions through markets to trade and exchange emissions credits, known as carbon trading, and the Clean Development Mechanism (CDM). The CDM provides financial incentives for the development of clean and renewable technologies for development, while carbon trading allows polluters to trade credits for cash. The protocol mandated that thirty-seven industrialized nations plus the European Union cut their greenhouse gas emissions back to 5 percent below 1990 levels between 2008 and 2012.

Over one hundred so-called developing countries, including China and India, were exempted from the treaty, although all developing countries were asked to voluntarily comply. The treaty finally came into effect in 2005 after the Russian Federation signed and over fifty-five countries, accounting for 55 percent of the world's emissions in 1990, had ratified it. The Kyoto Protocol suffered from a lack of global consensus, however. President Clinton signed on behalf of the United States in 1998, but the U.S. Senate refused to ratify it and in 2001 President George W. Bush withdrew from the treaty. In 2011, as the deadline for emissions reductions was looming, Canada renounced the treaty (having previously signed) as impossible to implement because the United States and China, the world's two largest producers of greenhouse gas emissions, did not commit to reducing their emissions. The Kyoto Protocol was extended to 2020 in December 2012 (and as of 2016 it was replaced by the Paris Agreement discussed below), but many were skeptical of the ability of the UNFCCC to hold the signatory countries accountable for their promised emissions reductions. The EU and many countries in Europe made concerted efforts to reduce their emissions, but many were faced with rather daunting prospects of needing to reduce their emissions by far more than 5 percent, the reduction required under Kyoto, since emissions have been increasing, rather than decreasing, since 1990.

The Kyoto Protocol was the first attempt to hold countries most responsible for greenhouse gas emissions accountable to the rest of the world. It acknowledged that relatively rich, industrialized countries need to act first and with more commitments to emissions reductions, and that they need to enable the Global South to take alternative, nonfossil development pathways while respecting their right to development at the same time. These nonfossil fuel development pathways continue to be framed as sustainable development. Yet, Kyoto failed to provide adequate financial flows and most developing countries felt their concerns were unaddressed.

The Paris Agreement of 2015 is the most recent international attempt at responding to climate change and was heavily influenced by developing country demands. To the surprise of many observers, it was rapidly adopted by most countries and thus came into effect in 2016 more quickly than anyone expected. The Paris Agreement, which allows countries to negotiate their own targets, replaces Kyoto and there are high hopes that it will prove more effective and easier to implement than Kyoto. In many respects, Paris takes several steps backwards from Kyoto in terms of addressing the biophysical aspects of climate change, but several steps forward in terms of addressing climate justice and the financial implications of climate change for poorer nations. Unlike Kyoto, the Paris Agreement does not bind countries to quantified emissions reductions. Rather, it requires countries to define their own pathways to reduce and stabilize emissions. China and developing countries also need to define such pathways, but there is more emphasis on mitigation efforts and clearer financial obligations for wealthier countries to provide support for emissions reductions and adaptation to developed countries. As such, several key climate justice demands by developing countries were satisfied, although observers remain concerned about the lack of quantified targets for carbon reduction overall.

Like Kyoto, the Paris Agreement was almost immediately confronted with the same political challenges of building global consensus for action. The U.S. President Donald Trump, who took office in 2017, stated that he will withdraw the United States from the accord and, as of this writing, the exact details of how this will affect the rest of the world's compliance is not known. It is precisely because of these unknowns that we advocate the approach taken in this book. Evaluating sustainability issues by asking questions about sustainability of what, for whom, and with what costs helps to unpack the challenges of Paris and allow democratic debate on the best ways forward. Unilateral decisions like those taken by the United States tend to close down such debates rather than opening them up.

As we have seen throughout this book, the approaches to sustainable development range widely—from deep ecology, *buen vivir*, and socionatures attempts to radically reframe society–environment dynamics, to green capitalism and "development as modernity," which are largely interested in maintaining the status quo. As a result, there is no agreed upon strategy and no common conceptualization of what meeting the challenges of climate change through sustainable development means. The lack of consensus over what sustainable development is and how to achieve it is aggravated by the fact that the UNFCCC has limited powers of enforcement, meaning that many believe the present mechanisms for global governance of climate change issues are insufficient.

Ultimately, the present governance mechanisms rest upon, first, the willingness of individual countries to comply with globally agreed treaties and, second, the ability of signature nation-states to regulate and impose sanctions upon their citizens in accordance with those treaties. As a result of these limitations, some scientists are calling for a supranational body of global governance that can agree on standards and impose penalties upon not only nation-states, but also private entities that fail to comply with climate change regulations. These proposals are highly controversial, however, and major questions have been raised over who will control them and whose perspectives will guide their operation.

There is no question, however, that climate change poses radical challenges for sustainable development and many of the aspects of sustainability addressed in this book. Emerging critiques of the UN's Sustainable Development Goals (see Chapters 3 and 4), for example, cite climate change as undermining the logic of the SDGs, which are still oriented around nation-states and which are also discrete, doing little to tackle the intertwined socio-natural character of many environmental challenges today.

## Energy and Technology

The final two topics in this chapter, energy and technology, are closely related to climate change. The burning of fossil fuels is the prime contributor to greenhouse gases, and technology is both blamed for as well as put forward as the solution to climate change and other sustainability issues. In this section we first look at energy itself and why it poses such a dilemma for sustainability. The section ends with a brief discussion of microtechnologies including genetic engineering and nanotechnology.

### Energy

Energy takes many forms but is most useful to humans when it has the ability to perform work. Today, the work of the global economy is powered by fossil fuels and nuclear energy. Industrialization would not have been possible without the exploitation of a new more powerful form of energy, namely coal, which replaced energy produced by draft animals, wood (sometimes converted to charcoal), and water mills in the eighteenth century. To explain one example above, coal replaced charcoal because its mining required less labor than cutting wood and converting it into charcoal. This new relatively easily obtained source of energy, along with other changes in the political economy that are outlined in Chapter 2, contributed to a change in human organization with more people working in factories and fewer on the land. The use of a more potent energy source, however, did not lead to reduced resource use. In fact, quite the opposite occurred. The adoption of coal led to an acceleration in the exploitation of other resources. Thus, while coal replaced wood, many forests were clear-cut, and industrialized countries sought to replace forests and other natural resources elsewhere. This quest for natural resources underpinned the age of merchant capitalism, also discussed in Chapter 2. Coal remained the principle source of energy until the 1950s, when oil took over as the dominant fuel in the global energy system.

All fossil fuels—coal, natural gas, oil, and petroleum—ultimately originate in energy from the sun. Plants take up the energy from the sun and, through photosynthesis, convert carbon from the atmosphere to grow. Fossil fuels are dead plant and animal material that has been decomposed and then compressed and transformed deep under the earth over very long periods of time. Most of the deposits of fossil fuels exploited today were laid down millions of years ago. In theory, fossil fuels are renewable, but on the timescales humans are extracting they are not. Yet even with this kind of profile, some companies are mobilizing sustainability narratives to justify their extraction practices. They often present their technologies as more sustainable than other forms of oil extraction, in a classic form of "greenwashing" (see Chapter 4). Fossil energy sources like oil and coal, however, do not reflect the environmental costs in their price, nor are the climatic effects of the release of carbon dioxide into the atmosphere reflected in their claims to sustainability.

Since the 1950s, the world has seen a fourfold increase in energy use, and despite everything that is known about the dangers of the overuse of fossil fuel, that trend is increasing. Energy use, however, is highly unevenly distributed. China is fueling its industrial boom on coal, India also consumes large amounts of coal for its power plants (although in 2017 made a sudden commitment to eliminate coal), and the United States shows few signs of curbing its excessive energy consumption. People in the United States consume the most fossil fuel-based energy per capita of any nation. The 2011 per capita energy use in the United States is around 313 million Btu per person, according to the U.S. Energy Information Administration website, International Energy Statistics, while the same figure for Eritrea is one million, revealing major global inequalities both in depleting fossil fuel resources and in contributing to global warming.

Such figures raise questions of justice concerning why some people are able to consume so much more fossil fuel energy than others. China, for example, uses its historical lack of industrialization as justification for dirty fuels now—fuels that are largely used to produce commercial goods consumed in the Global North. The Chinese government is beginning to rethink that logic, however: air pollution in Chinese cities is so bad that it is now making major investments in renewable energy sources and has made a strong commitment to the Paris Agreement. China is also well below other industrialized countries in terms of its emissions per capita: 78 million Btu per person, according to International Energy Statistics, versus 134 for the United Kingdom in 2011 (and the United States, as noted above, consumed 313 Btu per person in 2011). But the UK's per capita emissions has declined over a three-year period whereas China's has increased. Climate justice advocates are calling for more accountability from high emitters both now and historically. While China as well as India and some other developing nations are using polluting energy sources to drive economic development, it remains a fact that the most severe climate change impacts are predicted in the parts of the world that contribute least to global emissions. Climate justice movements have emerged around the world as a result of these inequalities. Such movements press for more accountability and more redistribution of both the benefits of energy consumption and the burden of climate change now (see also Chapter 5).

Observers from multiple quarters thus call for more investment into renewable energy. While fossil-based energy is best described as a finite resource, other types of energy sources are renewable and considered more sustainable. In line with the narrative first articulated in the Brundtland report, sustainable energy is a type of energy obtained from nonexhaustible sources, ensuring that the provision of energy can be sustained into the future. Sustainable energy utilizes processes and phenomena such as rivers, waves, wind, and sunlight and converts them into working energy (mainly electricity, but sometimes hot water or mechanical energy such as from a water mill). The idea is to exploit the earth's systems, rather than its nonrenewable substances, to obtain energy for human use. These types of energy include hydropower, solar power, wind energy, wave power, tidal power, bioenergy, and geothermal energy.

Many countries and individuals in Europe have invested in wind, solar, and geothermal power to complement their use of hydropower, nuclear energy and, of course, coal, gas, and oil, as part of meeting their commitments under the Kyoto Protocol, and later under the Paris Agreement. However, while sustainable energy forms are $CO_2$ neutral, their use raises other concerns. Renewables are not without their own set of complex sustainability issues.

A massive geothermal installation in Pomarance, Italy, a small town in Tuscany.
*Source:* Valerie Tomaselli.

Hydropower, or using the flow of a river to generate electricity, requires building dams and changing the properties of the river itself. Large-scale dams are now considered by many to be ecological disasters, as they disrupt the migration of fish and other aquatic species, flood homes and ecosystems upstream of the dam, and can be a major hazard if the dam should burst. Despite these well-known problems, they are still being built as solutions to renewable energy needs, particularly in developing countries. Smaller hydro schemes that do not rely on dams (but that use the natural "run of the river") create fewer problems for aquatic ecosystems, fish migrations, and risk of flooding, but nevertheless do cause transformations in river ecosystems. Similarly, wind power has been criticized for being aesthetically disruptive to the landscape or seascape—increasingly wind farms are proposed offshore—and are known to kill significant numbers of birds in the turbines. Offshore turbines present further disruption to marine habitats and conflicts are emerging between fishing communities and governments seeking to promote marine-based energy, including both wind and tidal turbines. Finally, biofuels, or fuels that are produced from crops, are criticized for creating nearly the same kinds of emissions as fossil fuels when burned, and competing with food crops where it is grown. Sugar cane and maize, for example, are some of the most important biofuel crops and not only can these crops be used as food, they also are planted on lands that could otherwise be used for growing food crops.

So while renewable energy sources are preferable to fossil fuels from many sustainability perspectives, they nevertheless are largely unsustainable from most so-called "strong sustainability" stances. The bottom line is that we humans need to try to reduce our energy consumption if our use of energy is to produce fewer sustainability quandaries.

At the same time that energy is necessary to sustain life, its overuse threatens to undermine the condition of life. The sustainability challenge, many believe, thus lies in finding ways to reconcile the necessity and demand for energy with its impact on life and environment. The UN's Sustainable Development Goal number 7 (SDG7) proposes to "ensure access to affordable, reliable, sustainable and modern energy for all." While these ambitions are laudable and the framing of them attempts to encompass the three pillars of sustainability, the SDGs themselves do not offer concrete pathways for achieving such ambitions (see Chapters 3 and 4). The SDGs try to outline sustainability of *what*, but fail to ask *for whom* and *through what means*—questions that are at the core of this text. Proposing access to energy for all as a goal may be admirable but the inevitable unevenness of the impacts of extraction and energy use persists.

### Technology

Technology is often brought into these debates on energy as a way to solve problems arising from the human use of unsustainable sources of energy. Technology can help to create efficiencies so that overall energy consumption is reduced. New light-emitting diode (LED) technologies, for example, drastically reduce electricity consumption for the same brightness of light, and have an added advantage of being free from mercury. Similarly, more fuel-efficient motor vehicles allow for the same amount of driving with less fuel use and lower emissions. New technologies are being developed for coal- and gas-fired power plants to capture carbon and other emissions before they reach the atmosphere. There are even attempts to convert some of these emissions back into usable energy. From an economic perspective, all these innovations also help contribute to economic growth, which is central to so many definitions of economic sustainability.

Like all the sustainability debates presented in this book, however, there is a spectrum of views on the ability of technology to address current issues with energy and sustainability. Those on the green modernization end of the spectrum support technological advances that are not only more environmentally sustainable, but more beneficial to economic sustainability and, by extension, social sustainability. Those towards the deep ecology and *buen vivir* end of the spectrum, however, reject the idea that new technologies can solve sustainability issues. Rather, they advocate for a radical transformation of present ways of life to reduce overall demand for energy and to create socioeconomic systems that are more in line with environmental processes. The case studies in Chapters 8, 12, and 13 discuss quite different attempts at putting such alternative lifestyles into practice.

For this discussion here, technology can be understood as any invention developed by humans that helps to transform natural resources into usable forms. As such, technology encompasses all our high-tech gadgets, but also agricultural tools such as horse-drawn plows and mechanical water mills. Technology is brought in as a solution to a variety of other sustainability problems in addition to energy. Here we briefly discuss two technologies to illustrate how human invention might be seen from various points of view in sustainability debates: genetic engineering and nanotechnology. (The case study on GMO crops in Chapter 11 gives a more in-depth treatment of how genetic engineering is applied to agriculture.)

Genetic engineering developed beginning in the 1970s as a way to speed up a process that humans had already engaged in for millennia, ever since they began manipulating the genes of other species through selective breeding. Individual animals or plants that express a desired trait are bred with similar individuals in order to try to make that trait stronger. The domestication of animals was done through this process as well as the development of crop plants. Genetic engineering, however, is profoundly different in that it uses biochemistry techniques to place genes into cells from other individuals and species. This process requires the use of viral and bacterial gene sequences which are good at replicating themselves. Genetically engineered organisms therefore carry not only new genetic material that contains the desired traits, but also other genetic material that helps ensure the genes will function.

Genetic engineering is employed primarily in medicine and agriculture today. In medicine, it is used to help develop new cures for cancer and genetic diseases, among other treatments. In agriculture, it is used to develop crops that are resistant to certain pest species and to particular kinds of herbicide, or to increase production, like the use of genetically engineered recombinant bovine growth hormone (rBST) to induce individual cows to produce more milk. Its proponents claim that genetic engineering can help to solve food security concerns by developing crops, supplements like rBST, and animals that produce more. Crops that are genetically engineered to be pest resistant require fewer pesticides. With increasing population and climate change, genetically engineered crops are being promoted as "climate smart" since they can help to compensate for changes in growing conditions.

Genetic engineering, unsurprisingly, also comes with its own set of sustainability issues. There are significant fears over the impact of genetically engineered species on original populations. Genes from genetically engineered crops have been found in crops grown in adjacent fields, leading the EU to ban the testing of genetically engineered plants in open fields. Because the viral and bacterial gene sequences used to help insert genes into the genome of the target individuals are so good at inserting themselves, genetically engineered gene sequences from crops have been found to move into other species relatively easily. As a result, many critics are calling for a halt to the use of genetically engineered crops altogether. However, they have been widely planted in the Americas and some African countries, making their sudden disappearance unlikely.

Genetically engineered crops are also considered unsustainable on justice grounds. Most GMOs are unable to replicate naturally, meaning that farmers, rather than being able to harvest seeds from one year's crop for planting the next year, have to buy new seeds each year from the multinational company that developed them. In addition, many of these crops are engineered to grow best when used in combination with particular chemical inputs. This locks farmers into buying expensive seeds and inputs rather than being able to use organic fertilizers and propagate their own seeds. While many large-scale farmers already use commercial seeds and inputs, smaller scale farmers, particularly those in developing countries, do not. GMOs lock them into commercial relationships that are not always profitable if growing conditions are poor in individual seasons. In addition, some genetically engineered crops have caused allergies and can even be fatal to some people as the case of soybeans genetically engineered with Brazil nut genes illustrates. Other sustainability and justice concerns with GMOs are covered in more detail in Chapter 11. But of most importance to this book is the way that GMOs reflect a particular logic of society–environment interactions. It is perhaps the best example of the Enlightenment divide between society and nature. It is only possible to imagine genetic engineering if organisms are conceived as machines with individual parts. While this does not necessarily make genetic engineering wrong, from a socionatures perspective, it is not very surprising that engineered genes are more footloose than scientists imagined possible and that they are creating a variety of unanticipated effects.

Nanotechnology also illustrates the complexity of using human intervention in the sustainability debate. Nanotechnology refers to technology and science that operates at the molecular or atomic scale, from 1 to 100 nanometers. It is applied across many scientific fields, including chemistry, biology, and physics. Nanotechnologies are also used in other technological fields, such as electronics, medicine, and energy. They can help to target individual cancer cells, for example, and have allowed for the development of smaller and more powerful computers, mobile phones, and other everyday electronics. Researchers are optimistic about the ability of nanotechnology to make significant contributions to sustainability by improving existing technologies and developing entirely new ones. As a result, it is attracting increasing international investment by businesses and governments.

Nanotechnology, of course, also has its critics and concerns from a sustainability perspective. Nanomaterials can appear in air, water, soil, or even organisms as a consequence of producing, using, and disposing of products that contain them. Not much is known about their impacts on environment and health, and it is yet to be a major topic of research. Yet, we know from naturally occurring nanoparticles such as asbestos that they might in high enough quantities produce a toxic response in the lungs, and because of their nano-size, could spread very widely and rapidly through ecosystems. We are therefore back at our original questions: who is making what decisions about where, when, and how new potentially viable and fair sustainable solutions are applied to current environmental quandaries?

For more on nanotechnology and the environment see:

- Sustainable Nanotechnology Organization. www.susnano.org/index.html.
- "Nanotechnology and the Environment – Potential Benefits and Sustainability Effects." *Nanowerk*. www.nanowerk.com/spotlight/spotid=25910.php.

## Conclusion

In this chapter we have introduced some of the environmental concerns—and solutions relating to them—that fall under a sustainability rubric. We have left out many that could have been discussed here, including the accumulation of microparticles of plastic in the oceans, persistent organic pollutants and their bioaccumulation properties, climate change adaptation and geoengineering for climate change mitigation and the security and justice questions that arise around them, the accumulation of waterborne toxins and hormones, and groundwater depletion, among many other issues. All of these are championed in different ways by different sustainability advocates, and indeed, all of them are real concerns for humanity.

Which narratives we use to explain sustainability concerns, the conceptualizations we have of socionatures, and whether we take an anthropocentric or ecocentric stance all shape not only the issues we prioritize and our responses to them, but also their very character. As the concerns discussed in this chapter show, how we address them transforms the problems themselves. If Kyoto had been more successful at the time it was proposed, we would be facing a different climate scenario today and the Paris Agreement would have had a different focus and impact as a result. It is therefore crucial that we approach sustainability not as a thing that can be known, or a state to be achieved, but rather as an evolving social–political process that is deeply shaped by knowledge, social relations, narratives, communities of practice, human–nonhuman dynamics, and, of course, politics. It is in this spirit that we have explored how sustainability travels around the world.

This book does not attempt to provide answers, and nor do any of the narratives we discuss here provide simple solutions for a viable future on the planet, but there are lessons that can be learned. Some of these lessons come from the insights possible through careful scientific inquiry, while some of the barriers to learning come from precisely the same source. A clear message emerges: more informed citizens should engage critically with deliberations about what sustainability is, who it is for, what it seeks to achieve, and how. We have attempted to provide a foundation for critical thinking on these issues, but our analysis should not be taken as comprehensive. Rather, we see it as a first step in what we hope will become a lifetime commitment to thinking about sustainability of what, for whom, by whom, and what it means for us to try to govern it as individuals and a global community.

The case studies in the second part of the book explore both the possibilities and limitations of very different attempts at producing sustainability. Many of the case studies feature examples of gross power inequalities making participation in sustainable outcomes unlikely. Others demonstrate the ways in which ideas for sustainability are on the move and taking root. Ordinary people in the rising powers of Brazil and India show greater concern for contributing to decision making over food security matters than they are given credit for by so-called experts, for example. Inhabitants of the Sahel continue to pursue self-organized, mixed livelihood, inter-communal patterns of adaptation to variable weather systems and territorial access rights, despite government or other outside attempts to encourage, or even force, inhabitants to settle in one place in the name of sustainability. Feminist contributions have normalized diversity thinking for elites, and promoted grassroots actions for equality of employment, sexuality, and reproductive rights that have made measurable differences for both justice and governance questions in relation to sustainability. And crucially, many look hopefully at the vibrant debates around more traditional views of sustainability and their alternatives (such as *buen vivir*) to open up sustainability thought and practice in a globalizing world.

## References and Further Research


Bennett, Jane. 2010. *Vibrant Matter: A Political Ecology of Things*. Durham, NC: Duke University Press.

Brander, James A., and M. Scott Taylor. 1998. "The Simple Economics of Easter Island: A Ricardo-Malthus Model of Renewable Resource Use." *American Economic Review* 881: 119–138.

Carpenter, Stephen R., Harold A. Mooney, John Agard, Doris Capistrano, Ruth S. DeFries, Sandra Díaz, et al. 2009. "Science for Managing Ecosystem Services: Beyond the Millennium Ecosystem Assessment." *Proceedings of the National Academy of Sciences* 1065: 1305–1312. doi:10.1073/pnas.0808772106.

Chadwick, Ruth F., and Doris Schroeder. 2002. *Applied Ethics: Critical Concepts in Philosophy.* Volume 6. London: Taylor & Francis.

Cronon, William, ed. 1996. *Uncommon Ground: Rethinking the Human Place in Nature*. New York: W. W. Norton.

Daily, Gretchen, ed. 1997. *Nature's Services: Societal Dependence on Natural Ecosystems.* Washington, DC: Island Press.

Douglas, Leo, and Gary Winkel. 2014. "The Flipside of the Flagship." *Biodiversity and Conservation* 234: 979–997. doi:10.1007/s10531-014-0647-0.

Duan Kequin, Tandong Yao, and Lonnie G. Thompson. 2006. "Response of Monsoon Precipitation in the Himalayas to Global Warming." *Journal of Geophysical Research-Atmospheres* 111.

Ellstrand, Norman C., and Loren H. Rieseberg. 2016. "When Gene Flow Really Matters: Gene Flow in Applied Evolutionary Biology." *Evolutionary Applications* 97: 833–836. doi:10.1111/eva.12402.

Evans, Jay D., Claude Saegerman, Chris Mullin, Eric Haubruge, Nguyen, Kim Bach, Maryann Frazier, et al. 2009. "Colony Collapse Disorder: A Descriptive Study." *PLoS ONE* 48: e6481.

Fisher, Brendan, R. Kerry Turner, and Paul Morling. 2009. "Defining and Classifying Ecosystem Services for Decision Making." *Ecological Economics* 683: 643–653.

Georgescu-Roegen, Nicholas. 2014. *Energy and Economic Myths: Institutional and Analytical Economic Essays*. Amsterdam: Elsevier.

Goodman, David, and Michael J.Watts, eds. 1997. *Globalising Food Agrarian Questions and Global Restructuring*. London: Routledge.

Gould, Stephen Jay. 1994. "The Evolution of Life on the Earth." *Scientific American* October: 85–91.

Haraway, Donna Jeanne. 2008. *When Species Meet*. Minneapolis: University of Minnesota Press.

Hecht, Susanna B., Kathleen D. Morrison, and Christine Padoch. 2014. *The Social Lives of Forests: Past, Present, and Future of Woodland Resurgence*. Chicago: University of Chicago Press.
Houghton, R. A. 2007. "Balancing the Global Carbon Budget." *Annual Review of Earth and Planetary Science* 35: 313–347.
Hubbell, Stephen P. 2001. *The Unified Neutral Theory of Biodiversity and Biogeography (MPB-32)*. Volume 32. Princeton, NJ: Princeton University Press.
Hunt, Terry L. 2006. "Rethinking the Fall of Easter Island." *American Scientist* 945: 412–419.
IPCC. 2014. "Summary for Policymakers." In *Climate Change 2014: Impacts, Adaptation, and Vulnerability. Part A: Global and Sectoral Aspects. Contribution of Working Group II to the Fifth Assessment Report of the Intergovernmental Panel on Climate Change,* edited by C. B. Field, V. R. Barros, D. J. Dokken, K. J. Mach, M. D. Mastrandrea, T. E. Bilir, et al., 1–32. Cambridge: Cambridge University Press.
Leach, Melissa, Kate Raworth, and Johan Rockström. 2013. "Between Social and Planetary Boundaries: Navigating Pathways in the Safe and Just Space for Humanity." *World Social Science Report* 2013: 84–89.
Lenton, Timothy M., Hermann Held, Elmar Kriegler, Jim W. Hall, Wolfgang Lucht, Stefan Rahmstorf, and Hans Joachim Schellnhuber. 2008. "Tipping Elements in the Earth's Climate System." *Proceedings of the National Academy of Sciences* 1056: 1786–1793. doi:10.1073/pnas.0705414105.
Loreau, Michel, Alfred Oteng-Yeboah, M. T. K. Arroyo, D. Babin, R. Barbault, M. Donoghue, et al. 2006. "Diversity Without Representation." *Nature* 442, no. 7100: 245–246.
Lorimer, Jamie. 2006. "What About the Nematodes? Taxonomic Partialities in the Scope of UK Biodiversity Conservation." *Social & Cultural Geography* 74: 540–557.
———. 2007. "Nonhuman Charisma." *Environment and Planning D: Society and Space* 255: 911–932. doi:10.1068/d71j.
Lovelock, James. 2000. *The Ages of Gaia: A Biography of Our Living Earth*. Oxford: Oxford University Press.
Luby, Claire H, Jack Kloppenburg, Thomas E. Michaels, and Irwin L. Goldman. 2015. "Enhancing Freedom to Operate for Plant Breeders and Farmers Through Open Source Plant Breeding." *Crop Science* 556: 2481–2488.
Matondi, Prosper B., Kjell Havnevik, and Atakilte Beyene. 2011. *Biofuels, Land Grabbing and Food Security in Africa*. London: Zed Books.
Mead, Margaret. 1976. *New Lives to Old: The Effects of New Communication on Old Cultures in the Pacific*. Honolulu: East–West Communication Institute.
Messier, François. 1991. "The Significance of Limiting and Regulating Factors on the Demography of Moose and White-Tailed Deer." *Journal of Animal Ecology* 602: 377–393. doi:10.2307/5285.
Millennium Ecosystem Assessment. 2005. "Ecosystems and Human Well-Being: Wetlands and Water." World Resources Institute: Washington, DC.
Moss, Richard H., Jae A. Edmonds, Kathy A. Hibbard, Martin R. Manning, Steven K. Rose, Detlef P. Van Vuuren, et al. 2010. "The Next Generation of Scenarios for Climate Change Research and Assessment." *Nature* 463, no. 7282: 747–756.
Newell, Peter. 2012. *Globalization and the Environment: Capitalism, Ecology & Power*. Cambridge: Polity Press.
Nordlee, J. A., S. L. Taylor, J. A. Townsend, L. A. Thomas, and R. K. Bush. 1996. "Identification of a Brazil-Nut Allergen in Transgenic Soybeans." *New England Journal of Medicine* 334, no. 11: 688–692. doi:10.1056/nejm199603143341103.
Parks, Bradley C., and J. Timmons Roberts. 2010. "Climate Change, Social Theory and Justice." *Theory, Culture & Society* 27, no. 2–3: 134–166. doi:10.1177/0263276409359018.
Plumwood, Val. 1991. "Nature, Self, and Gender: Feminism, Environmental Philosophy, and the Critique of Rationalism." *Hypatia* 61: 3–27.
Steffen, Will, Katherine Richardson, Johan Rockström, Sarah E. Cornell, Ingo Fetzer, Elena M. Bennett, et al. 2015. "Planetary Boundaries: Guiding Human Development on a Changing Planet." *Science* 347, no. 6223. doi:10.1126/science.1259855.
Strong, Donald R., and Kenneth T. Frank. 2010. "Human Involvement in Food Webs." *Annual Review of Environment and Resources* 351: 1–23. doi:10.1146/annurev-environ-031809–133103.

Tilman, David. 1998. "The Greening of the Green Revolution." *Nature* 396: 211–212.

Wackernagel, Mathis, Larry Onisto, Patricia Bello, Alejandro Callejas Linares, Ina Susana López Falfán, Jesus Méndez García, et al. 1999. "National Natural Capital Accounting with the Ecological Footprint Concept." *Ecological Economics* 293: 375–390.

Weart, Spencer R. 2008. *The Discovery of Global Warming*. Cambridge, MA: Harvard University Press.

PART TWO

# Case Studies

# 7

# Introduction to the Case Studies

Andrea J. Nightingale and Tom Böhler

This volume lays out background, history, and conceptual considerations in Part One and examines those foundations in specific localities or targeted trends in the case studies in Part Two. The goal is to help readers see theory in action.

As we have emphasized throughout, the basic questions of sustainability of what and for whom, and with what implications, undergird much of our analysis. The case studies are, therefore, starting points to think about other specific situations that can be explored by applying the foundations outlined earlier.

**Chapter 8: Alternative Food Movements and Transition Towns in the United Kingdom**, as author Ben Campbell states, "highlights the importance of fitting alternatives conceived at wider scales to local realities, so that initiatives aimed at producing beneficial change can work with and for local people." He does this by exploring Transition Towns, those that, recognizing the need for action on climate change at the local level, aim for zero or negative carbon emissions. Campbell explores various approaches to this strategy, focusing on two towns with different economic profiles. In his case study, local groups reject the idea that technology and factory farming can solve sustainability issues and the assumption that sustainability is something that can be achieved under the present capitalist system. Instead, the two examples illustrate localized attempts to radically reshape the use of resources and overall carbon emissions.

In pursuing this strategy, Campbell brings together several central themes of this book. How and why do ideas of sustainability show up in particular places, at specific moments in time? Who takes up the ideas, and for what purposes? What are the consequences socially, politically, and ecologically? In particular, Campbell explores how global concerns and narratives are reworked into local practices, and investigates attempts at putting alternative lifestyles into practice. Especially, when people strive to regain control over their own food system, new discourses of sustainability emerge that are more in line with localized environmental processes. These localized movements need to be understood as also part of globalization. On a much larger scale, agribusinesses are promoting the use of chemical inputs such as fertilizers and pesticides to try to increase production. Yet, as Campbell shows, communities experiment with organic agriculture, local food, and other initiatives to oppose these practices, instead trying to connect more strongly community and food production within their localities and across the globe. Both ends of this spectrum claim to be working for sustainable food production, although with very different methods and motivations, demonstrating how sustainable development takes very different pathways dependent upon the scale of intervention, and who articulates the goals.

In **Chapter 9: Community Forestry and Sustainability in Nepal**, Dinesh Paudel also investigates the way local communities engage in sustainability. Nepal's community forestry is no longer a local, community-centered forestry program; rather it has developed global connections and decision-making power, gradually giving community forestry actors global influence and coverage. As a model for sustainable development, community forestry aims to reestablish local peoples' control over forests to support the protection of the environment. Under ideal circumstances, it shifts power hierarchies so that local and marginalized people gain access to environmental governance. Paudel shows, however, that community forestry is not simply a decentralization of forest resource management; rather, it has opened up local environments to global market forces that are characterized by demands for economic growth. As such, community forestry has drawn local and often vulnerable populations into development processes and, in turn, the global market economy. Therefore, as Paudel suggests, community forestry is rather contradictory. On the one hand, it is probably among the best available development strategies in Nepal to achieve social and environmental sustainability; on the other hand, community forestry has allowed both the state and market to operate through local elites, coopting its progressive potential.

Paudel's case study clearly demonstrates that sustainability is an inherently political project, involving numerous actors with different sustainability agendas. Thus, while its advocates maintain that community forestry is well equipped to address the global goals of simultaneously protecting the environment and achieving economic growth, its critics claim that community forestry has become an instrument of the state and the market to exploit cheap labor and forest resources, and therefore exacerbates social inequalities. Perhaps of most concern, it sidesteps the real challenges of how to empower people to participate in the articulation of their own future, and replaces that with participation in economic development only. Nepal's community forestry is no longer a local, community-controlled forestry program; rather, decision-making power has been gradually transferred to global actors, and there is an increased realization that poor people are again losing control over forests because the market dictates how to manage the environment. The case study therefore shows how conservation often becomes a practice of ecological modernization, allowing governments to retract from environmental regulation and instead rely on market forces to determine how the environment is used, with significant consequences for social, ecological, and economic relations.

**Chapter 10: Sustainable Mining and the North** by Nancy Langston explores contested narratives of sustainability in mining practices in the boreal north—the Lake Superior Basin in North America and Sápmi in northern Scandinavia. In the case study, Langston investigates how decisions about sustainable mine developments are not purely scientific or technical decisions—at heart, they are social decisions—and how the meaning of sustainability is open for interpretations. For mining companies, mining stimulates economic activity and promotes development in poor regions of the world. According to this narrative, sustainability signals a commitment to economic growth, assuming that social and ecological sustainability will follow from that. In such corporate sustainability discourses, sustainable mining is framed as a harmonization of the "three pillars of sustainable development," claiming that it balances social, economic, and environmental factors. For others, however, mining operations are seen as destroying environments and displacing indigenous peoples who depend on the land for other activities, such as reindeer herding. From this perspective, sustainability revolves around questions of justice, and the protection of indigenous rights and socionatures, which is something that cannot be measured in monetary terms. Such alternative articulations of sustainability, or counternarratives, are often marginalized in dominant discourses of sustainable development, often deployed by large corporations.

Langston's case study reveals that frameworks of sustainable mining rarely recognize the fact that sustainability is a concept rather than an environmental state that can be known or achieved. Above all, Langston brings into clear view the fact that discourses of sustainability are often deployed for political purposes as competing actors articulate sustainability differently depending on their practical goals. The case illustrates a core theme of this book, namely that sustainability is not a neutral or value-free concept but rather involves competing claims over what sustainable development entails, for what, and for whom.

In **Chapter 11: Governing Genetically Modified Crops and Agricultural Sustainability**, author Phil Macnaghten examines the idea that concepts and models of agricultural sustainability need to embrace elements that are as much social and political as they are technical. He uses three spatial lenses in the case study: in Mexico, where genetically modified (GM) maize is conceived by the public as imposed globalization; in Brazil, where GM crops (sugar) represent a symbol of wider struggle against neoliberal polices and U.S. imperialism (by resisters of GM organisms, or GMOs, not by the government); and in India, where both positive and negative attitudes towards GM crops exist.

Macnaghten's case study is centered on how global discourses of sustainable agricultural development are encountered in different places, where they interplay with questions of social justice and environmental integrity. On a global scale, agribusinesses are promoting the use of chemical inputs and GMOs to try to intensify and increase food production. In this narrative, food security is directly linked to carrying capacity, which is assumed to be extended through the use of genetically engineered seeds that produce more food per hectare. This has been the major driving force of the Green Revolution and most forms of large-scale commercial farming. Indeed, some argue that without these technological innovations, agricultural systems will never be able to produce enough food for the planet's population. Technological and scientific advances are considered not only more environmentally sustainable, but also more beneficial to economic sustainability and, by extension, social sustainability. Critics, however, argue that chemical inputs are causing an overall decline in ecosystem health and therefore challenge long-term food production. Others fear that GMOs imply increased corporate power over food production, with seeds and other agricultural inputs being controlled by global actors rather than by the farmers themselves. Consequently, the introduction of GMO crops can close down more local sustainable development pathways.

Food production has inspired many sustainability attempts at different scales, a theme echoing throughout many sections of this book. As Macnaghten shows, peoples often resist corporate interventions into local or traditional food production in which farmers have control over the food system. In such places, GMO technologies are often seen as a threat to local cultures and practices. Both critics and advocates of GMO claim to be working for sustainable food production under the name of sustainability, highlighting the fact that sustainability means very different things dependent upon perspective and priority.

**Chapter 12: Rethinking Sustainable Community Through Permaculture Initiatives in Zimbabwe** by Pamela Richardson-Ngweyna focuses on three local projects in Zimbabwe to investigate how permaculture, a global movement, can reinvent and reshape community in a sustainable way. Within permaculture movements, the natural and the social are not seen as independent from one another; rather, permaculture fosters a narrative of mutually dependent and integrated socio-natural relations. In her case in Zimbabwe, humans are positioned as participants in a wider system wherein animals, plants, waterways, and other socio-natural beings also have significant rights and roles to play. The author's examinations involve how these projects have expanded community life, including readdressing unfair power relations and seeking connection with other communities across the world for mutual support and learning.

In this way, permaculture represents a change in sustainability narration, away from mainstream discourses of the three pillars of sustainable development, and toward the idea of local integration of community and food production. While not always explicitly a critique of global capitalism, these narratives are rooted in different conceptions of economy and society based on respect for and sharing kinship with the nonhuman world, rather than control over it. Its advocates argue for a radical transformation of present ways of life to reduce our overall demand for energy and to create integrated systems that are more in line with local environmental processes. It is crucial, however, to view these movements as part of globalization. Indeed, the most successful movements are those that connect people from disparate localities in order to leverage political power. In addition, such attempts increasingly link localities in the Global South with localities in the Global North to redistribute resources and make direct connections between people living very different lives.

As this case study shows, many possible visions for sustainable development exist, and no project is sustainable in and of itself. Indeed, mobilizing local knowledges and alternative ways of life does not necessarily lead to long-term sustainability, but can bring into view, or obscure, other inequalities and unfair practices.

In **Chapter 13: The Entanglements of Oil Extraction and Sustainability in the Ecuadorian Amazon**, authors Angus Lyall and Gabriela Valdivia also examine local dynamics in addressing sustainability issues. In this case, the focus is on oil extraction, often viewed through a global lens, in Playas del Cuyabeno, in the northern Ecuadorian Amazon. In their investigation into the ways the Kichwa community in Playas engage with oil extraction, the authors explore how "sustainability" is not an abstraction to be applied worldwide, but rather one enacted at the local level.

Lyall and Valdivia's case study makes evident the gap between sustainability as a normative project that seeks to address collective social and ecological well-being, and sustainability as a contested and politically charged practice that entangles groups in uneven power relations. They show that while governments support the expansion and intensification of resource extraction in the name of sustaining future national well-being—in their case oil extraction in the Amazon—these same practices continue to put the environments and lives of indigenous peoples at risk. Yet, at the same time, local residents also respond to and negotiate oil extraction as a sustainability practice. In the Ecuadorian case, the local community tried to form their own community-managed oil company to prevent ethnic discrimination and to sustain social, economic, and even environmental conditions, as part of their response to a long history of exploitation, environmental destruction, and ethnic discrimination. Both the government and indigenous groups positioned themselves in relation to social, political, and economic processes to create and negotiate spaces for what they perceived as sustainable development. While the local community's attempt to become oil producers failed, it nevertheless shows how sustainability is taken up by different actors, illustrating starkly how sustainability is a fundamentally politically and historically contingent concept that can never be neutral.

Sarah Bradshaw, the author of **Chapter 14: Sustainability and Gender Equality: Exploring the 2030 Agenda for Sustainable Development**, explores "how gender has been addressed within international initiatives for bringing development to so-called 'less developed' or 'Third World' countries, and how gender has been included within environmental debates." The case study shows how dominant discourses of sustainability are gendered, or how they position men and women differently in relation to social, ecological, and economic problems. The author argues that, even in the post–2015 sustainable development agenda, women continue to be viewed as a key to efficiency in achieving sustainability, as they were in earlier international sustainability regimes. And that this responsibility for women amounts to uncompensated work, which brings "positive collective benefits but few personal gains."

As argued throughout this book, sustainable development will usually create opportunities for some people in one place, and close down possibilities for others elsewhere. In this regard it is crucial to scrutinize how gender becomes silenced in dominant debates. Many historical, economic, and environmental issues have been defined in predominantly scientific or technical terms, and consequently assumed to be gender neutral. This has contributed to making women largely invisible within policy debates. Technical knowledge often fails to acknowledge, or even exacerbates, underlying or structural inequalities such as gender and class. In particular, locating sustainability issues on a global or planetary scale implies that inequalities at other scales remain hidden. Among these hidden inequalities have been the discrimination and increased burdens of women. Bradshaw's case study tackles these issues, exploring who is included and who is excluded from sustainable development programs and how these processes of inclusion and exclusion are gendered, both locally and globally.

**Chapter 15: Sustaining What and By What Means? The Importance of Local Social Relations in the Sahel** by Matthew D. Turner explores common misunderstandings of the human–environment relationships in the Sahel. Here the author argues that official misreadings of pastoralist livestock production as economically irrational and destructive to the environment have resulted in added hardship for Sahelian pastoralists. Turner's case study looks carefully at how humans live with their environments in a context where people follow the resource (livestock), rather than containing it for human use. Pastoralism has long been seen as economically irrational, inherently destructive to the environment, and obstructive to sustainability efforts. Historically, government interventions into such practices have tried to pin down who uses what resources in time and space in efforts to improve environmental governance. However, neither the community nor the resources they utilize are easily bound, making it difficult to define property and use rights and to determine responsibility for environmental governance as a result. This often leads to a mismatch between the scale of resource use and the scale of resource governance, contributing to increased vulnerability at the local level. Instead, Turner suggests, the mobility of livelihood strategies in the Sahel requires sophisticated forms governance, with responsibilities shared across different scales and associated institutions in order to promote sustainability. The case study investigates how social relations between pastoralists and farmers in the Fakara region of Niger, alongside institutions such as land tenure and herding contracts, can reveal a range of innovations—some not environmental at all—that would improve rural livelihoods.

This case shows clearly how sustainability problems might manifest themselves in the form of land degradation or soil erosion, but the roots of these problems are often social and political relations, highlighting the need for more complex understandings of sustainability based on grounded engagements with local realities. Scientific misunderstandings of environmental degradation can be intimately linked to interventions that increase local vulnerability. In contrast, embracing alternative and local knowledges can help conceive of and enact sustainabilities suited to particular places and peoples.

* * *

Taken together, these case studies illustrate the many forms, contradictions, and challenges that sustainable development projects face. They show clearly how one single solution to sustainability is impossible, but rather that sustainability itself—what it means, what actions are necessary, and by whom—is always embedded within historical and cultural relations.

The past informs what possibilities emerge for the future and preexisting social relations, injustices, and environmental destructions cannot be easily undone by projects to promote sustainability. These projects need to emerge within those relations and injustices, placing them as part of the core sustainability challenge with constant querying of "sustainability of what, for whom, with what consequences" if they hope to chart new pathways for the future.

# 8

# Alternative Food Movements and Transition Towns in the United Kingdom

Ben Campbell

Working towards more sustainable food systems presents challenges for researching and enacting alternatives to existing, global food systems. The case study in this chapter addresses examples of a Transition Town, and a more successful town nearby that took an alternative approach. It examines the way that messages about sustainable food systems can have distinct cultural audiences. It situates sustainable food initiatives within historical and cultural contexts, and considers the relative uptake of ideas, designs, and pathways for more sustainable food, according to locally appropriate narratives and projects for change. The adoption of such strategies makes sense to local communities for economic, political, cultural, and ecological reasons, and highlights how bottom-up actors have embraced notions of sustainability for collective organizing. The background landscape for these initiatives can be characterized as one of widespread cultural reinvention of food relationships: from how food is produced, right through to the British diet, which has long suffered a bad reputation among health experts and culinary writers.

The main contribution of this chapter is to highlight the importance of fitting alternatives conceived at wider scales to local realities, so that initiatives aimed at producing beneficial change can work with and for local people. Sustainability here refers to a collective transition that can transform relationships with the nonhuman environment, or our socionatures, recognizing that efforts in this direction follow innovative and unpredictable pathways. The case study gives a context for how the Transition Town movement emerged in many parts of the United Kingdom beginning in 2006. Transition Towns strive for zero or negative carbon emissions through collective and individual social, environmental, and economic initiatives. Yet for the global Transition Town movement to be successful, connections with local communities of people are vital for translating ideas about sustainable transitions into their everyday worlds. These worlds consist of relationships that are simultaneously local and global. Big ideas like "transition" need to be made tangible and meaningful in locally lived worlds, or they will not endure and be useful for grassroots people's projects of change.

Transition Town initiatives stemmed from accelerated concerns about the twin effects of climate change and peak oil. Rob Hopkins' 2008 widely read *Transition Handbook* advocates for adopting local resilience principles for reorganizing our living patterns and our food system and for engaging in an "energy descent" to remove the excessive reliance of our economy and society on fossil fuels. Resilience thinking, as discussed in Chapter 4, is based on self-organizing growth principles of biological systems, and ecologically appropriate zoning design. The Transition Town movement is also linked to the Permaculture

Association established by Bill Mollison. Permaculture rests on observations of how plants grow naturally in a given locality and, from those observations, designs a system that encourages edible species to grow in relatively permanent combinations.

The term "permaculture" was coined by Australian research professor Bill Mollison in 1978. Mollison states "Permaculture is a philosophy of working with, rather than against nature; of protracted and thoughtful observation rather than protracted and thoughtless labor; and of looking at plants and animals in all their functions, rather than treating any area as a single product system." The Permaculture Association was founded by Mollison in the UK in 1983 to promote theory and practice of permaculture. More information is available at www.permaculture.org.uk/.

Mollison's colleague David Holmgren expounded on twelve principles of permaculture in his 2002 book *Permaculture: Principles and Pathways Beyond Sustainability.* Holmgren Design International Permaculture conferences have taken place since the early 1980s. More information can be found at the Holmgren Design website: www.holmgren.com.au/.

Today, public and academic debates recognize that the transition movement can offer an appealingly innovative, boldly integrated, scientifically justified, and policy-aware vision for achieving goals of more sustainable living. This recognition is a reaction to concerns over purity of food products and industrial-based food systems, and reflects the search for more democratic sovereignty in relation to food. Transition Towns and related movements focus on forms of organization and everyday collaborations of people that are required to activate community-wide networks for growing food and pathways to climate-friendly livelihood change. These practices point not simply to useful technical and horticultural developments, but to more embedded relational approaches to food issues than the term "food security" conjures up. Yet, many observers recognize that Transition Towns have tended to take root in middle-class enclaves. Interestingly, the Transition Town communities in focus here, in the industrial north of England, have a notable history of mobilizing for consumer rights to quality food as a public good. Organizing as a Transition Town is therefore another example of local reactions to a global food market. However, the class–cultural baggage that goes with Transition labeling risks limiting the appeal of the idea to educated middle classes. Indeed, another nearby community also examined here, chose to organize around a different label, Incredible Edible.

To understand why food has provided a platform for alternative social movements, we need to see how food has been central to debates about how to live in an increasingly urban and industrial world. The complex chains that mediate modern food production, distribution, and consumption are never purely economic in character, but rather are foundational to how societies change.

## Historical Roots

The nineteenth century saw transformations in the food system designed to meet the demands of new urban-based livelihoods that were rapidly growing due to industrialization. Over a period of only two hundred years, a wholesale shift took place from a world in which landed property was the foundation of social hierarchy and power, to one in which other forms of ownership, capital (e.g., money, factory machinery), and value dominated. These transformations had global consequences as extractive and plantation colonialism came to replace more localized economies, changes described in Chapter 2 of this volume. Colonialism had far-reaching transformations in how food was grown, where it was grown, and indeed even the foods people came to prefer.

The circulation of food commodities to support growing urban populations in early modern globalization created a rift between urban industrial economies, their geographical location, and the food commodity system that sustained them. Adjustments to urban living habits, factory-based working practices, and commodified mass production systems required a de-linkage from old regional agricultural economies. This was made possible by train transport and colonial plantation and trade infrastructures. The diet of British industrial society was built upon many of the movable products designed to keep crews of colonial marine shipping enterprises fed and working. These products included sugar, biscuits, and preserved foods, and marked a shift from the nutritional norms of preindustrial diets (staples, legumes, and sauces).

Early on in the history of the new urban working classes, concerns over the quality of foods and the right to know the content of food commodities provided the basis for consumer citizenship movements. The co-operative movement was formed in Rochdale, Lancashire, in 1844 to ensure control over the purity of flour for factory workers, after supplies for breadmaking were regularly found to be adulterated with chalk and other materials. The Co-operative Wholesale Society, an offspring of the movement, went on to join the top league of global trading enterprises, maintaining its economic distinctiveness from other capitalist enterprises by providing a consumer dividend, and took a political stance by being a major supporter of the British Labour Party. There are therefore historical roots to concerns over dependability of food products, resistance to industrial-based food systems, and the search for more democratic sovereignty in relation to food.

Other responses to the rift of urban–industrial lifestyles from previous food economies included that of William Cobbett, who posed searching questions about the risks people were facing with internationally traded commodities. Cobbett published *Cottage Economy* in 1823, advocating for backyard food growing, in order to give people control over the quality of their everyday diet. Cobbett emphasized the central importance of reliable and accessible seeds, which he sold in South Kensington at the edge of London.

> Seed sovereignty continues to be at the core of contemporary struggles over corporate domination of the food system. In the UK a heritage seed exchange system was set up in the 1980s after European Union regulations prohibited the sale of seeds that have not been registered and licensed through an expensive scrutiny process. See more information on the exchange system on the "Heritage Seed Library" page, published by the Garden Organic website, at this URL: www.gardenorganic.org.uk/heritage.
>
> The Food Sovereignty movement has a similar emphasis on growers' rights to exchange seeds, and the Open Source Seed Initiative (http://osseeds.org/seeds/) was launched in the United States in 2012.

In the twentieth century, the viability of farming systems was again challenged, this time by experiences such as the dustbowl drought and the Great Depression in the United States (see Hyams 1952). Since World War II, technological fixes to meet gaps in food provision have led to phases of top-down, productivist solutions. Productivist agriculture depends on capital-intensive cultivation methods using heavy machinery and chemical fertilizers and pesticides to increase yields. The history of productivist farming began much earlier when traditional communities and their cultivation practices were displaced (referred to as the "enclosure of commons") from lands that have potential for high productivity through technical inputs, processes that are ongoing in parts of the world today (see also Chapter 2). These technical transformations range from landscapes engineered for irrigation or drainage, to the introduction of crop varieties that are bred to be high yielding when fertilizer is applied. High-yielding crop varieties are often grown in monocultures, which means planting the same crop over large acreages, year after year, so that pests, diseases, and nutrient deficiencies in the soil have to be treated with ever more costly applications of chemicals.

Cycles of productivism have led to concentrations of animals in small spaces (intensive livestock systems), effectively separating arable land from livestock. This separation means that animal manure presents waste management and carbon emissions problems. Productivist agriculture causes many other environmental problems of water and soil pollution and biodiversity loss (e.g., bees and other pollinators), which are not factored into agricultural profits, but have to be dealt with through public sources.

Productivist agriculture is highly dependent on fossil fuels for its technology and chemicals, thus profitability goes down when oil prices go up. The food calorie energy return compared to energy invested in productivist agriculture, especially with intensive livestock systems, is calculated by David Pimentel and colleagues (2008) to be inefficient by a factor of 30:1. This form of agriculture has also caused a significant decrease in the types of food crops, with a prevalence of just wheat, maize, rice, and soya on global markets. Market concentration of these commercial crops neglects the global abundance of potential species known by local food communities, which are suited to local conditions and tastes, and has contributed to food being transported over long distances (food miles).

By contrast, alternatives to productivist agriculture insist that people's food needs and preferences come before corporate interests. Many take a perspective that considers long-term sustainability of human and soil health as indicators of success in agriculture, rather than annual yields. Vandana Shiva's 2008 book entitled *Soil Not Oil* takes this position, and wraps it into arguments for reducing greenhouse gas emissions. Transition Towns and Incredible Edible similarly use food systems as the starting point for reimagining communities and economies less tied to fossil fuels and the global capitalist economy.

Food systems are about calories and nutrition, as well as a nexus of technical factors such as soils, compost, water, and energy. While these technical components are vital, they are co-produced and transformed through collective notions of security, well-being, identity, and cultural sovereignty. As such, food systems need to be understood as socio-natural systems, ones that have tangible nutritional, ecological, and distributive consequences, and that equally have potent social and cultural agency. For example, most cultures express ideas of abundance and progress through food symbolism, and food preferences can further mark distinctions of social identity in unequal societies. The corollary is that lack of food and poor-quality food affront human dignity. For these reasons, thinking about food systems requires attention not only to material human sustenance and long-term agricultural production, but also to narratives about equitable distribution and to the public culture of food; in other words, the nourishment of self and community under changing political–economic and environmental conditions.

## From Countryside in Crisis to Transition Towns

The United Kingdom's industrial society could only be possible on the back of an agricultural revolution. This consisted of a privatization of land, intensification of food production by mechanization, and the extension of global markets in food commodities under conditions of colonialism as described above. In the latter part of the twentieth century, further expansion of food production was achieved by fossil-fuel powered tractors, chemical fertilizers, pesticides, and growing new high-yielding crop varieties. This model of industrial agriculture became ingrained in market-based food systems, which were extended into rural societies in the Global South from the 1960s through the Green Revolution (see Chapter 2). These global transformations had negative effects on domestic agriculture, however. European Union agricultural policies reduced subsidies for overproduction after 1989 as a response to both the availability of cheaper food from international sources, and recognition of the risks to the environment and human health of industrial agriculture. Attention to heritage stewardship began to replace productivist agriculture in the UK as the consequences of degraded field ecologies from lost hedgerows, depleted soils, and disappearing cultural landscapes became evident.

A number of crises then impacted the politics of food in the UK in the 1990s and early 2000s. The scandal of bovine spongiform encephalopathy (BSE), or "mad cow disease," (caused by feeding "waste" animal parts from meat production as a protein source to herbivores) turned consumers away from burgers, and removed British meat from export markets for several years. Genetically modified (GM) foods arrived in Europe in the form of U.S. soya, maize, and Flavr Savr tomatoes spliced with fish DNA, the latter an example of what some observers refer to as "Frankenfood." GM crops provoked a storm of protest especially in the UK, Denmark, and France, which led to a ban on GM foods in the EU. In 2001 foot and mouth disease (hoof and mouth in the United States) was caused by imported meat sandwiches being fed as edible waste to pigs. Foot and mouth disease quickly spread, fueled by the massive increase in animal movements between livestock markets that were at least in part driven by traders using cell phones. Foot and mouth disease closed down the whole UK countryside to any animal movement, and the control of the disease through mass slaughter led to massive piles of burning carcasses that cast a plume of acrid smoke across rural England in particular. Two books provide important analyses of these crises. Marianne E. Lien and Brigitte Nerlich compare similar global food system crises in their 2004 book *The Politics of Food,* and Tim Lang and Micheal Heasman's widely read book *Food Wars,* also published in 2004, advocates transformation of the food system for public health and environmental interests.

These crises did not lead to a fundamental restructuring of the food economy, however. Government and corporate answers to questions about sustainable food systems in the UK tend to focus on efficiencies of markets, deregulation, removal of trade barriers, and the promises of new technologies such as GM. The result is that increasing control over the food system has been granted in a top-down manner to fewer and larger market actors. These actors are bent on delivering a narrowly conceived notion of value for money in cheap food for urban consumers. Meanwhile the amounts of processed foods have proliferated, the dominance of a limited range of global agri-commodities reduces choice, and obesity and other food-related health concerns are becoming matters of public attention. The role of fossil fuels and petroleum-derived chemical inputs in industrial farming has become a broader concern given peak-oil scenarios (see Chapter 4) and an upward fluctuation in oil and food prices after 2008.

In contrast, responses from outside of government and corporations have turned towards a smaller-scale, networked accountability of food systems to immediate consumers, emphasizing the diversity of food crops and varieties available outside monocultural management. The number of groups and organizations that have contributed to this movement for a more sustainable food system is large and diverse. It is a generational and cultural phenomenon that counters the tendency to see food issues as simply having to do with policy or economic problems needing resolution. Some of these groups have national profiles and lobbying influence, like the Soil Association. Many are regional in focus, and some appear as heritage-oriented and parochial, while others are explicitly seeking to create alternatives to neoliberal globalization. Several of these organizations have long-standing histories behind them to demonstrate the viability of alternative possible worlds of food. Community supported agriculture, for example, is an internationally widespread practice that attempts to rebuild this accountability. In the UK in recent years, the systemic rethinking of food within a pattern of low-carbon, climate-friendly production and distribution has been most closely associated with the Transition Town movement, inspired by Rob Hopkins' *Transition Handbook*.

Information on the Soil Association, one of the UK's leading organizations concerned with sustainable food systems, can be found at their website: www.soilassociation.org/.

Rob Hopkins' *Transition Handbook* can be accessed online at several sites, including this web page published by Transition/United States: www.transitionus.org/transition-handbook.

As stated earlier in this case study, Transition Town initiatives work with the properties of self-organizing growth in socio-natural systems. They aim to reduce inefficient technological dependence on fossil fuels in our food systems and residence patterns, helping to lessen global warming. Emphasis is placed on locally grown food and cultivating community relationships that help replace dependence on the global capitalist economy. This minimizes labor-intensive or energy-intensive interventions in agricultural landscapes that rely on monocultures. The permacultural alternatives usually increase productivity from ground level up to tree canopy (of fruit and nuts) by using vertical layers of vegetation, or through successional ecological companionship. Successional companionship introduces edible plants or plants that attract beneficial insects as part of annual crop rotation. Permaculture design methods (see also Chapter 12) emphasize efficient use of local environmental conditions, and meeting consumption needs by working with ecologically adaptive technologies (e.g., water harvesting, passive solar, multifunctional building design to use opportunities for growing spaces, and renewable energy). It minimizes the need for agriculturally engineered environments for desired crop species that require costly nurturing. Holmgren's term "energy descent" maps out a set of alternative pathways for weaning the existing food system away from fossil-fuel dependency, and towards an economy that can avoid contributing to further global warming, while being more resilient to climate change shocks.

## "It Were Grim Up North"

Hebden Bridge in England's Lake District is one of the shining examples of cultural change for sustainability in the north of England. Since its days of postindustrial decay after the textile industry collapsed in the 1970s, smallholders, artists, musicians, writers, academics, media professionals, gays and lesbians, and landscape gardeners found an attractive, but often rainy, respite from the rat race. (It is very wet; the local Leeds–Rochdale canal built in the 1830s, is one of the only canals in the UK that does not require a back-up water reservoir to maintain water levels and remain navigable year-round.) It is on the route of the Pennine Way trail running along the geological spine of northern England. It has an arts festival, a thriving folk-roots music scene, several organic cafés and bakeries, and the local council only permits the traditional Co-operative store rather than the mainstreet shopping superstores.

It celebrates a thriving boutique retail atmosphere of jewelers, ethnic handbag and wooly hat shops, real ale pubs, and heritage sweet and candy shops. Many local environmental groups have long been active in the area planting trees on the hills of the Calder Valley and hosting "potato day" events promoting the cultivation of rare and distinct varieties. An Alternative Technology Centre was formed there in 2000. Within this context it was not surprising that following the first UK initiative in Totnes, a Hebden Bridge Transition Town group was launched with a festival in May 2010.

More information on Hebden Bridge is available at the "HebdenbridgeWeb" site: www.hebdenbridge.co.uk/.
Information on the UK's first Transition Town, Totnes, can be found at the Transition Town Totnes website: www.transitiontowntotnes.org/.

A view of the canal that runs through Hebden Bridge.

*Source:* Tim Green/Wikimedia: https://commons.wikimedia.org/wiki/File:Hebden_Bridge_(27383325 672).jpg.

A view of Todmorden and the canal that runs through town.

*Source:* Tim Green/Wikimedia: https://commons.wikimedia.org/wiki/File:Another_view_from_Watty_Lane,_Todmorden_(4838711542).jpg.

The neighboring town of Todmorden, in contrast, has responded to the same processes of globalization and cultural change in a profoundly different manner. Todmorden's historically working-class population chose to focus on the value of food, in contrast to such a vague term as "transition." From the beginning, while the goals were similar to Hebden Bridge's, the suggestion that Todmorden should also declare itself a Transition Town was seen as a nonstarter. The area has a deep culture of local distinction and has retained a stronger working-class population and culture than the "yuppified" Hebden Bridge. Local activists knew that for their movement to be genuinely grassroots, it had to appeal across class and ethnic divides.

Thus an alternative rallying cry and motivational appeal was devised to address the same issues: climate change, food insecurity, unemployment, health and well-being, sustainability education, and improving participation in a low-carbon local economy including food. Incredible Edible Todmorden soon became associated with dramatic visual imagery. Vegetable beds bursting with leeks, cabbages, runner beans, artichokes, and sweetcorn sprang up in growing containers outside the health center, on the railway platform, beside the canal, in front of the police station, in car parks, community colleges, in church graveyards, and the grounds of old people's homes. Groups of curious tourists began arriving from elsewhere in the north of England and internationally. The celebrity chef Hugh Fearnley-Whittingstall, who attended the first Incredible Edible harvest festival in 2010, brought a camera crew and worked with the secondary school head cook to turn all the produce that local people could bring into appetizing dishes. The message was therefore broadcast that even in the stereotyped rainy, cold, and economically marginal "grim up north," postindustrial towns, it is possible to grow an enormous range of edible foods. Perhaps even more antithetical to the stereotypes was seeing that health, education, employment, and environmental (including climate change) issues can be addressed together by a community-based mobilization around the theme of food in postindustrial communities like Todmorden.

The movement succeeded in scaling up small-scale individual and community groups who have for decades been growing a patch of vegetables and soft fruit (which grows on bushes, such as berries) into a broad coalition. The public was encouraged to help themselves to the free produce growing in various locations, and encouraged to join in every third Sunday of the month to take care of an adopted vegetable bed. As a sign of the national profile of this nonpartisan, inclusive, broad-based phenomenon, a royal train brought Prince Charles to confer his green seal of approval with a visit to the town in September 2011. In subsequent years, Incredible Edible Todmorden has successfully tapped into funding for food-related and environmentally restorative projects.

Information on Todmorden, including news articles, up-to the minute blogposts, photos, and videos can be found at the town's website: www.incredible-edible-todmorden.co.uk/.

As a community-based movement, this does mean there is diversity and significant difference in emphasis among its various components and leaders. One project has been the promotion of beekeeping to address the severe decline in honeybee numbers, which are vital for pollinating fruit trees and many vegetables. The Incredible Edible Farm has been set up on a permaculture paradigm, using polytunnels, ponds, and hugel beds (raised growing areas created by amassing branches and woody prunings into heaps covered with compost). Apprentices can train, and volunteers learn, by practice and instruction at this farm. In contrast, a more high-tech aquagarden greenhouse has been built at the back of the secondary school. The aquagarden uses heated water tanks to grow tilapia fish in a water cycling system that uses bacteria to break down nutrients from fish excrement, and then circulates the water along rows of giant tomato and cucumber plants, before returning to the fish tanks. These are just three models for education in sustainable food growing for the future. They provide relevant and feasible possibilities to provide employment and to supply restaurants, school kitchens, and food outlets with produce in a manner that reduces the carbon footprint in food miles. Instead of waiting for government to act, people in communities are taking up initiatives and achieving demonstrable, practical change.

## Global–Local Connections

The Incredible Edible name has not only helped to transform Todmorden, it has also inspired other such groups to form, to merge with transition groups, and to take off in translation around the world. At the time of writing, for example, France has four hundred citizens' groups self-described under the label "Incroyables Comestibles"—French for Incredible Edibles. Annual network conferences of Incredible Edible have extended the model to other communities and provided training in cultivation techniques, cooking skills, and financial management, as well as involving schools and putting sustainable food system issues into the curriculum.

Across the Global North, years of economic austerity precipitated by neoliberal economic policies have brought greater audibility and prominence to urban and peri-urban food growing as a sign of positivity flourishing in communities. In the UK, the anticipation of future rounds of welfare cuts by all the main political parties in 2016 has led to predictions of many more people who will be in need of food banks in order to survive. As Pat Caplan pointed out in a 2016 article in *Anthropology Today*, levels of food poverty that were once considered only the unfortunate fate of Third World countries have become normalized in the economy of austerity in affluent nations. The expanding scenarios of food poverty and poor health among the populations who have come to rely on the system of cheap food has come under intense critique in the UK (see Thirsk 2000). In this space there are some new collective processes of societal reflexivity underway, including movements like Incredible Edible.

## Sustainable Food for Cities

The initiatives in the northern English towns described here are part of a disruption to culturally prescribed patterns for the normal order of things in urban landscapes. Something of their transgressive character can be gauged in the breaking up of place-bound taboos, such as growing food in old cemeteries. The steps towards achieving a low-carbon food economy, however, require a much more ambitious striding into the unknown. Robert Biel outlines some features of what this scenario will look like in his article "Visioning a Sustainable Energy Future: The Case of Urban Food-Growing" (2014), which uses the technique of back-casting. Back-casting helps to identify measures that can deliver substantive reorientations of urban locations for food growing. Biel sees urban rather than rural environments as better suited to innovations in food-growing practices due to their proximity to consumers, while much of the rural landscape is locked in to vested interests in chemical agribusiness. A significant contribution to a future of food sustainability lies within the grasp of motivated groups of urban-based citizens prepared to capitalize on the increasing enthusiasm for allotment gardens, fruit and nut yielding parklands, and the potential of high-tech aquaculture providing vegetables and protein in more intensive systems.

There is no doubt that this pathway will disturb a multitude of folk who have formed attachments to a style of urban gentility and tidy straight lines of clipped hedges that mark rustic landscapes from unruly patches of growth. The town and country separation historically paralleled the professionalization of agriculture. This separation produced a notion of a distinctly cultural sphere of life and meant that townspeople could deliberately ignore the mundane provisioning of everyday food requirements. Instead, urban designs with public parks for aesthetic and moral education emerged, regulating food provision services to environments organized in entirely different places and aesthetics. These orders of place and purpose are intentionally and playfully reimagined by many of the urban food grower movements, while undertaking the serious task of reconnecting with culinary socionatures.

## Conclusion

Looking back over the decades since the 1992 Earth Summit's Agenda 21 (discussed in Chapter 2), alternative food movements in the UK have steadily grown at the same time as huge industrial food crises hit the headlines (including, as discussed earlier in this chapter, mad cow disease, genetically modified "Frankenfoods," and others). The Transition Town movement emerged as highly localized and nondirective practices of learning and mobilization for resilience. The rapidity of its spread among community groups from the first UK example in Totnes in 2006 has been impressive, but it is difficult to quantify that spread because "transition," true to the sense of the term, and with self-organizing systems at the fore, always has an impetus to become something else. Several transition initiatives have been grouped together under that label, only to disperse into other self-described categories and applied purposes. One example at the heart of this chapter, the town of Todmorden, was home to a number of people and groups with transition in mind who strategically decided that the term "transition" risked putting off the local working-class populace. They were keen to assert their own local distinctiveness and distance themselves from the middle-class enclave of Hebdon Bridge down the road. By focusing on food, the Incredible Edible initiative more effectively reached into local popular culture. This came with the allure of celebrity chefs and the message some of them have campaigned about. Namely, that people deserve better food and should reclaim the right to the kinds of foods and food-growing opportunities they would like to see.

Using a positive platform for encouraging possibilities for action in the present, rather than getting lost in doomsday scenarios, back-casting has been important to the practitioners of transition in envisaging pathways to an achievable low-carbon livelihood system. A radical direction is required for genuinely transforming the locales of our intensive contemporary habitats into flourishing environments for novel kinds of food growing systems that are not reliant on fossil fuels. The relationships among places, people, water, soil, animals, and plants need to be reinvented.

Many actors are stepping forward in the present moment to take responsibility for producing evidence about how to live more sustainably. The local level is where smaller scale, ecologically, and socially networked food systems accountable to immediate consumers can be carried forward.

The prospect of a multitude of people taking food systems out of the fossil-fuel regime requires more evidence than this one chapter written by a local resident and active grower and experimenter. There are no off-the-shelf solutions, but the case study here points to the potential of new collaborations that remake humans and nonhuman relationships, even those with the odd neighbors five miles up the road. As discussed, food systems bring together calories and nutrition, community, and visions for long-term sustainability. For these reasons, thinking about food system sustainability requires attention not only to material human sustenance and long-term agricultural productivity, but also to narratives about fairness and the public culture of food.

## References and Further Research


Barry, John, and Stephen Quilley. "The Transition to Sustainability: Transition Towns and Sustainable Communities." In *The Transition to Sustainable Living and Practice*. Published online: March 9, 2015, 1–28. doi:http://dx.doi.org/10.1108/S2041-806X(2009)0000004004.

Beck, Ulrich. 1992. *The Risk Society: Towards a New Modernity*. London: Sage.

Berry, Wendel. 1981. *The Gift of Good Land*. Berkeley, CA: Counterpoint.

Biel, Robert. 2014. "Visioning a Sustainable Energy Future: The Case of Urban Food-Growing." *Theory, Culture & Society* 31, no. 5: 183–202.

Boulianne, Manon. 2006. "The Movement for an Economy of Solidarity: Urban Agriculture and Local Exchange Trading Systems in Quebec." In *Markets and Market Liberalization: Ethnographic Reflections*. Published online. 261–279. doi:http://dx.doi.org/10.1016/S0190-1281(05)24009-X.

Campbell, Ben. 2008. "Fields of Posthuman Kinship." In *Kinship Matters: European Cultures of Kinship in the Age of Biotechnology*. Edited by Jeanette Edwards and Carles Salazar, 162–178. Oxford: Berghahn.

Caplan, Pat. 2016. "Big Society or Broken Society? Food Banks in the UK." *Anthropology Today* 32, no. 1: 5–9.
Cobbett, William. [1823]. 2013. *Cottage Economy.* Internet Archive/Amazon Digital Publishing.
Dirlik, Arif. 2000. "Place-Based Imagination: Globalism and the Politics of Place." In *Places and Politics in the Age of Globalization*. Edited by Arif Dirlik. New York: Rowman and Littlefield.
Drayton, Richard. 2000. *Nature's Government: Science, Imperial Britain, and the "Improvement" of the World.* New Haven, CT: Yale University Press.
Fara, Patricia. 2003. *Sex, Botany and Empire: The Story of Carl Linnaeus and Joseph Banks*. London: Icon Books.
Hajer, Martin. 1996. "Ecological Modernisation as Cultural Politics." In *Risk, Environment and Modernity: Towards a New Ecology*. Edited by Scott Lash, Bron Szerszynski, and Brian Wynne. London: Sage.
Holmgren, David. 2002. *Permaculture: Principles and Pathways Beyond Sustainability.* Hepburn, Victoria, Australia: Holmgren Design Services.
Hopkins, Rob. 2014 reprint. *Transition Handbook*. Cambridge: UIT Cambridge.
Hyams, Edward. 1952. *Soil and Civilization.* London: Thames and Hudson.
Kloppenburg, Jack. 1988. *First the Seed: The Political Economy of Plant Biotechnology, 1492–2000.* New York, NY: Cambridge University Press.
Lang, Tim, and Michael Heasman. 2004. *Food Wars: The Global Battle for Mouths, Minds and Markets*. London: Earthscan.
Lien, Marianne E., and Brigitte Nerlich. 2004. *The Politics of Food.* Oxford: Bloomsbury Academic.
Malthus, Thomas R. 1798. *An Essay on the Principle of Population, as it Affects the Future Improvement of Society*. London.
Marx Karl, 1954 [1867]. *Capital*, Vol. I. London: Lawrence and Wishart.
Mintz, Sydney. 1986. *Sweetness and Power: The Place of Sugar in Modern History.* New York: Penguin Books.
———. 2008. "Introduction." In *Food and Globalization: Consumption, Markets and Politics in the Modern World.* Edited by Alexander Nützenadel and Frank Trentmann. Oxford: Berg.
Mollison, Bill. 1991. *Introduction to Permaculture*. Tasmania, Australia: Tagari.
Pimbert, Michael. 2006. *Transforming Knowledge and Ways of Knowing for Food Sovereignty.* London: International Institute for Environment and Development.
Pimentel, David, Sean Williamson, Courtney E. Alexander, Omar Gonzalez-Pagan, Caitlin Kontak, and Steven Mulkey. 2008. "Reducing Energy Inputs in the US Food System." *Human Ecology* 36: 459–471.
Pollan, Michael. 2008. *In Defense of Food: An Eater's Manifesto*. London: Penguin.
Prebble, John. 1969. *The Highland Clearances.* London: Penguin.
Pretty, Jules. 2002. *Agri-Culture: Reconnecting People, Land and Nature*. London: Routledge.
Schumacher, Ernst F. 1976. *Small is Beautiful: A Study of Economics as if People Mattered.* London: Abacus.
Scott, James. 1998. *Seeing Like the State: How Certain Schemes to Improve the Human Condition Have Failed.* New Haven, CT: Yale University Press.
Scott-Cato, Molly, and Jean Hiller. 2010. "How Could We Study Climate-Related Social Innovation? Applying Deleuzean Philosophy to Transition Towns." *Environmental Politics* 19, no. 6: 869–887.
Sen, Amartya. 1981. *Poverty and Famines: An Essay on Entitlement and Deprivation.* Oxford: Oxford University Press.
———2008. *Soil Not Oil: Climate Change, Peak Oil and Food Insecurity*. London: Zed Books.
Shiva, Vandana. 2008. *Soil Not Oil: Environmental Justice in an Age of Climate Crisis*. Berkeley, CA: North Atlantic Books.
Stern, Nicholas. 2006. *The Economics of Climate Change: The "Stern Review."* Cambridge: Cambridge University Press.
Thirsk, Joan. 2000. *Alternative Agriculture: A History from the Black Death to the Present Day.* New York: Oxford University Press.
Thoreau, Henry. 1948 [1854]. *Walden, or, Life in the Woods, and On the Duty of Civil Disobedience*. New York: Rinehart & Co.
Williams, Raymond. 1959. *Culture and Society, 1780–1950*. London: Chatto and Windus.

# 9

# Community Forestry and Sustainability in Nepal

Dinesh Paudel

This chapter explores the theory and practice of community forestry and the various ways that it connects to and contradicts the idea of sustainable development in rural Nepal. Community forestry is often upheld as one of the leading environment and development programs in the Global South. It came to prominence after 1970, when international development agencies conceptualized various forestry programs to counter the increased alienation of communities from the conservation of forest resources. Among them, community forestry was the highest priority program implemented throughout the developing world with the original goals of conserving forest ecosystems and improving rural livelihoods. Nepal was one of the pioneer countries to implement community forestry, and over the past few decades Nepal's community forestry has been celebrated as a successful model for re-establishing communities' ownership of forests and addressing some of the growing social, economic, and environmental problems in rural areas. By 2014, over seventeen thousand Community Forestry User Groups (CFUGs) had been formed throughout the country, formally establishing more than one-third of the country's population as members of community forestry, which has been largely successful in reforesting Nepal's hillsides and redistributing forest resources (Department of Forests 2014). Nepal's community forestry has clearly demonstrated that guaranteeing communities' collective entitlement over the forest resources is the fundamental precondition for effective conservation and equitable livelihoods (see Ojha 2014; Malla 2001). Over time, however, community forestry innovations have become vulnerable to state and market interests and encroachments, increasingly undermining the possibility of achieving social and ecological sustainability in Nepal's forest sector in the present moment. This situation has been further exacerbated by existing community hierarchies and power differences in managing and utilizing forest resources.

As described by Donald Gilmour, R. J. Fisher, and other analysts, community forestry is a decentralized and participatory system of governing common forestlands in which authority for forest management and use is formally handed over to local communities (see Gilmour and Fisher 1991; Charnley and Poe 2007; Agarwal 2010). The concept of community is foundational for conceptualizing and operationalizing community forestry. Communities are often imagined as a collectively functional group of people living in a particular geography with common interests and shared socio-cultural practices. In reality, however, communities can be heterogeneous, geographically scattered, and differentiated in terms of size, interest, institutional structure, and everyday interaction. Despite this heterogeneity, historically, communities have successfully engaged in collective actions toward common goals related especially to their livelihoods and natural resources. Balancing these multiple definitions, community forestry defines a community group as those people who have been living in and utilizing a particular forestland for a long period of time, developing specific cultural and economic ties to that forest, and who are willing to take responsibility for collective stewardship and sustainable management of forest resources. Community forestry formalizes these community groups to legally transfer ownership of forest resources (but not the land itself) and responsibility for implementing forest management plans. In recent years, however, community forestry is moving beyond a locally based, everyday interaction between the forest and communities, and emerging as a complex web of relations among local people, bureaucrats, contractors, donors, environmentalists, and other interested parties in appropriating forest resources at different levels. This shift in interaction and control has redirected community forestry from its original indicators of sustainability as being locally managed, equity based, subsistence driven, and integral to ecological processes toward multi-stakeholder, externally influenced, and commercially oriented forestry practices.

Community forestry has gained momentum since the 1990s as a model of sustainable development in Nepal. Its proponents argue that community forestry is very well equipped to address the global goals of simultaneously protecting the environment and achieving economic growth. Critics claim that community forestry has not only exacerbated social inequalities and hierarchies, but has become an instrument of the state and the market to exploit cheap labor and forest resources. Mobilizing ideas from both critics and supporters, sustainability scholars have comprehensively explored parallels between principles of sustainable development and community forestry practices at the local level. The emphasis has been on the potential of community forestry in addressing social and environmental issues and achieving sustainability goals at the local level. In response, community forestry programs have identified and incorporated activities informed by the three pillars of sustainable development: ecological conservation, equitable distribution, and economic growth. These new initiatives, however, are also driven by the ideas of economic growth, market linkages, and infrastructure development rather than on the fundamental aspects of sustainability, including social and environmental justice.

Does community forestry challenge the existing status quo and provide alternative pathways toward nonhierarchical, nonexploitative forestry governance, or is it simply a new name for the same processes of extending market mechanisms and state control to enroll communities and common forests into destructive capitalist production of forest resources? How does community forestry compare with contemporary understandings of sustainable development? This case study from Nepal will help us to examine critically how community forestry parallels sustainable development and in what ways it can become an alternative method of forest–society interaction. The remainder of this chapter will be structured as follows: a history of community forestry, a description of the origins of community forestry as a sustainable development initiative in Nepal, and evolving trends.

## History of Community Forestry

To better understand community forestry, we need to look at its origins. The idea gained traction in the 1970s as an urgent response to the increased degradation of forest ecosystems resulting from the inability of governments and markets to halt deforestation, soil erosion, and habitat loss. Forest management in the 1950s and 1960s was dominated by complementary ideas of the "tragedy of the commons" (see Chapter 3; Hardin 1968) and timber production for economic growth. The idea of the tragedy of the commons was informed by Cold War capitalism (where natural resources owned collectively by communities were seen as an irrational and wasteful property regime, as well as a potential seedbed of communism) and hardcore economics of private property. Collectively managed forestlands in the developing world were therefore viewed as open-access resources that would inevitably be overexploited, leading to eventual collapse of forest ecosystems and of forest-dependent communities and economies. The solution, based on this conceptualization, was to transfer forest ownership to either the state or the private sector.

See "The Tragedy of the Commons" by Garrett Hardin at this Garrett Hardin Society web page: www.garretthardinsociety.org/articles/art_tragedy_of_the_commons.html.

During this same period, international development agencies, especially the World Bank, were investing massive loans in the recently independent postcolonial countries in the Global South to promote industrial extraction of natural resources as a main strategy of economic growth. The governments in these countries perceived this development as an integral part of state-building processes and as a way to expand their access to the global market. Post-colonial countries with ties to the West and the World Bank (such as Brazil, Indonesia, etc.) typically pursued large-scale natural resource extraction through privatization, while those countries with ties to centrally planned socialist countries (such as India) typically nationalized natural resources and industries. For both capitalist and socialist countries, however, large-scale, export-oriented, and non-community-controlled natural resource extraction was seen as essential for the development of these newly independent countries.

Influenced by the combination of these multiple factors, countries in the Global South nationalized most of the common forestlands and made them available for commercial extraction and revenue collection in the 1950s and 1960s. For example, in Nepal all of the country's forestlands were nationalized in 1957 by establishing a national forest bureaucracy to control and utilize forest resources. Communities were denied collective ownership and legally banned from continuing traditional forest-use practices. This left them no option but to act as free riders or bribe state officials to get access to the forest resources. Ironically, this community pressure, paired with the government's priority of extracting timber from the southern plains of Nepal, transformed the entire forestlands into a de facto open-access regime, triggering massive deforestation, degradation, and landslides throughout the country.

The alarming rate of deforestation in the 1960s and 1970s and the growing discontent among forest-dependent communities against top-down forestry systems prompted policy-makers, development agencies, economists, and environmentalists to consider inclusive and community-based forest management policies throughout the Global South. Community forestry emerged not as a political innovation embracing social justice and the authority of indigenous practices. Rather, it grew out of an urgent need to address the failure of state- and market-based policies, to halt deforestation, and to develop coping strategies for forest-dependent communities.

Two evolving ideas in the 1970s provided a conceptual framework for the implementation of community forestry in the Global South. First, the idea of alternative paths of development against market-centric development programs became popular in the 1970s among critical thinkers, activists, and indigenous communities. Mainstream development was criticized for being an exploitative, environmentally destructive, and top-down imposition of market rules that led to further social inequality and marginalization. In its place, communities and activists demanded bottom-up, participatory, and locally controlled development programs that would ensure community livelihoods and ecological integrity. The idea of community forestry fulfilled those demands and emerged as a model of sustainable and participatory development in the Global South. Some considered community forestry a radical departure in the development paradigm and evidence for the arrival of a new era of development.

Second, revising the earlier understanding of the tragedy of the commons, natural resource economists came up with the idea of Common Property Resource Management (CPRM) in the late 1970s, arguing that communities could manage forest resources rationally and that community institutions were compatible with state and market rules (see Ostrom 1990 for the institutional mechanism of managing the commons). The concept of CPRM allowed scholars and development practitioners to see the commons as a legitimate property regime outside the structure of state or private ownership. They found that communities could provide effective stewardship to maximize the production and conservation of commonly held natural resources given a handful of criteria: they needed to have identifiable boundaries, well-defined membership criteria, robust institutional arrangements (whether formal or informal), clear rules of resource allocation, and mechanisms for punishing rule breakers. Informed by this rational understanding of group dynamics and common property, community forestry was initiated to involve the local people who were living in the vicinity of the forests in designing and implementing forest protection and management activities. In this conceptualization, the common property regime is ensured by legally recognizing community groups as institutionally organized, collective owners of common forestlands where communities acquire exclusive property rights so that they can enforce property rules to produce goods and services for both subsistence and commercial utilization. Emerging from these two different contexts, community forestry includes elements of participatory development, representation, markets, and the common property management regime.

## Origin and Expansion in Nepal

Beginning in the early 1980s, international development donors conceptualized and experimented with Nepal's community forestry to involve local communities in controlling the increased rate of deforestation in the Nepali mountains (see Thompson and Warburton 1985; Gilmour and Fisher 1991). The donor-funded development projects introduced community-based forest governance and management practices by blending together indigenous knowledge systems and forestry practices with the discourses of participatory development and environmental sustainability. The move from the state-controlled extraction of forest resources for industrial utilization during the 1950s and 1960s to a more participatory management of forests by local people received overwhelming support across scales. Development agencies found community forestry to be a good alternative strategy to define property rights of common forestlands, which could be mobilized for economic growth in rural areas (see Agrawal 1996). Similarly, local communities, especially the elites, believed this to be an opportunity for gaining control over forests that were under their control before the nationalization of the forests in 1957 (see Malla 2001). Initially forest officials were reluctant to support this initiative. Community forestry garnered their support especially with the motive of mobilizing rural people for the protection of forests in areas where forest authority was practically absent.

The emergence of community forestry and its subsequent expansion in Nepal were grounded in four prominent national and international contexts. First, the state-led industrial utilization of forests with the emphasis on revenue generation was portrayed as a barrier for the promotion of free markets (see Thompson and Warburton 1985). States' increased regulation of the forest sector was criticized for being top-down by both pro-market development agencies and the proponents of bottom-up, participatory development. Nepal's nationalization of forests in 1957 was seen as a move toward a centralized economic system in line with the growing socialist economies in Asia, including India and China. The nationalization of the forest exacerbated already rampant deforestation for new settlements, timber smuggling, bureaucratic corruption, and alienation of communities from forestry activities. In remote villages, especially in the hilly areas, communities were unaware of nationalization and continued to provide informal stewardship of the forests, in many places successfully avoiding free-rider problems and maintaining ecological integrity. In line with the shifting paradigm in international development from state-led industrialization to fulfilling basic needs of the people in the 1970s, state control in forestry was discredited in development and understood as a barrier for sustainability and a reason for the increased degradation of the forests and growing poverty in rural areas.

Second, the idea of participatory development that became popular since the 1970s provided the conceptual framework for community forestry in Nepal. Practicing participatory forestry for the involvement of communities who traditionally had been using forests for their subsistence requirements was conceptualized as liberation from repressive and corrupt state bureaucracy (see Agrawal and Gibson 1999; Malla 2001). Proponents of participatory development viewed community forestry as an anti-capitalist form of development. Giving power to communities who would collectively own local resources was a victory for them. The idea of bottom-up development was popular among the donors. Community forestry fulfilled the desire for an alternative development model and received direct support from critical development thinkers, progressive NGOs, leftist political forces, and social activists. They considered this evolving shift in the forest sector as a necessary precondition for sustainable development where community forestry was understood as a locally controlled and ecologically harmonious developmental change. Generally, community forestry was celebrated as a success against widespread corruption in the forest sector. It was a movement against the relentless extraction of timber and non-timber products by the state and private companies, and against the continued extralegal coercion of forest bureaucracy against the communities.

Access Agrawal and Gibson's 1999 article, "Enchantment and Disenchantment: The Role of Community in Natural Resource Conservation" published in the journal *World Development*, here:

www.sciencedirect.com/science/article/pii/S0305750X98001612.

Third, the Theory of Himalayan Environmental Degradation (THED) provided an ecological justification for the emergence of community forestry in Nepal. This theory was developed by Western environmentalists in the 1970s to explain the "inevitable" ecological catastrophe in the mountains of Nepal. According to this theory, the Himalayas were going through a serious environmental crisis leading to soil erosion, catastrophic natural disaster, and loss of biodiversity. The theory projected that the continuous mismanagement of the Himalayan landscape would lead the people of the region, especially rural farmers, to rise up in protest, sending all of South Asia into social chaos and political instability (see Guthman 1997; Ives and Messerli 1989). Projections of Himalayan degradation as a serious environmental crisis shifted development priorities away from infrastructure and industry to conservation of the degrading Himalayas. This new theory encouraged participation from inhabitants of the Himalayas out of the belief that it would change their land-use practices, which THED adherents thought were irrational, unscientific, inhuman, and therefore the causes of deforestation, soil erosion, and flooding (see Metz 1991 and Eckholm 1976). International environmental organizations identified local inhabitants as threats to the forests. The idea of community forestry was the perfect model of development to counter these environmental problems, allowing local people to participate in these conservation efforts. Conservation interests of environmental programs seeking involvement of local people in protecting Himalayan environments culminated as community forestry in Nepal.

**A THED-Induced Forestry since the 1970s**

According to a retired government official, "We used to send hundreds of extension workers to villages with posters to show people good land use systems, and seedlings for plantation to control soil erosion. We did that for more than fifteen years during the 1970s and 1980s." These programs were informed by the idea of "environmental crisis" in the Himalayas as developed by THED. Many donor agencies including the World Bank, USAID, UNICEF, and the German Organization for Technical Cooperation organized a workshop to talk about the growing environmental problem in the Himalayas in 1974. Eric Eckholm—a reporter for the *New York Times*—was hired to publish the proceedings. Eckholm later published a book called *Losing Ground* in 1976, calling for an urgent intervention to halt environmental degradation in the Himalayas, which was elaborated by other organizations as THED. The book became famous especially among donor agencies and completely transformed the priorities of international development agencies in Nepal. For details read Eckholm 1976; Ives and Messerli 1989; Metz 1991; Guthman 1997.

Fourth, the existence of indigenous resource appropriation practices, mostly controlled by feudal elites, which was later conceptualized as "indigenous systems of common property resource management," served as a theoretical justification for the emergence of community forestry in Nepal (see Malla 2001 and Ostrom 1990). Feudal elites were extremely unhappy about nationalization of forests, which were under their control before 1957. They saw community forestry as an opportunity to regain their control over forestlands. Similarly, some influential forestry officials also endorsed the idea of indigenous community forestry mainly to serve interests of feudal families, which was where they came from, and to make international donors happy so that foreign scholarships for higher study were ensured. These new interests created an opportunity for the emergence of community forestry in Nepal.

Those who were suspicious about community forestry included landless peasants, especially Dalits (members of the lowest caste), and poor households with small landholdings. These groups saw community forestry as an imposed control of local elites over forests and people. The open-access situation before community forestry was beneficial for them as they could access forest resources with few restrictions. But new community-enforced regulations would affect them more than other groups, as they had to rely on forests for fulfilling everyday needs. For them community control over forests became a site for restriction, compliance, and punishment, and at the same time it reinforced elite domination. Forest officials who were interested in timber trade were not a threat to poor households as they rarely harvested timber for their own use.

Despite these multiple interests and ideas behind the emergence of community forestry, the program showed success as local-level sustainable development: it led to the devolution of power to forest-dependent communities, the prioritization of subsistence utilization over commercial exploitation, and the natural regeneration of species and ecological balance. Community forestry priorities and programs shifted massively especially since the 1990s, when the commercial and political interest of the elites became dominant over genuine possibilities of achieving social equity and ecological integrity.

## Evolving Trends

Nepal's community forestry has gone through a substantial shift from its original goals of securing community livelihoods and preserving forest ecosystems to the contemporary priority of producing commercial goods for the market, including timber, non-timber products, and environmental services. The idea of sustainable development understood as a method of achieving simultaneous advancement in economic growth, ecological conservation, and social equity has driven these shifts in community forestry policies and priorities. As elaborated more fully in this section, community forestry has mobilized the concept of sustainability to highlight the importance of fulfilling everyday subsistence requirements of the people and also of satisfying market demands for income generation.

With these changes, community forestry has become an intersection of two simultaneous but contradictory ideas. First, some people consider community forestry as a commons that reproduces collective spaces by nurturing nonalienating, nonhierarchical, and nonexploitative relationships between the commoners (communities) and the commons (forests) outside the market. In this understanding, community forestry is about reproducing indigenous forest–people relationships and securing subsistence livelihoods of forest-dependent communities and ecological integrity of the forests as an ecosystem of coexistence. Second, community forestry is a way of enrolling communities and the commons into market relations in order to produce commodities that otherwise were inaccessible to the market and/or to avoid losing the stock of commercial value because of the free-riding problem. Community forestry reproduces both these conditions simultaneously, though it is increasingly integrated with market mechanisms prioritizing commodity production instead of everyday subsistence. Communities as a group have a difficult time gaining a fair share in the market, while the middlemen and contractors extract forest resources as local agents of the market value chain. To arrive at this point, community forestry evolved through various changes over the last few decades in Nepal.

Nepal's Forest Act of 1993 was a major force in these new changes, including the establishment of Community Forestry User Groups (CFUGs). Members of a CFUG have: (a) the right to self-organize, (b) entitlement to forest resources, and (c) the right to benefits generated from the community forest.

A CFUG comprises about 100 households on average in Nepal. These households are often divided into interest-based subgroups such as non-timber harvesters, blacksmiths, women-, and grazing-only groups. Members participate in *tole* (hamlet) meetings, executive committee, and the general assembly.

To read the full text of the Forest Act of 1993, see this document from FAOLEX, the database of the UN's Food and Agriculture Organization: www.fao.org/faolex/results/details/en/c/LEX-FAOC004527/.

As the initial objective of community forestry during the late 1970s was to fulfill daily forestry needs of rural people, the authority to control forest protection was given to local village leaders. The main idea behind this approach was based on a development concept of the 1970s that rural poverty could be reduced by providing people with the opportunity to fulfill their own basic needs (see Gilmour and Fisher 1991). Subsistence farmers were asked to plant trees on their private land as well as common forests (see Metz 1991). The idea of community as a group of people to manage local forest resources became appealing in the early 1980s when community forestry started to create local institutions of community members. A formal community institution was required to be able to hand over a defined area of forestland for community protection and management. Community groups were found to be very heterogeneous, so community forestry focused on developing a common institutional mechanism by which forestry activities could be harmonized throughout the country (see Thompson and Warburton 1985). Community forestry was still criticized by NGOs and development projects on the grounds that state control was still prominent. They wanted increased local control and full autonomy of community groups. To ensure local institutional robustness and shifting of authority to local communities, the Master Plan for the Forest Sector was prepared by donor projects, and the national community forestry act was also legislated in the early 1990s. The Forest Act of 1993 recognized community groups as self-governed institutions with full entitlement over the forest resources.

The focus of the community forestry programs in the early 1990s was on institutional strengthening and empowerment of community groups. Community groups needed extra skills and new systems of governance for the effective implementation of community forestry programs. Local institutions were dominated by informal practices, lack of transparency, and elite domination. Creation of awareness to empower community members to take part in community forestry processes became a priority. Household visits, group-level training, community meetings, and other on-the-ground initiatives were organized to change people's attitudes and behaviors so that they might practice norms and values as developed by community forestry. Forest officials, NGO workers, community facilitators, development projects, and local consultants were mobilized intensively as a campaign of community forestry. NGOs were very supportive of this change as they were actively involved in community activities. Community leaders, benefiting from various empowerment and training programs, could establish themselves as development leaders locally. Donor projects mobilized forest officials for the implementation of local empowerment programs. These officials saw the opportunity to be part of empowerment programs as a way to improve relationships with people as well as with other social activists and leaders.

Community forestry programs focused on enhancing the networking abilities and social capital of community groups in the mid-1990s. The focus on local empowerment and formal institutional initiatives in communities was criticized for being too narrow to promote overall economic growth and development. A new multi-partnership approach was based on the emerging developmental concepts of "social capital" and "multi-stakeholder" ownership (see Malla 2001). This approach was a clear move to allow civil society and private sector actors to take an active role and ownership in community forestry. Communities were unable to commercialize forest products without the active involvement of private sector investors and middlemen. NGOs' facilitation was sought to help establish commercial networks and transfer forest management skills. Elite members of the community took a lead in establishing forestry-related networks. Community forestry programs conducted various study tours, joint workshops, and networking campaigns to facilitate the process of developing multi-partnerships at the local level. Alliances between these newly developed stakeholders started to emerge at the local as well as national levels.

The private sector started to play a strong role in utilizing forest products. Community groups were encouraged to develop collaborations with the private sector. Contractors were the most effective in translating their interests into community operational plans by developing alliances with community leaders. Community elites, who acted as representatives of an entire community, often sought to build alliances with officials and contractors that ultimately benefited themselves. This was a major shift in community forestry when the state's role was minimized and previously unknown private sector actors started to control community activities.

The late 1990s was an era of good governance in community forestry in response to criticisms that, through its multi-stakeholder approach, existing power relations were reinforced by excluding women and other marginalized populations. Critical researchers highlighted the increased elite domination in benefit sharing (see Nightingale 2005). Community forestry responded with new programs of gender inclusion, equitable distribution, transparency, and productive use of resources, income generation, and efficiency in decision making. Wrongly, the exclusion of community members from forestry activities was thought to be the result of the lack of individual capacity to participate rather than institutional barriers established by community forestry itself. As a result, community forestry programs focused on changing community dynamics through the ideas of good governance and encouragement to participate in community activities. These programs rested on the belief that institutionalization of good governance would be sufficient to deal with the problems of power differences, inefficiency, and exclusion, and a formal governing structure would provide a comprehensive and sustainable development model at the local level. The irony was that those community members who were not taking part in the community forestry process were identified as incapable to participate when in fact their actions were often a form of resistance to these changes in community forestry.

The beginning of the millennium was a major turning point in community forestry with the introduction of the market as a main instrument of conservation and economic growth. Entrepreneurship, income generation, self-help community development, and scientific forestry were some of the new programs in Nepal's community forestry during the early 2000s. Communities' continued subsistence utilization of forests was explained as a reason for their increased poverty and lack of development (see Paudel 2016). Those communities that could finance local development activities were considered sustainable. Entrepreneurial ideas were taught to community members to develop commercial linkages. The argument was that communities were inclined to protect forests if they would get sufficient economic incentive from their collective actions. This notion of economic growth became popular also among those who were advocating for local and autonomous development. These entrepreneurial ideas were successful in mobilizing all forces that were opposed to market-led development. A wide range of community members and development facilitators considered that community involvement in entrepreneurship and commercial production was a path toward local prosperity and autonomy. Very radical-sounding development strategies like poverty reduction, inclusion, and income generation to poor people became so powerful that community forestry received broad support from every sector of society including the poor, marginalized, and working people in rural areas.

The most recent shift in community forestry concept is its focus on pro-poor economic growth and peace building. This shift is a response to the criticisms of entrepreneurialism in widening the gap between rich and poor people and the disproportionate flow of benefits to private investors. The reasons identified for such skewed distribution of benefits were lack of capacity of poor people in taking part in market systems. The solution then was to train poor people to take part in entrepreneurial production processes. Loan provisions, microcredit mechanisms, and plantations of commercial species have been the main activities in recent years. Community leaders are very appreciative of this shift, and since then they have been developing commercial relationships with private-sector investors.

This trend shows that Nepal's community forestry has already become an integral part of commercial production moving away from subsistence-based forestry programs. Forest resources are measured and managed for their economic value rather than their ecological importance. Old trees, slow-growing species, crooked branches, tree hollows, and noncommercial species are now categorized as non-values or waste. Communities are encouraged to collect these low-value products for subsistence and to prioritize fast-growing, young forests with marketable timber and non-timber species. These changes have naturalized the technical measurement of forest products and calculation of cost and benefit as a basis for community decision making. One can argue that community forestry is following private property relations by introducing commercial and entrepreneurial principles. Local actors of community forestry, who had not necessarily intended to achieve such market-based outcomes, are being mobilized for the commercial production of forest resources.

In its few decades of development, Nepal's community forestry has become part of a truly international agenda. Moving beyond forestry, international development agencies found community forestry more conducive to overall rural development goals. Transforming the local agenda of development, community forestry groups provide leadership for wider community development activities, and they have actively organized themselves into networks and federations. Community forestry has become a movement encompassing a variety of interests ranging from local autonomy to an engine of economic sustainability.

## Shifting Community Dynamics

Even though commercial extraction has dominated much contemporary forest management in Nepal, everyday workings of community forestry have brought in tremendous socio-environmental changes to rural areas, changes that belie simple positive or negative classifications. Co-optation of local potential and resistance against exploitation and domination are happening simultaneously. In spite of exemplary participation of local communities, forest officials have never welcomed handing over forests that they used to control. Furthermore, external actors such as forest officials, timber contractors, middlemen, carbon traders, NGO workers, and development facilitators are increasingly dominating local decision making and everyday activities. At the user-group level, growing internal inequalities and the distribution of benefits from community forestry favor elites, male and rich. However, communities are also radically empowered, organized along the lines of different interest groups and able to politicize issues of inequalities, marginalization, and access to resources. Bottom-up resistance is enormous where communities have been persistent in organizing protests and public campaigns against the state's interest of curtailing communities' authority and responsibility. These actions have demonstrated an enormous potential to create a truly sustainable space as the following description of the unfolding of community forestry practices reveals.

The process of forming community forestry generally follows three main steps in establishing forest user groups and implementing forestry activities with the communities. The first step includes identification of the forestland to be managed as community forest and the empowerment of communities who have historical ties with that particular forest area and are willing to take responsibility for its management. The crucial part of this step is the implementation of rigorous and comprehensive activities to empower community members about their rights and responsibilities in managing forest resources. Issues of inequalities, marginalization, poverty, decentralization, and rights to access forest resources are discussed individually, within neighborhoods, and collectively as a larger community. The belief is that empowered community members would be institutionally organized to avoid conflicts and implement forestry activities effectively and rationally. It should be noted that this part of the process is intended to diffuse contentious political questions relating to forest ownership, control, and access, and to generate confidence among community members toward social justice and community rights. However, the unintended consequences of these activities have often been a re-politicization of forest questions by supporting elites' interests of control and domination while at the same time bolstering demands for equality by other group members.

Following identification and empowerment, the second step focuses on the establishment of a formal institution as an organized Community Forestry User Group. The group Constitution defines the institutional aspects of the CFUG, including membership, subgroups, committee representation, decision making, forestry rights, individual responsibilities, group participation, and mechanisms of distributing benefits. An Operational Plan outlines the ecological details about forest resources and yearly plan of operation to be carried out in the forest including protection, planting, cleaning, harvesting, and so forth. There are certain standards and guidelines used to develop these documents, but each CFUG is independent in defining the social arrangement and the responsibilities based on the local tradition, culture, and practices. This step is sometimes liberating for communities, where they can formalize their indigenous practices and organize themselves as an autonomous, collective owner and manager of forest resources. A number of external actors are involved in this process including forest officials, contractors, NGO activists, and community elites with their own agendas and interests. Communities usually compromise with external interests

Community Forestry User-Group handmade paper factory, eastern Nepal.

*Source:* Andrea J. Nightingale.

while developing these documents, but local autonomy and people's livelihoods remain central in this process of formalizing community forestry at the local level. Nevertheless, possibilities for revitalizing historically coexisting relationships between the forest and people are strong at this step.

Community forestry becomes most vulnerable to exploitation by external actors during the production stage. Depending on the ecological conditions and management practices of a particular community, the stock of the forest grows and becomes available for harvesting after some years under community management. The market promotion programs act as a seductive force to make commercial extraction of timber and non-timber products seem like a natural and inevitable stage of community forestry for development and progress. Not every community commercializes forest products, but there is increased pressure on communities to generate financial resources locally to implement community development activities. Once commercialization is institutionalized locally, community forestry crosses community boundaries and the contractors, middlemen, and forest officials become dominant in decision making by developing alliances with community leaders. As commercialization grows, economic valuation techniques are used for categorization, enumeration, and priority setting of species in Nepal, deploying precisely the same principles used in private commercial endeavors. These tools minimize the government roles in local development, but open up spaces for the market and NGOs.

In order to systematize these changes, development tools such as well-being ranking, microcredits, and individual livelihood plans are implemented in rural Nepal. Well-being ranking is a tool to rank community members based on economic assets they own, and identify poor households with fewer economic opportunities. When community forestry was criticized for not being sufficiently committed to the reduction of poverty, a method for well-being ranking was introduced to identify households for income-generating activities. Once the poor households are identified, community funds are established and financial institutions appear to provide them with microloans. The outcomes of well-being ranking and the processes of microcredits are the basis of individuals' livelihood improvement plans, another development tool, used for promoting family-based entrepreneurship. This idea of individual planning is introduced to systematically change indigenous practices that are seen as behavioral limitations to commercial production.

As a final step, community forestry has recently introduced a tool, linked with the market, to establish commercial connections between the communities and the contractors and buyers. This intervention has helped to promote the public–private partnership model in community forestry. These changed priorities and practices have reshaped the composition and distribution of forest species. As forest species are identified, categorized, and measured using economic valuation techniques, those species that are not economically significant are of lower priority for protection and management. Communities remove economically less valuable species to facilitate the growth of productive ones. Such changes in species composition is not only generating a less biodiverse forest with the domination of economically profitable species, but buyers located outside the community actually acquire power to make decisions about forest management plans in the community.

These commercial possibilities are sustained through the emerging multi-scalar alliances between community elites, contractors, and government officials, institutionalizing the market as a decisive force in community forestry. This shift has satisfied the interests of all dominant forces at the local level: (1) for community elites, taking leadership and earning additional income; (2) for government officials, maintaining control over commercial distribution and gaining personal financial benefits (bribes) with less legal and social risks; and (3) for contractors, increasing profits in less time. Such invisible but powerful alliances are sustained by establishing informal benefit-sharing agreements. The involved actors interact informally making it hard to recognize how they operate in everyday activities. These alliances are dangerous to equitable and sustainable community forestry. They threaten the means of subsistence livelihood especially of forest-dependent and marginalized people, putting them at risk of being dispossessed of their resources. However, the possibilities of reversing this trend are always present as such alliance building is not monolithic or straightforward. The process of building alliances must reproduce legitimacy and acceptance among the involved actors in every transaction. To increase the share of the total benefit, each actor mobilizes its constituency as a force for renegotiation of its position. For example, community elites mobilize community members as a collective force if they feel that their share (bribe) is not appropriate. They may obstruct, delay, produce low quality, or reduce quantity to reassert their position for getting more benefits. This moment of renegotiation, however, can be an opportunity for instigating sustainability practices, as local communities can reorganize the production and distribution processes to dismantle these alliances using the existing legal and institutional provisions in community forestry.

## Resistance and Control

The contemporary discourse around natural resource sustainability revolves around eco-friendly production, consumption, and lifestyle at individual, family, and to some extent community levels. Ideas, methods, and tools are rarely available when natural resource management programs such as community forestry move beyond communities and involve large-scale actors and institutions. Nepal's community forestry is no longer a local, community-controlled forestry program; rather, it has developed a global connectivity and the decision-making power has been gradually transferred to the actors with global influence and coverage. This transformation from a local activity to a global agenda happened because of the persistent involvement of international development agencies in conceptualization and implementation of constantly shifting community forestry policies. The emerging networks of civil society and intellectuals have been playing a central role in creating a discursive space to expand community forestry connectivity from local to global. This involves multiple functions including research, activism, knowledge production and articulation, advocacy, mediation, facilitation, reflections, and multi-actor collaborations. These changes have created the conditions of possibility for external actors to be influential at the local level in generating change so that the local communities transcend community boundaries and participate in market rules.

Local communities, however, are not just passive recipients of external ideas. At present, community resistance against state encroachment of community rights and market exploitation of their resources has grown at an unprecedented level throughout the country. Communities are putting aside their differences and emerging as a united force against this organized assault on their autonomy and rights over forest resources.

In Nepal, aid projects have contributed substantially to sustain inequalities at multiple levels, and they have cultivated development ideas away from indigenous conceptions of change. Relentless reproduction of techno-bureaucratic hegemony in the forest sector and limited livelihood impacts from community forestry are the results of structural constraints created by the aid industry in Nepal (see Ojha 2014). In recent years, donor-supported techno-bureaucratic forest policies have promoted sustainability as a means of turning community forests into de facto conservation areas instead of focusing on the livelihoods of the people.

Nepal's community forestry has a double life of creating opportunities for autonomous, sustainable, and livelihood-oriented forest and people relationships, and simultaneously expanding market mechanisms of resource extraction in the rural hinterlands by enrolling community forests and communities into commercialization. Increased community participation in forest management and production are the indicators of the growing potential for creating politically conscious alternative and environmentally friendly practices. The potential of community forestry to create alternative meanings of sustainability could be realized if we conceptualize and mobilize alternative practices that community forestry has been generating in Nepal.

## Conclusion: Alternative Imagination

Community forestry has the potential to provide alternative development strategies in Nepal that can create radical transformations in current socio-environmental practices and relations in a manner that addresses the needs of most people. Decentralized local management of forests instead of control by state bureaucracies is a progressive step forward. But the emerging question is how such devolution of development power to communities is actually achieving such transformations. In fact, community forestry has allowed both the state and market to operate through local elites, seriously threatening opportunities for transformation. There is an increased realization that poor people are losing control over forests because the market decides what to grow and how to extract it. Resources are being produced to sell into markets rather than for consumption at the local level. People are losing their means of subsistence and have become cheap laborers for enterprises to earn "cash" for their daily requirements. They are paid less and in return they pay more for buying goods and services they need, leading to further poverty and vulnerability. Private sector and elite members of the community own the forest-based enterprises, and poor people are their laborers in community forestry in Nepal.

It is however clear that community forestry has also provided innovative spaces for revitalizing the people and forest relationship. These spaces have proved to be effective in promoting community entitlement and stewardship, the conditions of possibility for forging collective action, social equity and justice, and diverse livelihood opportunities. How useful community forestry is to achieve sustainability depends upon how we define sustainability. In Nepal we see how sustainability carries different definitions and meanings for different discourses and mindsets; it has been used by those promoting market integration and by those rallying for radical redistribution of resources alike. This chapter has explored how, despite this ambiguity, community forestry has helped foster powerful discourses for equity and justice. Likewise, community forestry has provided an important space for resistance and struggle for "real" decentralization: decentralization that frees people and forests from external exploitation. The present threats facing community forestry can perhaps be counteracted by building community networks across the region to develop counterstrategies and struggles for resistance at every level. Community forestry has a huge emancipatory potential if elements of expropriation are blocked and progressive aspects are mobilized to create alternative and sustainable forestry.

## References and Further Research

Agarwal, Bina. 2010. *Gender and Green Governance: The Political Economy of Women's Presence Within and Beyond Community Forestry*. New York: Oxford University Press.

Agrawal, Arun. 1996. "The Community vs. the Market and the State: Forest Use in Uttarakhand in the Indian Hiamalyas." *Journal of Agricultural and Environmental Ethics* 9, no. 1: 1–15.

Agrawal, Arun, and Clark Gibson. 1999. "Enchantment and Disenchantment: The Role of Community in Natural Resource Conservation." *World Development* 27 no. 4: 629–649.

Charnley, Susan, and Melissa Poe. 2007. "Community Forestry in Theory and Practice: Where Are We Now?" *Annual Review of Anthropology* 36: 301–336.

Department of Forests. 2014. *Community Forestry Status Database.* Kathmandu, Nepal: Department of Forests.

Eckholm, Erik. 1976. *Losing Ground: Environmental Stress and World Food Prospects.* New York: W. W. Norton.

Gilmour, Donald, and R. J. Fisher. 1991. *Villagers, Forests and Foresters: The Philosophy, Process and Practice of Community Forestry in Nepal.* Kathmandu, Nepal: Sahayogi Press.

Guthman, Julia. 1997. "Representing Crisis: The Theory of Himalayan Degradation and the Project of Development in Post-Rana Nepal." *Development and Change* 28: 45–89.

Hardin, Garrett. 1968. "The Tragedy of the Commons." *Science* 162: 1243–1248.

Ives, Jack, and Bruno Messerli. 1989. *The Himalayan Dilemma: Reconciling Development and Conservation.* New York: Routledge.

Malla, Yam. 2001. "Changing Policies and the Persistence of Patron–Client Relations in Nepal: Stakeholders' Response to Changes in Forest Policies." *Environmental History* 6, no. 2: 287–307.

Metz, John. 1991. "A Reassessment of the Causes and Severity of Nepal's Environmental Crisis." *World Development* 19, no. 7: 805–820.

Nightingale, Andrea J. 2002. "Participating or Just Sitting In? The Dynamics of Gender and Caste in Community Forestry." *Journal of Forest and Livelihood* 2, no. 1: 17–24.

———. 2005. " 'The Expert Taught Us All We Know': Professionalization and Knowledge in Nepalese Community Forestry." *Antipode* 37, no. 3: 581–604.

Ojha, Hemant. 2014. "Beyond the 'Local Community': The Evolution of Multi-Scale Politics in Nepal's Community Forestry Regimes." *International Forestry Review* 16, no. 3: 339–353.

Ostrom, Elinor. 1990. *Governing the Commons: The Evolution of Institutions for Collective Actions.* Cambridge: Cambridge University Press.

Paudel, Dinesh. 2016. "Re-Inventing the Common: Community Forestry as Accumulation Without Dispossession in Nepal." *Journal of Peasant Studies* 43, no. 5: 989–1009.

Thompson, Michael, and Michael Warburton. 1985. "Knowing Where to Hit: A Conceptual Framework for the Sustainable Development of the Himalaya." *Mountain Research and Development* 5, no. 3: 203–220.

# 10

# Sustainable Mining and the North

Nancy Langston

This chapter explores contested narratives of sustainability in mining, focusing on two mining regions in the boreal north: the Lake Superior basin in North America and Sápmi in northern Scandinavia. Both regions are witnessing new mining booms, with dozens of companies actively exploring the regions. State governments are supporting the mining companies, promising new revenues to economically depressed northern regions with few steady industrial jobs. Yet Lake Superior and northern Scandinavia have long histories of mining, which different communities interpret in quite different ways. Mining advocates hope to return to an imagined past when thriving mines brought good jobs, schools, roads, and other forms of economic development. For indigenous communities, however, mining has left a legacy of displacement and dispossession. Indigenous peoples in both regions lost their lands and their sovereignty, and toxic wastes from mining have devastated the watersheds that once sustained their cultures. Conflicts over the meanings of these differing views of the past continue to play an important role in current controversies over the sustainability of possible future mines. Decisions about sustainable mine planning are not purely scientific or technical decisions; at heart, they are social decisions. This chapter examines various narratives of mining sustainability to explore why different stakeholders in the north hold such different positions on new mines. How do indigenous groups, mining companies, and state governments deploy narratives (and counternarratives) of sustainable development in order to gain traction in current debates over the future of northern regions?

Mining companies now exploring the two regions argue that while past mining practices may have caused some damage, the future will be different because the industry has embraced sustainability. New mines, they argue, will bear little resemblance to past mines. Furthermore, mining companies insist, mining is an essential part of sustainable development because it helps to fund economic development and environmental protection. A few sacrifice zones in remote regions will enable nations to experience the full benefits of neoliberal development.

> Neoliberalism as a political and economic project has attracted both staunch supporters and critics alike. Some argue that neoliberal policies are necessary to ensure healthy domestic and global economies, while others blame these same policies for creating one of the biggest gaps in wealth ever recorded in history. For more on neoliberalism see Anup Shah's "Primer on Neoliberism" on this Global Issues web page: www.globalissues.org/article/39/a-primer-on-neoliberalism.

Indigenous groups view such claims with skepticism. They point out that they have historically borne the brunt of sacrifices made in the name of economic development. Current mining industry discourse, they say, only replicates assumptions that places remote from urban centers are essentially barren and empty, rather than landscapes peopled with indigenous communities. Indigenous communities fear that their land tenure rights will once again be ignored in the rush to mine the new north—yet this time it will be justified by the discourse of sustainability.

These two case studies reveal that sustainability questions in mining are not purely technical disputes. Rather, they encompass environmental, economic, and social domains. Resolving mining disputes is not simply a question of evaluating trade-offs between jobs and clean water. Indigenous communities, though often bearing the greatest burden from the toxic wastes and social instability fostered by mining projects, have rarely had much decision-making power in the planning process. Important dilemmas are posed when governments must decide about permitting new mines. Who decides when the benefits and consequences of new mines are not equitably shared? If minerals are central to sustainable development, as mining proponents argue, what should be sacrificed to further those developments? Different groups approach these questions in quite different ways, their perspectives shaped by their specific historical experiences in their respective regions. No simple answers are possible, because, as the introduction to this volume argues, sustainability is not a single environmental state that we can easily measure or know. Rather, it is a cultural construction that emerges in the negotiations and relationships that develop within particular places.

While numerous frameworks for sustainable mining have appeared in United Nations, corporate, and environmental NGO literature, these rarely recognize the fact that sustainability is a concept rather than a stable, measurable standard or goal. One framework, for example, sets maximum limits for twenty-two contaminants downstream of a mine. If the mine water contains less of those particular contaminants than those standards require, then the mine is deemed sustainable. Yet, as the initial chapters of this volume show, such technical measures cannot tell us what is sustainable and what is not. Given that sustainability is a *concept* (rather than a thing), this chapter asks: What work do different conceptions of sustainable mining accomplish for mining corporations, for governments, and for indigenous communities? How do different stakeholders involved in mining use the concept of sustainable development in their efforts to gain a voice and power in debates over proposed developments?

## Mining and Sustainability

Many scholars argue that sustainable mining is impossible because minerals are not renewable on timescales that matter to people. Recovery from mineral extraction involves geological processes extending over millions of years. Mining removes minerals from the ground forever, and technological solutions cannot make the ore grow back. Yet demand for many minerals is increasing, especially in the growing economies of Asia. Existing mines cannot meet those demands. While conservation and recycling has the potential to *decrease* the demand for new mines, in the foreseeable future it will not *eliminate* that demand. Because demand for minerals persists, denying all mine permits in a particular region does not solve the problem of environmental consequences from mining; it simply shifts those consequences to other regions that may have weaker capacities for environmental and social protection. Reducing global environmental consequences from mining ultimately requires reduced consumption as well as more sustainable mining practices.

Questions regarding water and waste are particularly critical for mining. Mines are often among the largest water users for their regions. When mine pits extend beneath the groundwater, those pits need to be dewatered, with pumps extracting groundwater as it seeps into the pits. Such dewatering can affect the movement of groundwater into local surface springs and water bodies throughout a region much larger than the mine area itself, affecting many water users. As one report noted, "The depletion of groundwater by dewatering operations and the presence of large mine facilities can take decades to replenish after mining ceases, and in some instances, groundwater levels and flow directions can be altered indefinitely" (Initiative for Responsible Mining Assurance [IRMA] 2014). As water moves through mining sites, it often becomes contaminated with mercury, arsenic, sulfates, and other toxic chemicals. Acid drainage from mines can leave legacies that last for centuries. Technological treatments can reduce the risk of contamination, but those technologies are extremely expensive, and local communities and governments have often not had the capacity to insist on such treatment.

Waste is another core issue in responsible mining, because only a small fraction of what is mined can be sold at a profit. The rest becomes waste. Two forms of waste are typical: the overburden, or the rock that is removed to get at the ore-bearing rock; and the tailings, or the waste left after processing the ore into the final product. Both the overburden and the tailings can contain extremely toxic materials, and containing those materials is still not possible, either financially or technologically. Controlling and managing waste materials remains a major challenge for the mining industry.

While all mining operations must address the problem of waste, different mines mobilize toxics in different ways. Some mines use extremely toxic materials in the refining process, adding chemicals such as mercury to systems that did not contain those chemicals. Other mines, however, do not use particularly toxic chemicals in the refining process, leading industry advocates to argue that such mines are safe and sustainable because they only rearrange natural elements. Yet it is essential to question simplistic categories of natural versus synthetic. Even in the absence of added synthetic chemicals, some geological formations contain natural deposits of heavy metals that—even though they are natural—would be quite toxic if they were mobilized into biological systems. Typically, however, such metals are bound in stable formations, where they do not move into the atmosphere or the water supply on timescales that matter for biological life. (Over millions of years, of course, they do mobilize, so spatial and temporal scales are critical.) But when mining exposes those deposits to air and water, natural minerals can move into biological systems in ways that make them toxic to biological life. Whether or not a particular mine has waste and water problems depends partly—but not entirely—on the geologic context. Certain minerals, for example, are more likely to create acid drainage problems than other minerals. Geology alone does not lead to toxic consequences; rather, human *interaction* with that geology creates toxic waters.

## Dominant Discourses of Sustainable Mining

In recent decades, a dominant discourse of sustainable development has emerged within certain sectors such as the United Nations (UN). This discourse frames sustainable development as a balance between capitalist expansion and nature preservation. Sustainable development requires nature protection, the discourse assumes, but protecting nature requires money, which can only come from economic development. For example, in Rio+20, the UN team wrote a statement on sustainable mining that places neoliberal economic development at the core of sustainable development.

Rio+20 is the short name for the United Nations Conference on Sustainable Development that took place in June 2012 in Rio de Janeiro, Brazil. In 1992, the UN had held the Earth Summit in Rio, where countries adopted Agenda 21, a blueprint to combine economic growth, social equity, and environmental protection. Twenty years later, the UN brought together governments, international institutions, and other groups to agree on measures that might "reduce poverty while promoting decent jobs, clean energy and a more sustainable and fair use of resources" (from the "United Nations Conference on Sustainable Development, Rio+20" web page; see URL below).

For more on the United Nations Conference on Sustainable Development, see: "Sustainable Development Knowledge Platform." United Nations Department of Economic and Social Affairs (https://sustainabledevelopment.un.org/rio20).

The Rio+20 statement on mining affirms that because mining brings money to poor regions, and money is necessary for sustainable development, mining is therefore central to sustainable development. The statement reads:

> We acknowledge that minerals and metals make a major contribution to the world economy and modern societies. We note that mining industries are important to all countries with mineral resources, in particular developing countries. We also note that mining offers the opportunity to catalyze broad-based economic development, reduce poverty and assist countries in meeting internationally agreed development goals, including the Millennium Development Goals, when managed effectively and properly. We acknowledge that countries have the sovereign right to develop their mineral resources according to their national priorities and responsibility regarding the exploitation of resources described in the Rio Principles.
>
> (United Nations [UN] 2012)

Future mining, in other words, is a necessary tool for sustainability, even if past mining was completely unsustainable, having left long-lasting social and environmental scars.

The mining sector has been quick to adopt these sustainability premises, even though after the first Rio Conference on Sustainable Development in 1992, the mining industry appeared to be anything but sustainable by the definition that emerged from Rio. In the early 1990s, concerns about the social legacies of past mining activities in poorer nations shaped the concept of "resource curse" among economists. Countries richest in minerals (and other nonrenewable resources such as oil) tend to have greater extremes of poverty and inequity than other countries (Auty 1993). A growing concentration of wealth in the hands of a few, in addition to conflict, corruption, and the long-lasting legacies of toxic contamination to both land and water, had all contributed to mark mining as a troubled industry.

Concerned about growing criticism of the sustainability of the mining sector, in the late 1990s some of the world's largest mining companies formed a group called the International Council on Mining and Metals (ICMM) to articulate a vision of mining's contribution to sustainable development and to "reposition the industry in terms of both performance and perception" (International Council on Mining and Metals [ICMM] 2012).

### Goals of the International Council on Mining and Metals

As ICMM writes in its promotional literature, ICMM was formed:

> to catalyze improved performance and enhance the contribution of mining, minerals and metals to sustainable development. Today, it brings together 22 mining

and metals companies as well as 34 national and regional mining associations and global commodity associations. ICMM's member companies employ close to one million of the 2.5 million people working in the sector worldwide. These companies have some 800 operations in over sixty countries producing 30–40 percent of the world's hard mineral commodities including iron ore, gold, platinum and nickel. We engage with a broad range of stakeholders—governments, international organizations, communities and indigenous peoples organizations, investors, civil society and academia—in order to build meaningful relationships. Our vision is one of leading companies working together and with others to strengthen the contribution of mining, minerals and metals to sustainable development.

("The Role of Mining in National Economies" 2)

ICMM gives the mining industry a platform to promote a corporate narrative of sustainable mining. This narrative reframes the conversation about sustainability, insisting that the story should not dwell on the problems associated with *past* mining. Instead, sustainable mining narratives must focus on the economic benefits of *future* mining. As the ICMM document on sustainable mining states,

> The focus is not on how mining can be sustainable (recognizing that any individual mining operation has a finite lifespan) but on how mining, minerals and metals can contribute to sustainable development. This is a conceptual shift away from a singular analysis and mitigation of impacts to a more comprehensive analysis that looks at the wider contribution of the industry and its products.
>
> (ICMM 2012, 5)

In other words, the ICMM statement encourages us to focus less on toxic performance of individual mines and more on the industry's contributions to economic development.

For more on the International Council on Mining and Metals, see the Council's website: www.icmm.com/minings-contribution. Also, "The Role of Mining in National Economies," quoted in the text box, can be found at this URL: www.icmm.com/website/publications/pdfs/social-and-economic-development/romine_1st-edition.

In this document, ICMM notes the importance of mining for colonialism and exploitation of indigenous regions, but reframes this narrative as a good thing: "Throughout history, mineral discoveries have catalyzed the development of sparsely populated and remote areas. The discovery and development of mineral resources has provided an opportunity for addressing poverty reduction and helping countries in achieving the Millennium Development Goals" (ICMM 2012, 10). Sustainable development, in this corporate narrative, equates to modern lifestyles, which in turn require money that mining can provide:

> Importantly, while minerals and metals provide a material foundation for contemporary society, they are also a means for transforming current society into one marked by greater efficiencies, lower environmental stresses and more effective public services. . . .

> More and more developing and emerging nations are finding that the role of mining and metals in their national economies can be a positive factor, that their dependency on mining and metals for economic strength is growing and that mining and metals can play a key role in addressing poverty reduction and other developmental issues.
>
> (ICMM 2012, 11–12)

In the late 1990s, indigenous groups began to challenge the assumptions about sustainability inherent in the industry and UN formulations. They asked: Did economic development from mining ever benefit indigenous communities? Did nature conservation ever benefit indigenous communities? Mining, these indigenous communities argued, is not necessarily an integral part of sustainable development, because it ignores key questions: Sustainable for whom and for what? What is lost and by whom in the push toward sustainable development? Who decides which places become sacrifice zones? Who decides which places become nature conservation areas, and who decides what long-standing cultural activities should continue to occur in either mining or conservation areas?

Indigenous communities have worked to articulate alternative definitions of sustainability to counter the dominant sustainability rhetoric of mining companies. While these vary by indigenous group, one analysis by indigenous scholar Michael Dockry (2012) found that the concepts of sovereignty and territory were shared across tribes in their emerging discourse of sustainability. Economic opportunity was important in indigenous counternarratives, according to Dockry, but no sharp boundary could be drawn between nature preservation and economic development, with one justifying the other, as in the dominant narrative of sustainable development. Examining mining in two northern regions reveals the fact that mining companies and indigenous communities have contested claims over northern regions, justified by different concepts of sustainability and socionatures.

## Mining the Boreal North: Northern Sweden and Lake Superior

Much of the world's current mining boom is happening in the boreal north, in some of the world's wealthiest regions: Canada, United States, Norway, Sweden, and Finland. Indigenous peoples have inhabited the boreal north for millennia, but as they have adapted to modern state regulations they have struggled to retain sovereignty over their land. For centuries, governments have used the idea of the boreal north as uninhabited and remote to promote its colonization and exploitation. The abundance of space combined with rich natural resources and relatively few people translated into a particular kind of spatial and representational logic: as long as we put the extractive activities out there, away from where most people live, it is acceptable. Open-pit iron mines proposed for indigenous territory in the Lake Superior basin and in northern Sweden continue to be justified by similar logic (see Langston 2013).

An open-pit mine is defined as "an excavation or cut made at the surface of the ground for the purpose of extracting ore and which is open to the surface for the duration of the mine's life." For more on open-pit mines, see:

"Basics of an Open Pit Mine" at the Mine-Engineer.Com website at this URL: www.mine-engineer.com/mining/open_pit.htm.

## Mining in Northern Sweden

Northern Sweden contains some of the largest iron ore deposits in the world; 90 percent of all iron ore used in Europe comes from the region. From the Swedish government's perspective, expansion of mining in the north is inevitable because the world needs iron ore for steel, and the government needs mining profits to fund the social programs integral to Swedish society. But from the perspective of the Sámi, the indigenous populations living in Sweden's northern regions, the proposed mines would make it impossible to continue reindeer herding, ending thousands of years of cultural adaptations to boreal ecosystems.

### The Sámi

The Sámi are the indigenous peoples of northern Europe, and their territory is called Sápmi. As the inland ice retreated from Sápmi some ten thousand years ago, reindeer *(Rangifer tarandus)* followed the edge of the ice as it retreated northwards, and the ancestors of the Sámi people went with them. The anthropologist Piers Vitebsky has argued that without reindeer, human cultures could not have thrived in the boreal forest (Vitebsky 2006). Originally hunters of reindeer rather than herders, the Sámi shifted to herding semi-domesticated reindeer after European incursions into their lands reduced their subsistence resources.

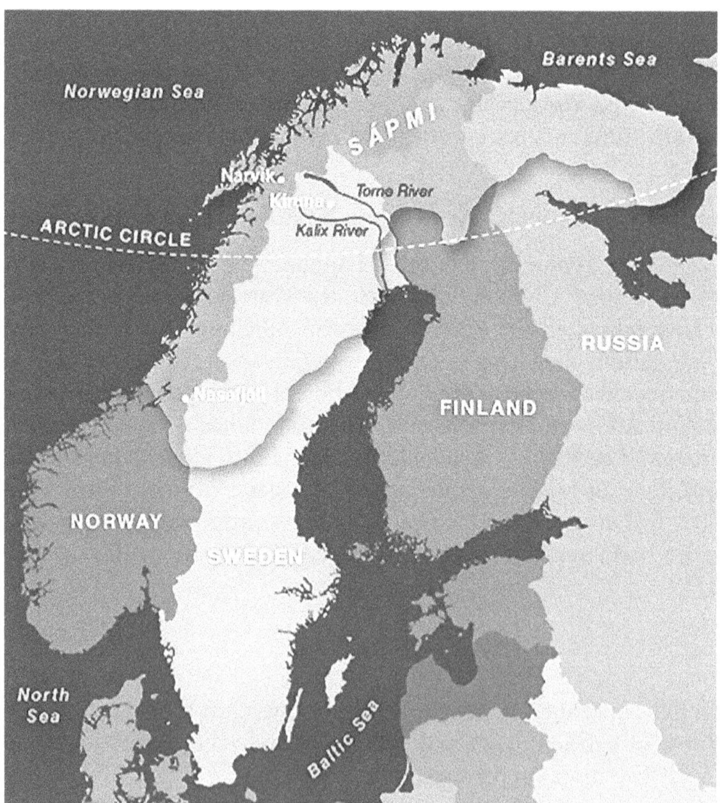


The Sápmi territory, home of the Sámi indigenous group.

*Source:* Map from *American Scientist*, used with permission.

For more on reindeer peoples, see:

Vitebsky, Piers. 2006. *The Reindeer People: Living with Animals and Spirits in Siberia.* Boston, MA: Houghton Mifflin Harcourt.

For more on the Sámi, see the website of the Sámi parliament:

"Sámediggi." Sametinget. www.sametinget.se/lang/english.

In the ninth century, Norse chieftains moved into Sápmi and began taxing the Sámi. Trappers and traders from southern Scandinavia soon came to Sápmi to capitalize on the growing European market for furs, selling firearms to the Sámi and receiving furs in return. The Sámi intensified hunting, depleting the boreal forests of fur-bearing animals and reindeer. As subsistence resources dwindled, the Sámi began to domesticate wild reindeer, and by the seventeenth century, herding of semi-domesticated reindeer had largely replaced hunting of wild reindeer (see Beach 1993).

The north, in European eyes, seemed essentially empty of people (even though the Sámi provided essential taxes), and that emptiness justified it becoming Sweden's colony for resource extraction—first for furs, then for minerals, timber, and energy. In 1542, the Swedish king Gustav Vasa proclaimed that "all permanently uninhabited land belongs to God, Us and the Swedish Crown, and nobody else." In 1635, cabinet minister Carl Bonde wrote to the Swedish statesman Axel Oxenstierna describing mineral deposits in the north, expressing his belief that the boreal north could become Sweden's own "West Indies." The year before, a Sámi man named Peder Olofsson had mentioned to some Swedes that he had found silver ore in Nasafjäll in Pite Lappmark, close to the Norwegian border. Silver revenues might help fund war expenses that burdened the Swedish state, so in 1635, mining began at the Nasafjäll works (Langston 2013).

Transport of ore to smelters and then to markets was vital for mining development. Believing the Sámi to be too weak and lazy to dig ore, but useful for transporting it, the Swedes ordered Sámi peoples to use their reindeer to carry ore sixty kilometers between the Nasafjäll mine and the smelting works. The Sámi refused, fearing the work would interfere with reindeer herding cycles. Swedish workers described how they forced the Sámi to comply: "We tied them [the Sámi] to a couple of timbers, pushed them down into the rapids a few times and then pulled them up to allow the water to run out of their mouths again, so all the people working in the mine were aware of the situation" (Samer.se n.d.). Contemporary accounts speak of the road "lined with bleached reindeer skeletons long after mining finished" (Samer.se n.d.). While the company's deployment of violence persuaded many Sámi to obey, many others refused what they saw as enslavement and instead fled to Norway, even though the Swedish government sent troops to prevent their escape. Much of their territory near the mine became depopulated.

In 1673, the Swedish king encouraged Swedish citizens from the south to move north and colonize the Sámi area, believing that privately owned farms in the lower elevation forests would not interfere with Sámi communal ownership of reindeer pasturage at higher elevations. Few Swedes, however, were interested in moving to a place they saw as remote, barren, and inhospitable, even when the king promised relief from military service and taxation. The Swedes who moved into the forest found settled farming nearly impossible in the harsh climate, so they began extracting resources from Sámi lands, hunting, fishing, and cutting hay in winter reindeer pastures. The Sámi herding territories and fishing grounds were squeezed by the new immigrants, but when conflicts arose, the Sámis were able to assert their rights to communal reindeer pasture in Swedish courts. Even though the monarchy had encouraged emigrants to move into Sápmi, courts recognized that the Sámi had never given up their traditional rights to common lands necessary for reindeer pasturage (Langston 2013).

With nineteenth-century industrialism, colonial development of the north's resources exploded, transforming Sámi relations with reindeer and territory once again. Although a series of earlier court decisions had affirmed Sámi rights to communal ownership of reindeer pastures, the Swedish state now saw Sámi communal landownership as an obstacle to the exploitation of boreal resources. To justify stripping the Sámi of their property rights, the Swedish government turned to the contemporary science of racial difference. Eugenists and Social Darwinists argued that the Sámi were an inferior race, unable to make adult decisions, so others would have to decide what would happen on Sámi lands. Swedish government policy towards the Sámi was defended with a racist ideology that claimed to have found "scientific verification" of Darwin's theory of natural selection and the survival of the fittest within the human community.

In 1886, Sweden passed the Reindeer Husbandry Act, which reduced Sámi ownership of land and water to mere rights of use (or usufruct). The Sámi could no longer own the land; they only retained limited rights to graze, fish, and hunt. Those Sámi who did not herd reindeer and instead made their living primarily from fishing lost their land- and forest-use rights. Only Sámi who met certain ideas of traditional reindeer culture were protected, and only if they remained frozen within static notions of primitiveness (Beach 1993; Borchert 2001).

**The 1886 Reindeer Husbandry Act**

The 1886 Reindeer Husbandry Act, also known as the Reindeer Herding Act, was the first reindeer herding law in Sweden. The intent was to reduce conflicts between Sámi herders and colonizing farmers. Farmers had been complaining that Sámi reindeer were ruining their fields, while Sámi had argued the farmers were ruining forests and catching too many fish. "By passing this law, the Swedish government aimed to promote more intensive herding among the Sami and to establish liability for damages. This law also divided the Sami into villages. This was an attempt to make the Sami, a traditionally nomadic people, more stationary" (Korsmo 1993).

For more on the 1886 Reindeer Husbandry Act, see: Dirk Sonniksen's "Reindeer Herding in Sweden" at the Sami Culture website: www.laits.utexas.edu/sami/diehtu/siida/herding/herding-sw.htm.

Swedes had long suspected that the north contained some of the richest iron ore deposits in the world, but exploitation could not begin until the transportation problem was solved. In 1903, Sweden completed a railroad running from the Baltic port of Luleå, through the iron ore region of Kiruna, down to the ice-free port of Narvik on the Norwegian coast. The mining boom at Kiruna was on. Initially open-pit mines, the Kiruna mines transformed the landscape. Historic photos show that mountaintops were removed to reach the rich ore deposits, a precursor to today's controversial mountaintop removal practices in Appalachia. The work was dangerous, and pollution from the pit and tailings piles was notorious. Sulfates contributed to acidification both of local watersheds and larger regional watersheds. Acids leached toxic heavy metals from rocks, where they had been safely locked up, into river systems where they contaminated food chains that eventually included people. Smoke from smelters contained heavy metals such as cadmium and lead, which were captured by lichen and then accumulated in reindeer and the people who ate them. Mine tailings were stacked in enormous terraces, and winds blew toxic dust over reindeer pastures. Sediments washed over fish-spawning beds.

In the 1920s, an agreement between the Norwegian and Swedish governments prevented the Sámi in Sweden from grazing their herds over the border into Norway. This, along with mining and timber projects, reduced the pastures available for the Sámi, and overgrazing resulted. The Swedish state believed that overgrazing resulted from the traditional land tenure systems (a "tragedy of the commons"), rather than from the undermining of those traditional land tenure systems by state policy decisions. The Swedes decided that they needed to rationalize Sámi grazing and ordered massive reduction of herds and herders.

As the Sámi lost their land tenure rights, timber harvests increased across Sápmi, further reducing grazing habitat for reindeer. Clear-cutting removed nearly all the older spruce forest, critical habitat for the pendulous lichen that forms important winter food for the reindeer. Modern forestry involved plowing, drainage, and soil scarification, which reduced the ground-growing lichens that sustained reindeer during fall and spring. A series of court cases in the 1990s, in which forest companies argued that reindeer were damaging replanted trees, further undermined the rights of Sámi to graze their herds in traditional winter forest pastures.

Hydropower development also undermined reindeer husbandry critical to Sámi economic sustainability. In the 1950s and 1960s, Sweden turned to intensive hydroelectric development to fuel modernization. The government envisioned a system whereby each river that flowed

Reindeer herding in Sápmi.

*Source:* Mats Andersson/Wikimedia; https://commons.wikimedia.rg/wiki/File:Reindeer_herding.jpg.

to the Baltic would be dammed, the power of the flow fully utilized for the nation's industrial development. The state formed a hydroelectric company, Vattenfall, which set out to rationalize the river systems of the north. Between 1957 and 1961, Vattenfall turned to the Kalix and Torne river system in Sápmi. In what would have been the largest hydro project in Swedish history, the Atlantprojekt, Vattenfall proposed reversing the flow of the rivers, so they would fall westward to the Atlantic Ocean instead of eastward to the Baltic. With an intricate system of dams, canals, and tunnels, half the flow of the rivers would have been diverted into Norway to power an enormous hydropower plant. Commercial fishing interests and Sámi reindeer herding villages organized, halting the project. These battles created alliances between Sámi activists and environmental groups, and they brought attention to the issues of Sámi sovereignty and land tenure rights, helping to set the stage for opposition to new mining expansions a generation later.

Global markets continue to affect local conditions in Sápmi, remote as the region seems to many outsiders. A global steel boom in the first decade of the twenty-first century increased Asian demand for iron. LKAB, the Swedish-owned mining company that operates the Kiruna iron mine, decided to expand the mine, which means that Kiruna's city center must be moved, an enormous logistical undertaking. Local communities typically accept the need to move Kiruna, but new open-pit-mining proposals in the Kalix River valley have met with more opposition. In 2010, an Australian mining company named Kiruna Iron proposed to create three new open-pit mines in the Kalix River valley. These mines would operate for ten to twenty years, depending on steel prices; Sweden would agree to collect no royalties or taxes on the profits.

The Swedish government approved the open-pit-mining proposal, even though the proposed mines lie in the heart of two Sámi reindeer villages. The Swedish government contended that the Sámi have no rights to exclude competing uses from Sámi territories. From the government's perspective, mineral development is an inevitable part of developing a sustainable society and the Sámi must make way for it. If the good ore happens to interrupt a migration route, then one can move the reindeer somewhere else. If tailings piles eliminate lichen, then one can feed the reindeer something else. In this view, if the Sámi or their reindeer cannot coexist with mining, they must give way to the need for economic development of the north to benefit all of Sweden. Sámi communities, however, have a different perspective on sustainable development, wondering how twenty years of mining can take priority over thousands of years of Sámi culture (for more, see Fowler 2012). While the Swedish government agrees with the mining industry that economic development of the north is necessary for sustainable development, the Sámi have appealed to the UN, arguing that sustainable development should not lead to the erosion of their culture and sovereignty.

## Lake Superior Mining

Similar processes are at work in the Lake Superior region on the border between Canada and the United States, where the Anishinabeg (Ojibwe) tribes have lived for many generations. In 2011 a company named Gogebic Taconite (GTAC) proposed to develop the largest open-pit mine in the world just upstream of the Bad River Band's reservation on Lake Superior. The mine would exploit the region's low-grade iron ore deposits (known as taconite), which are manufactured into the steel needed for the world's building and steel commodities boom.

The United States federal government made three major land cession treaties in the nineteenth century with the Anishinabeg in the Lake Superior basin. This proposal to mine so close to native lands has implications for the historic treaties that established reservations controlled by the tribes. In addition, the tribes were careful to retain the right to hunt, fish, and gather on ceded territories, and this meant that they retained rights over management

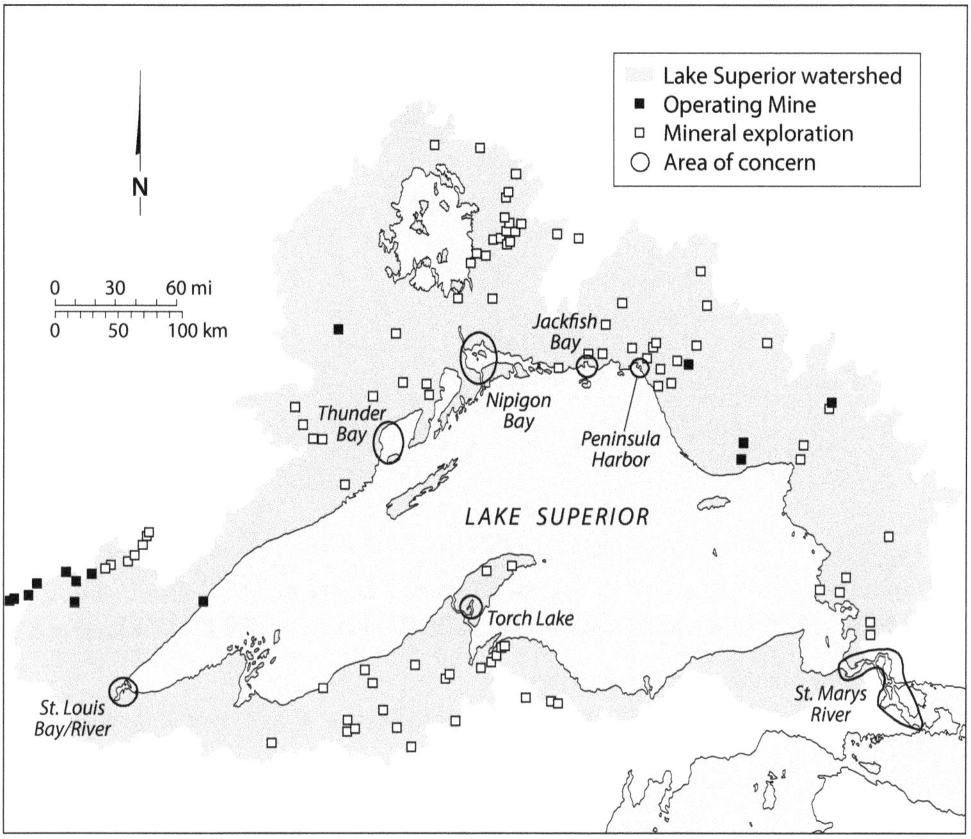


Mining boom around Lake Superior.

*Source:* Nancy Langston; map drawn by Bill Nelson.

of natural resources, including water (for more, see Reynolds 2003). While the proposed GTAC mine would lie a few kilometers outside the legal boundaries of the Indian reservation, it would be within ceded territories, giving the tribe the rights to contest the mine on sovereignty grounds.

The iron ranges of Minnesota, Wisconsin, and Michigan—collectively known as the Lake Superior District—long supplied much of North America's iron ore. In 1890 more than half of all the iron ore used by the American iron and steel industry came from the Lake Superior District. By the end of World War II, the region supplied 85 percent of the United States' supply of iron ore. For most Americans, Lake Superior seems extremely remote. It lies at the northern edge of the country, far from urban industrial centers. The largest American city on the lake, Duluth, Minnesota, has fewer than 85,000 people. No major highways run along the lake, and few jet routes pass over. Yet the iron mines of the Lake Superior District lay at the heart of American industrialization. As the key ingredient in manufactured iron and steel, iron ore was essential for building automobiles, railroads, skyscrapers, barbed wire, tanks, bombs, and bullets (Thistle and Langston 2015).

Open-pit iron mining in the Lake Superior basin mobilized toxic chemicals that moved into local bodies—particularly the lungs of miners—and into distant watersheds and the atmosphere. Iron mining typically creates tremendous amounts of dust containing silica, asbestos, and other toxins that mine workers and local people inhale, causing fluid buildup and scar tissue to form in the lungs. Some of the mining dust contained a form of asbestos that could lead to a particularly virulent form of lung cancer known as mesothelioma.

Wild-rice growing in the Lake Superior region.

*Source:* Superior National Forest/Wikimedia; https://commons.wikimedia.org/wiki/File:Wild_rice_1_(6129843903).jpg.

Workers on the iron range were three times as likely to get mesothelioma as the general population, according to the 2013 *Annual Report to the Legislature: Minnesota Taconite Workers Health Study* by Lourey and colleagues. When tailings were dumped into water, those toxic fibers could move out of mining spaces and into much more dispersed spaces, particularly when overburden and tailings were dumped into water bodies.

The Bad River runs through the proposed GTAC mine site before entering the 16,000-acre Kakagon–Bad River Sloughs—the largest undeveloped wetland complex in the upper Great Lakes. In 2012, this was designated as a Ramsar Site, recognizing it as a wetland of international importance. The Ramsar designation noted that as the only remaining extensive coastal wild-rice bed in the Great Lakes region, the Kakagon–Bad River Sloughs are critical to ensuring the persistence of Lake Superior wild rice. The sloughs make up 40 percent of the remaining wetlands on Lake Superior's coast, and they contain the largest natural wild-rice beds in the world.

**The Ramsar Convention**

The Convention on Wetlands, called the Ramsar Convention, is an intergovernmental treaty providing a framework for conservation of wetlands. The Convention was adopted in the Iranian city of Ramsar in 1971, hence the name. Designation as a Ramsar Site means that a wetland has been formally recognized as having international importance and "significant value not only for the country or the countries in which they are located, but for humanity as a whole. . . . The inclusion of a wetland on the List embodies the government's commitment to take the steps necessary to ensure that its ecological character is maintained" (www.ramsar.org/about/wetlands-of-international-importance-ramsar-sites).

For more on the Ramsar Convention and the criteria used to determine qualification as a Ramsar site, see the Ramsar website: www.ramsar.org.

For members of the Bad River Band, these wild-rice beds are central to their cultural history and sovereignty. When the Anishinabeg migrated westward from the Saint Lawrence River Valley, the ancestors of the Bad River Band chose to make their homes along the Kakagon Sloughs because the wild-rice beds they found there had figured heavily in their visions. In the sloughs, they found wild rice or *manoomin*, "the food that grows on water," which they continue to see as a "sacred gift from the Creator" (Reynolds 2003, 146). Wild rice is sensitive to sulfates in the watershed, which can be released by taconite mining and processing. Band members argue that stopping the mine is essential for their survival, not just to ensure thriving wild-rice beds and fisheries, but also to sustain the connection with the past that is at the core of their cultural identity.

The proposed mine would target an iron formation in the Gogebic Range called the Ironwood Formation, which is a significant ore deposit in the national and indeed global context. Best estimates suggest that the Ironwood Formation contains at least 3.7 billion tons of taconite ore, or 20 percent of known iron ore deposits in the United States. This translates into one billion tons of steel, or sixty-six years of domestic supply (Fitz 2012). This is such an enormous deposit, GTAC argues, that it is inevitable that it will be mined. The language of inevitability figures heavily into the pro-mining discussions, even though the company's own projections show that the ore body would rapidly be depleted.

In the planning process, GTAC has done its best to present taconite as a safe ore to mine with minimal toxicity concerns. It is so safe and so pure, GTAC insisted to the Wisconsin legislature, that the state should quickly pass a new law exempting taconite mining from many of Wisconsin's existing environmental regulations (Gedicks and Blouin 2013). The argument had two parts: first, because taconite itself rarely contains iron sulfides or pyrites—which can produce acid mine drainage—it is pure. Second, the extraction process uses magnets and clean water, not hydrosulfuric acid. Technically, both these details may be correct, but the conclusion that follows—i.e., therefore taconite mining has no toxicity concerns—ignores the larger geological and biological context of taconite within a watershed. First, taconite is a low-concentration iron ore, and to extract that valuable part, the rest of the rock (the tailings) must be crushed to a fine dust, mixed with water, then dewatered and stacked in piles or dumped into water. These tailings particles are quite easily eroded by wind and water, and from there they become mobilized into the watershed. At least 70 percent of the volume of the Ironwood Formation would be turned into waste and fine tailings, and the waste for the initial four-mile stage of mining alone would create a pile "600 million cubic yards" in size, stretching one-mile square, "600 feet high" (Gedicks and Blouin 2013). Bill Williams, the President of GTAC, promises that all water leaving the mining site will be treated, so that it will not harm downstream users. He promotes his own sustainability credentials, promising that his mining experiences in other locations have given him the skills needed to mine responsibly in the Lake Superior basin while providing much-needed economic development.

Since January 2011, when news of the possible GTAC mine first spread, the issue has become extremely polarized. Because the proposed mine lies in ceded territories, the Anishinabeg tribes in the basin have vowed to elevate the fight against the mine to a federal level, claiming violation of their tribal sovereignty. These debates have an important historic context in the region. Wisconsin rarely recognized tribal sovereignty rights on ceded territories until a series of conflicts took place in the late twentieth century. In the 1990s, Exxon proposed to build a copper and zinc mine in sulfide-containing ore bodies within the Wolf River watershed (a National Wild and Scenic River). Like the currently proposed iron mine, the Exxon mine would have been located just upstream of an Anishinabeg reservation (the Mole Lake Sokaogon reservation), within ceded territories (Grossman 2005). The state of Wisconsin had promoted the Exxon mine, just as it currently promotes the GTAC mine, arguing that both would bring economic development to a poor, northern region. The Wisconsin Department of Natural Resources decided to set standards for water leaving the Exxon mine at industrial water quality levels, allowing for forty million tons of tailings and acidic mining waste that would have degraded wild-rice beds on the reservation. To block the mine near Crandon, Michigan, the Mole Lake Sokaogon worked with the Environmental Protection Agency to win the right to create its own clean water standards. Eventually the United States Supreme Court upheld the authority of the tribe to set water standards that protected reservation waters. These decisions established important precedents for current mining conflicts, for they allow the tribe to challenge the dominant corporate and state discourse of sustainable development (Reynolds 2003).

When the GTAC proposal became public in 2011, tribal sovereignty again became the focus of debate. Within the state permitting agencies, discussions focused on technical aspects of sustainability, without ever addressing concerns about sovereignty. The Department of Natural Resources argued with the company about the composition of the ore body, the appropriate sampling techniques for water quality, and the storage of tailings from the bulk sampling process. The assumption underlying these discussions appeared to be that, if these details could be corrected, the mine would be permitted. But the tribe challenged these premises, arguing that a sustainable development framework which ignored tribal sovereignty and cultural meanings of pure water for the indigenous community was useless.

While the various groups seemed to have little common ground, what they all shared was a tendency to base their readings of sustainable futures on their interpretations of the past—interpretations that dramatically differed. Mining advocates believed that they had once created prosperity through regional mining, so any sustainable future needed to begin with the economic development that had once sustained their community. Tribal mining opponents, on the other hand, interpreted the same past as a time of brutal dispossession and toxic pollution, so their vision of a sustainable future insisted upon a future very different from that past. Conflicts over the meanings of the region's mining history, in other words, continued to play a key role in current controversies over future planning.

## Recent Sustainable Mining Frameworks

In recent years, groups of mining companies have tried to deflect criticism of mining by creating extra-governmental, voluntary standards to improve the perception of their industry and promote sustainable mining. ICMM has created one such set of standards, while the Mining Association of Canada (MAC) has created a similar program called "Towards Sustainable Mining" or TSM. The TSM program requires participating companies to prepare a set of sustainable mining plans, including ones focused on tailings management, emergency response, and energy use. No plans or assessments are required that concern water in these voluntary standards (quality, pollution, disposal, or quantity)—a striking omission. Not only are standards voluntary, each company assesses its own performance and rates itself (Mining Association of Canada [MAC] 2007).

Two examples illustrate problems with such voluntary self-assessments. Early on August 4, 2014, the tailings dyke failed at Mount Polley, an open-pit copper and gold mine in British Columbia. Ten million cubic liters of contaminated mining water and nearly five million cubic meters of tailings were released into the upper watershed of the Fraser River—a world-class salmon river. Even though engineers and foremen associated with the mine had been reporting flaws in the tailings dyke well before the disaster, company management ignored the warnings from staff. According to news reports, this was the worst mine tailings disaster in Canadian history and perhaps the nation's worst environmental disaster ever (Hutchinson 2014; Judd 2014; Woo and Posadzki 2014).

Imperial Minerals—the Canadian mining company that operated the mine—is part of the Canadian Mining Association and involved in the Towards Sustainable Mining program of self-assessment. In 2014, Imperial Minerals released its report on their sustainable mining performance for the year that included the tailings breach. While recognizing the breach, the company nevertheless awarded themselves an AAA rating (the highest possible) on nearly all sustainability measures. By every reasonable standard, Imperial Minerals had a most "unsustainable" year—yet the Towards Sustainable Mining self-assessment did not reflect that (MAC 2014). Every three years, an independent, external verification takes place, and that external verification dropped Imperial Mineral's "tailings management assessment" grade down to an A—a decrease, to be sure, but not exactly reassuring (Mining Association of Canada, and Mining Association of British Colombia 2015).

A second example of problems with industry-led sustainable mining frameworks hits even closer to home for critics of the proposed GTAC mine near Lake Superior. Between 2005 and 2014, a Canadian mining company named Inmet Mining Corporate operated the Cobre Las Cruces mine in Spain, an open-pit copper mine that is the largest open-pit mine in Europe. Bill Williams—now the President of GTAC—was associated with Cobre Las Cruces from 2006 to 2011. He was director of water operations until August 2008, and then he became general mine director, leaving Spain to become GTAC's President in 2011. GTAC argues that its proposed open-pit mine will be sustainable because his years in Spain gave Bill Williams the experience to operate a successful mine.

The Spanish government, however, has pressed charges against Cobre Las Cruces and Williams for polluting local drinking water with arsenic from the mine. According to the Spanish prosecutors, when Williams was in charge of surface and groundwater at the Las Cruces mine, he knowingly and illegally violated the mining permits by pumping arsenic-contaminated mine water into a nearby aquifer that serves as drinking-water supply for surrounding communities (Bergquist 2014; Rodríguez 2013, 2014). In addition to eight civil charges, Williams has been charged with two criminal violations by the Spanish government, with penalties of five and a half years in prison if found guilty. Inmet, like Imperial, was part of the Canadian Mining Association's "Towards Sustainable Mining" program during those years. Perhaps not surprisingly, the company gave itself very high rankings for sustainability even after the arsenic contamination (MAC 2011).

Scholars have critiqued voluntary industry standards, arguing that "they have not consistently ensured responsible behavior in mining operations, and negative environmental and social impacts occur more frequently than they should" (Miranda, Chambers, and Coumans 2005, 2). Voluntary frameworks are insufficient for many reasons. First, they fail to create the social or governmental capacity to ensure compliance, transparency, and independent verification. This translates into the curious situation that, even after well-documented mining disasters, the two companies discussed above still assessed themselves as doing an excellent job at sustainable mining—illustrating the point that sustainable development can be used for different political and economic justifications.

For the Anishinabeg and the Sámi, the corporate sustainability discourse misses the core point: clean water and healthy watersheds have values that go beyond what can be measured in the downspout of a water treatment facility. Like other indigenous communities faced with new mining projects, the Anishinabeg and the Sámi both argue that they have the right to "Free, Prior, and Informed Consent" before mining (or other development) projects should proceed. Both groups call for stakeholder processes that will set aside certain areas of special environmental and cultural concerns as "no-go" zones for mining—but not as areas for nature conservation that exclude customary uses of the landscape. Just as most Roman Catholics would probably agree that the Sistine Chapel is too sacred to be mined, indigenous groups argue that they too should have the right to define sites too special to be exploited. The key point is not that one perspective is correct and the other is wrong. Rather, the tension among communities stems, in part, from the stark differences between how different people define sustainability. One thing is clear: defining sustainability in some culturally neutral, scientifically proven manner is impossible.

ICMM objects to the establishment of "no-go" zones that would exclude mining. Rather than simply stating that such zones would limit the ability to develop profitable ore deposits, ICMM borrows the discourse of sustainability, claiming that no-mine zones would violate core principles of sustainable development. For example, in one 2005 memorandum, ICMM objected to an environmental NGO's call for setting up a stakeholder process that would have allowed communities to identify sites that should not be mined, including Ramsar Wetlands, protected wilderness sites such as International Union for Conservation of Nature (IUCN) sites, and World Heritage Sites. ICMM argued that "This approach is not consistent with sustainable development principles which require a balancing of social, environmental and economic factors in decision making. The implicit assumption in prohibiting all mining from IUCN Protected Area Categories I to IV is that the conservation values of such lands always outweigh potential socio-economic benefits" (ICMM 2005, 4). Few sustainable development proponents believe that balancing social, environmental, and economic factors means that every single site on earth should be open to mining. Yet ICMM interprets the discourse of sustainable development to argue just that.

ICMM also objects to requirements that their members should obtain "Free, Prior, and Informed Consent" from indigenous communities before mining in their territories. They argue such consent is "inconsistent with the fundamental principle of international law of territorial sovereignty and a state's right to manage its own natural resources" (ICMM 2005, 3). In 2013, ICMM added that "neither Indigenous Peoples nor any other population group have the right to veto development projects that affect them" (ICMM 2013, 3). Yet for the indigenous communities, the entire point is that their sovereignty is *not* subordinate to the state.

Industry groups such as ICMM use the language of sustainable development to argue that mining can bring an underdeveloped region into a modern condition. In the Lake Superior basin and northern Sweden, this modernization discourse rests on a deeper historic narrative that situates the north as a colonial warehouse, a source of resources to make the central state rich. Industry arguments about the benefits of mining for a sustainable society assume, first, that mining operations *do* lead to economic development for impoverished communities, rather than the resource curse of profits for the few and poverty for the many. Second, sustainable mining arguments assume that a few sacrifice zones—areas damaged by mining—are necessary for the rest of the country to benefit from economic development. Mines do involve short-term sacrifices to environmental quality, ICMM and other proponents of sustainable mining development admit. Rather than degradation, they use the neutral term "change" to represent this: "Mining and metals operations bring significant change—social, environmental, and economic. Change is inevitable" (ICMM 2012, 8). As environmental historians John Thistle and Nancy Langston point out,

> Mines are often located far from urban centers, creating a sense of remoteness (at least in the eyes of decisionmakers), which reinforces the sense that the environmental effects

> of mines are fairly limited. . . . Mining interests continue to promote this view of their activities—that the effects of digging a massive hole in the ground and extracting ore might seem severe, but in the global context, those effects are quite local and brief.
>
> (Thistle and Langston 2015)

The corporate narrative of sustainable mining obscures the fact that, rather than being remote and self-contained, mining sites are inhabited by people who have the legal right to claim the same rights of self-determination as people in urban cores.

## Conclusion

As these two northern cases have shown, in the late nineteenth and early twentieth centuries, Swedish and North American capital and state interests brought mineral resources into production. From one perspective, that of the mining interests and industrial colonizers involved, this was a story of progress and human improvement: the advance of enterprising capitalists, civilization, and modernity into lands considered empty and unproductive. Urban governments long envisioned the north as a remote hinterland best suited for resource extraction. Yet the north is intimately connected to sites of industrial activity by animal migrations, by toxic mobilizations, and by historical legacies. Mining conflicts in the north continue to revolve around what kinds of relationships to socio-natural communities will be supported by development—and who has the right to decide. Conflicts over sustainable mining show that there is no single, correct meaning of sustainability. Rather, we always need to ask: sustainable for whom and for what?

## Acknowledgment

Some material in this chapter has been previously published in Nancy Langston, "Mining in the Boreal North," *American Scientist* 101, 2 (March–April 2013), 98. doi:10.1511/2013.101.98.

## References and Further Research

Ali, Saleem H. 2009. *Mining, the Environment, and Indigenous Development Conflicts*. Tucson: University of Arizona Press.

Auty, Richard. 1993. *Sustaining Development in Mineral Economies: The Resource Curse Thesis*. London: Routledge.

Beach, Hugh. 1993. *A Year in Lapland: Guest of the Reindeer Herders*. Washington, DC: Smithsonian Institution Press.

Bergquist, Lee. 2014. "Gogebic Mining President Accused in Spain of Environmental Crimes." *Milwaukee Journal-Sentinel*, February 28. http://archive.jsonline.com/news/gogebic-mining-president-accused-in-spain-of-environmental-crimes-b99215710z1-247895871.html.

Borchert, Nanna. 2001. *Land Is Life: Traditional Sámi Reindeer Grazing Threatened in Northern Sweden*. St. Leon-Rot, Germany: Nussbaum Medien.

Brännlund, Isabelle, and Per Axelsson. 2011. "Reindeer Management During the Colonization of Sami Lands: A Long-Term Perspective of Vulnerability and Adaptation Strategies." *Global Environmental Change* 21, no. 3: 1095–1105.

Dockry, Michael J. 2012. "Indigenous Forestry in the Americas: Comparative Environmental Histories in Bolivia and Wisconsin." PhD diss., University of Wisconsin-Madison.

Fitz, Tom. 2012. "The Ironwood Iron Formation of the Penokee Range." *Wisconsin People & Ideas*: 33–39.

Fowler, Marilyn Ellen. 2012. "Wisdom from the North: Sami Perspectives on Sustainable Leadership." PhD diss., California Institute of Integral Studies.

Gedicks, Al. 2001. *Resource Rebels: Native Challenges to Mining and Oil Corporations.* Cambridge, MA: South End Press.

Gedicks, Al, and Dave Blouin. 2013. "Science and Facts Show a Need for Tight Regulation of Taconite Mining." *Duluth News Tribune*, February 13.

Grossman, Zoltán. 2005. "Unlikely Alliances: Treaty Conflicts and Environmental Cooperation between Native American and Rural White Communities." *American Indian Culture and Research Journal* 29, no. 4: 21–43.

Howkins, Adrian. 2015. *The Polar Regions: An Environmental History*. Cambridge: John Wiley.

Hutchinson, Brian. 2014. "Anger and Confusion After Worst Disaster in Canadian Mining History Darkens B.C. Town." *National Post*, September 12. http://news.nationalpost.com/news/canada/anger-and-confusion-after-worst-disaster-in-canadian-mining-history-darkens-prosperous-b-c-town.

Initiative for Responsible Mining Assurance (IRMA). 2014. *Standard for Responsible Mining Draft v1.0*. Washington, DC: Initiative for Responsible Mining Assurance. www.responsiblemining.net/images/uploads/IRMA_Standard_Draft_v1.0(07-14).pdf.

International Council on Mining and Metals (ICMM). 2005. *ICMM Preliminary Comments on the Draft Framework for Responsible Mining*. London: International Council on Mining and Metals. www.icmm.com/document/37.

———. 2012. *Mining's Contribution to Sustainable Development: An Overview*. London: International Council on Mining and Metals. www.icmm.com/minings-contribution.

———. 2013. Indigenous Peoples and Mining: Position Statement. London: International Council on Mining and Metals. www.icmm.com/publications/icmm-position-statement-on-indigenous-peoples-and-mining.

Judd, Amy. 2014. "Mount Polley Mine Tailings Pond Breach Is One of the Worst in the World." *Global News Canada*, August 7. http://globalnews.ca/news/1498222/mount-polley-mine-tailings-pond-breach-is-one-of-the-worst-in-the-world-experts/.

Korsmo, Fae. 1993. "Swedish Policy and Saami Rights." *Northern Review* 11: 32–55.

Langston, Nancy. 2013. "Mining the Boreal North." *American Scientist* 101, no. 2: 98.

LeCain, Timothy. 2009. *Mass Destruction: The Men and Giant Mines that Wired America and Scarred the Planet.* Piscataway, NJ: Rutgers University Press.

Lourey, Tony, Kathy Sheran, Tim Mahoney, Sheldon Johnson, Tom Huntley, and Rep. Tina Liebling. 2013. *Annual Report to the Legislature: Minnesota Taconite Workers Health Study*. Minneapolis: Minnesota Taconite Workers. http://taconiteworkers.umn.edu/resources/reports.html.

MacNeil, Kristina Sehlin. 2015. "A Case of Cultural and Structural Violence in the Power Relations Between a Sami Community and a Mining Company in Northern Sweden." *Ethnologia Scandinavica* 45.

Mining Association of Canada (MAC). 2007. *Towards Sustainable Mining Framework*. Ottawa, ON: The Mining Association of Canada. http://mining.ca/towards-sustainable-mining.

———. 2011. *TSM Assessment Protocol*. Ottawa, ON: The Mining Association of Canada. http://mining.ca/towards-sustainable-mining.

———. 2014. *Towards Sustainable Mining Progress Report*. Ottawa, ON: The Mining Association of Canada. http://mining.ca/towards-sustainable-mining/tsm-progress-report-2016

Mining Association of Canada, and Mining Association of British Colombia. 2015. *TSM Progress Report: Imperial Metals*. Ottawa, ON: The Mining Association of Canada. http://mining.ca/towards-sustainable-mining.

Miranda, Marta, David Chambers, and Catherine Coumans. 2005. *Framework for Responsible Mining: A Guide to Evolving Standards.* Executive summary. Bozeman, MT: Center for Science in Public Participation. www.frameworkforresponsiblemining.org/pubs/Framework_ES_20051018.pdf. O'Faircheallaigh, Ciaran. 2013. "Extractive Industries and Indigenous Peoples: A Changing Dynamic?" *Journal of Rural Studies* 30: 20–30.

Owen, John R., and Deanna Kemp. 2013. "Social License and Mining: A Critical Perspective." *Resources Policy* 38, no. 1: 29–35.

Poelzer, Gregory A. 2015. "A Stake in Mining: Participatory Elements in Swedish Mine Development." *Northern Review* 39.

Reynolds, Glenn C. 2003. "A Native American Water Ethic." *Transactions of the Wisconsin Academy of Sciences, Arts and Letters* 90: 143–161.

Rodríguez, Chema. 2013. "La Audiencia Procesa a Tres Directivos de Cobre las Cruces." *El Mundo*, November 24. www.elmundo.es/andalucia/2013/11/24/5292232b61fd3d95778b4573.html.
———. 2014. "La Fiscalía Pide Cárcel para Tres Directivos de Cobre las Cruce." *El Mundo*, February 20. www.elmundo.es/andalucia/2014/02/20/530515fce2704ea0528b456a.html.
Samer.se. *Sápmi.* Accessed November 9. www.samer.se/4549.
Thistle, John, and Nancy Langston. 2015. "Entangled Histories: Iron Ore Mining in Canada and the United States." *Extractive Industries and Society* 3, no. 2: 269–277.
United Nations (UN). 2012. *Sustainable Development in Mining Activities*. New York: United Nations. https://sustainabledevelopment.un.org/partnership/?p=1636.
Vitebsky, Piers. 2006. *The Reindeer People: Living with Animals and Spirits in Siberia.* Boston, MA: Houghton Mifflin Harcourt.
Woo, Andrea, and Alexandra Posadzki. 2014. "Red Flags Raised Years before B.C. Mine-Tailings Spill, Consultant Says." *The Globe and Mail*, August 5. www.theglobeandmail.com/news/british-columbia/bc-mine-had-issues-with-rising-waste-water-ahead-of-breach-consultant-says/article19920040/.

11

# Governing Genetically Modified Crops and Agricultural Sustainability

Phil Macnaghten

Agricultural sustainability—and the parallel imperative of food security—is at the heart of debates about environment and sustainability in a globalizing world. With rising world populations, persistent hunger, and a growing demand for food globally, the need to protect land for biodiversity and ecosystem services and the mounting threats associated with climate change, it is unsurprising that agricultural sustainability is fast becoming recognized as one of this century's most critical challenges.

Yet it remains unclear what precisely agricultural sustainability means and how it is best pursued. On the one hand, as neatly characterized in a 2009 report from the Royal Society, there is the policy discourse arguing for the "sustainable intensification of global agriculture." Arguing from a "globalist" perspective, the challenge is presented as the largely technical one of how to meet demand for food, feed, fiber, and fuel on an area of land that is unlikely to increase in the future. Within this frame, science and technology are presented as having a primary role to play in meeting these challenges. Without radical advances, particularly at the molecular level, it is hard to imagine at least within current conceptual frameworks how yields can be increased without adverse environmental impact or the cultivation of new land. Novel research methods have, so it is commonly claimed, the potential to contribute to food production through forms of genetic improvement, including the genetic modification (GM) of crops that have been altered to introduce new and desirable traits.

Yet, on the other hand, a number of advances in the agricultural sciences that have been seen as promising by agricultural scientists and policymakers have met with considerable social and political resistance. The introduction of GM crops, in particular, has been deeply mired in controversy and polemic. Although the rise of GM crops has been dramatic in recent years, their uptake has not seen the smooth, universal transition predicted by advocates of GM. Controversy has been marked even in those countries where approvals have been impressively rapid. The regulation of GM crops has been challenged as inadequate, even biased, and in some settings, such as India and Mexico, planting of certain crops has been legally suspended. Over the last two decades, the promotion and regulation of GM crops have precipitated acute political and scientific controversy arguably on a scale without precedent in modern times, resulting in substantial disruption to some of the world's largest companies and to the plans of incumbent governments. Unless we understand why this issue continues to evade policy resolution, attempts aimed at enhanced genetic improvement of crops will generate further controversy, misunderstanding, and polemic. For this reason, GM crops and foods represent a fascinating case to examine the contested and historically contingent character of sustainability discourses. Whereas advocates promote GM crops as an integral element of the future of sustainable agricultural, opponents see them as opposed to the basic principles of sustainability. As explored throughout this volume, concepts of sustainability represent a site at which different visions of justice and social progress often collide.

**Innovations in Genetic Modification**

A range of new plant breeding techniques is being developed by agricultural scientists that offer radical prospects of genetic "improvement" that extend beyond genetic modification involving the transfer of genes across species boundaries. These range from cisgenesis, where GM is achieved using only genes from the species itself, to intragenesis, where genetic modification is achieved through a new combination of functional elements of different genes. There has been much excitement in the scientific community associated with new gene editing techniques—such as CRISPR/Cas9, TALENS, and "zinc fingers"—and in their promise for delivering more precise and targeted interventions.

More information on new GM practices are available at these links:

- "Opportunities of New Plant Breeding Techniques." Wageningen UR. http://edepot.wur.nl/357723.
- Ledford, Heidi. "CRISPR: Gene Editing is Just the Beginning." *Nature*. Springer Nature. www.nature.com/news/crispr-gene-editing-is-just-the-beginning-1.19510.

In this chapter I examine the impacts of GM crops and offer an analysis of the dynamics of controversy that could contribute to the development of a particular pathway towards agricultural sustainability. Following a section setting out what GM crops are and the context in which they have been developed, I then draw on qualitative and ethnographic research from three rising global powers—Brazil, India, and Mexico—exploring the views of farmers, publics, and scientists with regard to GM crops. Following an analysis of the dynamics that have underpinned the controversy in these three diverse geopolitical contexts, I then set out a governance framework for the responsible innovation of agricultural GM technologies. The argument promoted in this chapter is as follows: that concepts and models of agricultural sustainability need to embrace elements that are as much social and political as they are technical; that current governance frameworks of agricultural sustainability, including of GM crops and foods, rest upon primarily technical considerations (e.g., through the use of risk assessments) that tend to ignore wider social and political considerations; and that new governance architectures are required if agricultural sustainability is to be pursued in ways that are just, socially inclusive, and respectful of traditions and livelihoods.

## Background and Context

Since the end of World War II, global crop yields have increased dramatically through three routes: improved hybrid varieties as a result of intensive selection breeding; the use of more effective agrochemicals; and improved agronomic practices. The development of hybrid high-yielding varieties of cereal grains was a central plank of what became known as the "Green Revolution," a set of initiatives that took place between the 1940s and 1960s that increased agricultural production worldwide, particularly in the developing world. Hybrid varieties of rice, wheat, and maize (corn) were researched and developed mainly by public agricultural research centers. The widespread adoption of hybrid crops during this period led to steady increases in cereal production and to the general transformation of agriculture, principally in Asia and Latin America. Even though the Green Revolution has had its critics, such developments undoubtedly increased yields, helped feed growing global populations, and arguably assisted in avoiding widespread famine in places such as India. However, in the last twenty years or so, this increase in crop productivity has leveled off, posing serious problems for the near future as the human population increases to a predicted 9 to 10 billion by 2050 from its current level of 7.2 billion. Populations in developing countries are increasingly demanding more meat in their diet, which requires more arable land and feed to sustain livestock, while arable land availability is set to decline through adverse effects of climate change, soil salinization, and urbanization of rural areas. Therefore enhanced yield is a key sustainability target for crops grown in increasingly difficult arable conditions.

Genetically modified crops are plants whose genetic material has been modified using genetic engineering techniques to introduce new and targeted traits. Genetic modification techniques offer the potential to "speed up" traditional plant breeding by introducing selected novel genes into a crop plant in the laboratory. The use of GM technology is seen as potentially valuable because it allows the transfer of genes underpinning yield (and other characteristics) from wild relatives of crops that cannot naturally cross-fertilize, and so are not accessible to conventional breeding programs. The two most common key GM traits that are commercially available are resistance against particular insect pests through the use of Bt toxins and herbicide-tolerant genes to protect crops against herbicides that kill weeds.

Since 1996 the cultivation of GM crops has increased dramatically. Of 1.5 billion hectares (3.7 billion acres) of arable land worldwide, approximately 175 million hectares (4.32 million acres) or around 12 percent are cultivated with GM crops—with GM soya bean, maize (corn), cotton, and canola accounting for nearly all GM crops. Of the twenty-seven countries that planted GM crops in 2013, most were grown in five countries: the United States, Brazil, Argentina, Canada, and India. The pattern of adoption is also shifting: since 2011 the developing world (or what we term the "Global South") overtook the developed world in the area of GM crops under cultivation (*Nature* 2013). According to figures produced for the International Service for the Acquisition of Agri-Biotech Applications (ISAAA), 81 percent of the global crop yield for soya is planted with GM varieties, 35 percent for maize, 81 percent for cotton, and 30 percent for canola (James 2013).

Current approaches to the regulation and governance of GM crops have been dominated by risk-based assessment methodologies, the assumption being that the key criterion mediating the release of genetically modified organisms (GMOs) into the environment should be based on independent case-by-case risk assessments of their impacts on human health and the environment. In the United States a regulatory regime emerged that considered genetic engineering as a process that presented no special risks that could not be addressed by existing product-oriented legislation. This led to the principle of substantial equivalence—meaning that the safety of a GMO food may be assessed by comparing it with a similar traditional food that has proven safe in normal use over time—that came to govern international trade policy, including that of the World Trade Organization (WTO). In Britain a more expansive view of the potential for GM technologies to generate harm developed, including those surrounding the industrial production of GMOs and their deliberative release. These were seen to require control and regulation that could not be accommodated with existing regulatory structures. A regulatory system developed in which the process of genetic modification became an appropriate basis for determining policy. This was complemented at the European level, starting in 1990, where GMOs were regulated to ensure that they would not cause "adverse effects." In Mexico, India, and Brazil, regulation followed a blend of European and North American models. In addition, regulatory regimes were driven in part by the need to implement the Cartagena Protocol on Biosafety, an international agreement that sought to protect biological diversity from the potential risks posed by GMOs.

> The Cartagena Protocol on Biosafety is an international agreement that aims to contribute to the safe handling, transport, and use of living modified organisms (LMOs) resulting from modern biotechnology. It aims to mitigate against potential adverse effects on the conservation and sustainable use of biological diversity, and on human health, focusing specifically on transboundary movements. It was adopted on January 29, 2000, and entered into force on September 11, 2003. For more information see the "Cartagena Protocol on Biosafety to the Convention on Biological Diversity" on the website of Secretariat of the Convention on Biological Diversity at this web address: www.cbd.int/doc/legal/cartagena-protocol-en.pdf.

One consequence of this regulatory framing is that the public debate surrounding GM crops has boiled down to one of safety: whether they are safe to eat, and whether they are safe for the environment. Yet, this framing does not exhaust the issue. In the Global South, for example, the uptake of GM crops has been associated with the widespread adoption of neoliberal policies aimed at the institutional reform of agriculture, where social need is equated to that of the free choice of consumers in the marketplace, and where the scope of regulation is restricted to cover solely the scientific evidence of harms rather than the appraisal of social and cultural contexts. In the Global North, public concerns have mobilized around a broader range of arguments: that GM foods would lead to an inevitable loss of consumer choice, that decisions had already been taken outside the public sphere, that GM crops would lead to the corporate control of food systems, that GM crops and foods would benefit only multinationals and large-scale farmers, that the technology was "unnatural," and that there would be unpredicted effects beyond the reach of risk science. Restricting public regulation and deliberation to a science-based analysis of harms, where wider issues and their associated framings and voices are effectively hidden from public accountability and influence, thus employs science to generate standardized procedures and processes that serve a particular model of global capitalism and interests.

Such critique formed the background for the GMFuturos project. The strategic question behind the project was to examine why GM crops have not been universally accepted as a public good, since if we do not address this, we will fail to understand the conditions under which GM crops may contribute to global agricultural sustainability in an inclusive manner.

> **The GMFuturos Project**
>
> The GMFuturos project ran from February 2012 to November 2014 with funding from the John Templeton Foundation. It included an interdisciplinary team of anthropologists, human geographers, philosophers, physicists, plant scientists, political scientists, and theologians at Durham University in the UK, alongside local teams of researchers working in Brazil, India, and Mexico.

The broad analytical framework guiding the project's research is the "pathways to sustainability" approach pioneered by the UK STEPS (Social, Technological, and Environmental Pathways to Sustainability) Centre. This approach attempts to link technological innovation with questions of social justice and environmental integrity. Focusing on the needs of the poor is an explicit theme of the approach, both in terms of seeking greater recognition of the everyday realities of marginalized peoples and in terms of developing policies aimed at addressing their needs. Borrowing from science and technology studies, the pathways approach emphasizes the importance of framing. Different ways of framing an issue or a problem by different kinds of actors will lead to different ways of understanding and representing the system and thus to different pathways for problem resolution. Framing involves choice—about which elements of the system to include and exclude, about whose knowledge counts and whose does not, about what scale is appropriate, about what constitutes policy resolution—and the pathways approach attempts to open up new choices, particularly those oriented towards the needs of the poor, as well as making existing choices more accountable. Given this focus on multiplicity, the pathways approach seeks continuously to challenge dominant pathways—and their associated framings and narratives—and to open up alternative, hidden, and otherwise marginalized pathways to sustainability.

The STEPS Centre is an interdisciplinary global research and policy engagement centre uniting development studies with science and technology studies. Its cross-cutting work covers agriculture and food; energy and climate change; health and disease; and water and sanitation. For more information, see their website at this URL http://steps-centre.org/.

## The GMFuturos Research

The central argument underpinning the GMFuturos research is that if we are to govern GM crops in a socially and scientifically sustainable manner, debate over relevant issues needs to include a broad range of actors and stakeholders. At the core of the project was fieldwork undertaken in three of the rising global powers, namely in Mexico (on GM maize), Brazil (on GM soya), and India (on GM cotton), and involving ethnographic, interview, and focus group research with farmers, scientists, and publics. The choice of these case studies is deliberate. Although the majority of scholarship on GM crops has focused on the Global North, it will be in countries such as Brazil, Mexico, and India where sustainable agricultural innovation is most needed, where the bulk of food provision is expected to come from, and where debates over GM agricultural technologies are likely to be most intense.

### An Analysis of the GM Debate in Mexico

Reviewing the debate on GM maize in Mexico, we found that the controversy over GM maize came to prominence after a highly publicized article in the journal *Nature* written by David Quist and Ignacio Chapela (2001) reported the flow of transgenes into wild maize populations. Although widespread protests from the academic community and beyond led to the paper's retraction, subsequent protest from civil society actors and the wider population over the issue continued. Maize has high cultural resonance in Mexico, and protests against GM maize came to signify the defense of Mexican culture and identity in the face of unwanted forms of globalization. In such charged circumstances, decisions by regulatory bodies were seen as compromised and lacking in transparency and contested vocally by NGOs and other actors. Notwithstanding the drama at the level of civil society, there was little evidence of sustained effort by institutional actors, including the Mexican state, to engage or inform the public (Carro-Ripalda, Astier, and Artia 2016).

In our fieldwork research with smallholder farmers and rural stakeholders in the Pátzcuaro Lake region in the state of Michoacán, we found that debates on GM maize were situated within the context of an ongoing crisis in rural agriculture. Within this context we found strong and enduring social relations around maize agriculture, reproduced by systems of local community exchange and day-to-day food and religious practices. Within this context GM maize was felt as an intrusion into traditional practices, with unknown and possibly negative impacts. Suspicion was exacerbated by deep patterns of mistrust expressed concerning the motivations of key actors, including the government and the seed companies. Smallholder farmers were also fundamentally opposed to GM maize, seeing it as artificial, unnecessary, and a threat to traditional smallholder agriculture.

Reporting on interviews with a variety of local stakeholders involved in the debate on GM crops, we found a clear division between the views of some of the respondents (smallholders, consumers, environmental NGOs, and social scientists) and those of others (large producers and seed companies). For the former, traditional maize agriculture was seen as highly significant for Mexican history and culture, and GM maize perceived as an imposition and transgression. Regulatory bodies and laws (including the Biosafety Law) also tended to be seen as compromised and ineffective. For the latter, GM maize was viewed as part of a modernity that would transform the Mexican countryside from its current malaise.

Next, we conducted a laboratory ethnography at the National Laboratory of Genomics for Biodiversity (LANGEBIO) in Guanajuato. Here, we found a clear distinction within the laboratory between senior researchers who were more avowedly in favor of the application of GM agricultural technologies, including GM maize, and younger and more junior researchers who were more cautious and nuanced. For the latter, extreme care was advocated in any attempt to restructure the maize genome, with a strong preference not to use genetic material from other species, and to minimize any risk that could affect maize's integrity.

Among predominantly middle-class urban Mexicans, we found an appreciation of traditional maize products and cooking as a part of Mexican identity and as a medium in the maintenance of diverse social practices. We identified a negative reaction to GM foods and crops generally, but especially to GM maize. This negative perception was compounded by various dynamics: by the lack of labeling and reliable information, by mistrust in the motives of those producing and selling them, by the unknown dangers GM foods may pose, and by the lack of proven necessity. The government, generally, was seen as in collusion with the large corporations at the expense of the public interest. Regulatory bodies were similarly lacking in credibility.

At the end of the fieldwork, we organized a workshop including a range of national stakeholders (smallholder and large-holder farmers, seed company representatives, natural and social scientists, and environmental groups) to explore our research findings and to make recommendations on how to advance the debate on GM crops. With some exceptions (mainly among some natural scientists and representatives of seed companies who believed that the argument in favor of GM agriculture had already been won), participants voiced the need to open up the public debate. Particular calls were made to develop more rigorous policies on maize, to give more voice to smallholder farmers, to develop agricultural research that aims to secure genuine sustainable development, and to reconsider all the issues surrounding the production and conservation of native maize and food security. These could be seen as some of the conditions necessary for GM crops to contribute to inclusive agricultural sustainability in Mexico.

## An Analysis of the GM Soya Debate in Brazil

Our report on findings from the Brazil case study focused in particular on debates and practices around GM soya (Guivant and Macnaghten 2016). First, we reviewed the debate on GM crops in Brazil, identifying the trajectory of the debate and the factors that led both to the resistance and to the widespread adoption and take-up of GM crops in Brazil. The story of GM crops in Brazil has been deeply polemical, with strong and heterogeneous coalitions of actors both for and against them. As with the Mexican case, GM crops came to be represented as a symbol of a wider struggle against neoliberal policies and U.S. imperialism, leading to a ban that was sustained up until 2003. However, following widespread cultivation of illegal GM soya in the southern states, President Luiz Inácio Lula da Silva issued a presidential decree permitting the temporary sale and distribution of illegal GM soya. This was followed by the passage of a Biosafety Law in 2005, creating the context for GM cultivation, and leading to subsequent and widespread cultivation of GM crops, particularly GM soya and maize. By 2012, the coverage of GM crops in Brazil had risen to 36.6 million hectares (90.4 million acres) or 21 percent of the global biotechnology crop, with Brazil becoming the world's second largest producer of GM crops, behind the United States.

### The Reach of Neoliberalism

Neoliberalism is a political philosophy, developed in the 1930s, that supports extensive economic liberalization policies including those of privatization, fiscal austerity, deregulation, free trade, and reductions in government spending in order to enhance the role of the private sector in the economy. The wide acceptance and implementation of neoliberal policies and economic theories began in the 1970s, through the International Monetary Fund, the World Bank, the Maastricht Treaty, and the World Trade Organization.

In our fieldwork research with family farmers and rural actors from the southern state of Santa Catarina, we identified the various ways in which GM crop technologies had been adopted into local agricultural practices. While GM crops were perceived to have certain technical advantages, such as ease of working the land and some initial improvements in yields, albeit at cost, we found that GM crops had tended to be accepted mainly on pragmatic terms, often as a survival tactic for smallholder farmers. Especially women and those working in organic agriculture felt marginalized from debates on GM crops, which in many cases were affecting their livelihoods but which tended to be presented as an inevitable part of Brazil's agricultural future. Here, multiple and competing sustainabilities were being played out at a practical level, with different sets of actors imagining their own practices as best contributing to long-term agricultural sustainability. We also found evidence of a conflict between farmers and technical experts from the seed companies, each blaming each other for the growing problem of weed resistance to glyphosate. Again, we can see how the sustainability promises of GM crops (e.g., in the promises of reduced use of agrochemicals) had not been met in practice, and with the idealized prescriptions designed into the technology proving unable to cope with backlashes from nature (in this case, weed resistance to the herbicide that is part of the GM technology package).

Next, we undertook a survey and interviews with a variety of stakeholders involved in the debate on GM agriculture. Notwithstanding a diversity of views, we found a clear alignment of responses with negative claims, chiefly: the propensity of GM agriculture to create dependency on seed companies, to cause potential problems with human health and/or the environment, and to threaten traditional forms of life. Most respondents agreed that while the debate had abated, it had been thus far dominated by a few powerful voices (large farmers, scientists, and corporate interests), and that involvement of the wider public or access to quality information had been limited.

Subsequently, we undertook a laboratory ethnography at the soya research division (CNPSO) of the state-owned agricultural research organization Embrapa, located in the southern state of Paraná. We found clear and unqualified optimism amongst scientists on the role of GM crop technologies to provide significant future agricultural improvement, and to create more productive and resilient cultivars with genetic resistance to pests and major diseases. The arguments deployed tended to emphasize economic benefits, the apparent unparalleled ability of GM crop technologies to provide "improvements," and the necessity for agricultural GM research to have a strong national base. We also found little evidence of a structured and sustained debate between scientists and the wider society, which the scientific community tended to see as uninformed and thus unqualified for entering the debate on GM crops.

Then, through a series of focus group discussions, we undertook research with urban publics on Brazilian responses to GM crops and foods. We found that food quality and safety were important topics for urban Brazilians, who expressed concern about the industrialization of foods, and, for at least the better-off, a desire to consume foods as organic and locally sourced as possible. Most of the participants displayed little awareness and genuine surprise about the extent of GM crop adoption in their country. Notwithstanding a general trust in expert systems, including science, participants adopted largely negative opinions about GM crops and foods in the discussions, not least because the technology was seen as benefiting the producer rather than the consumer and because they had not been consulted or clearly informed. They were concerned that the public debate had so far been largely restricted to academic scientists, government actors, and seed companies at the expense of wider civil society; some expressed a suspicion that these actors may have "manipulated" the debate to promote their own interests. As a response, participants called for wider government responsibility for assuring more robust regulation and oversight, for raising consciousness, and for promoting the public interest.

Finally, we organized a deliberative workshop, conducted with a range of national stakeholders, to explore research findings and how to develop the public debate on GM crops. We found that the clear majority of participants agreed on the need to reopen the debate on GM crops, on its regulation and oversight, on the need for concerted action to communicate reliable information, and for proper channels of citizen participation in strategic decisions. We found also a widespread feeling of impotence in confronting the power of the current alliance between scientists, the seed companies, and politicians. Again, we see addressing these concerns as critical conditions for fostering a socially inclusive model of agricultural sustainability.

## An Analysis of the GM Debate in India

The findings from the Indian case study focused in particular on debates and practices around GM *Bt* cotton (Egorova, Raina, and Mantuong 2016). Here, we also started by reviewing the debate on GM crops in India. We identified the trajectory of the debate and the factors that led to the ongoing resistance to the adoption and take-up of GM crops in India, culminating in the 2013 ten-year moratorium. We found that the issue has been hotly debated by parliament, by the Union government, by NGOs, and in the media, much of the debate revolving around regulation of GM trials and crops. We found that the promotion of GM crops has been encouraged by the Indian government, by the Ministry of Agriculture, and by the wider science and technology establishment. Within this context, even though there have been a few within these organizations who voiced concerns about the social, political, and ecological implications of GM crops, it has been the NGOs who have spearheaded the articulation of these concerns, with some success.

In our fieldwork research with smallholder family farmers in the Kalahandi district in the west of Odissa, we found an increasing prevalence toward the cultivation of GM *Bt* cotton, especially in upland areas, due to the perception of higher yields and increased incomes. We saw that *Bt* cotton, now accounting for over 90 percent of cotton cultivation in India, was being aggressively marketed by seed companies, through demonstration projects, the provision of advantageous loans, and the delivery of inclusive technological packages. This was perceived as having both positive and negative effects. As well as increasing incomes, the increased coverage of *Bt* cotton was making it difficult for the farming community to access traditional varieties of food crops, including millet, jawar (sorghum), and lentils. Increased incomes came with increased cultivation costs, seed costs especially, and increasing dependence on external agents. Moreover, in recent years *Bt* cotton crops had become affected by increased pest attacks and have led farmers to consider previously available seed varieties. However, farmers find themselves often "locked-in" to using *Bt* cotton with indigenous seeds no longer so easily available. Thus again we see how the sustainability promises of a GM crop were at best unevenly realized in everyday agricultural practice, with GM crops closing down alternative pathways to sustainability and reducing resilience.

In interviews with a variety of local stakeholders involved in the debate on GM agriculture, we found that both stakeholders in favor of GM crops (e.g., representatives from seed companies) and those against (e.g., environmental NGOs) tended to use a similar argument from science: what was required in both cases was more rigorous science to help settle the issue. For the environmental groups independent science would bolster their claims that GM crops had not been proved to be safe, while for the seed companies, science would prove that GM crops were safe and that public concerns were unreasonable.

Reporting on findings from a laboratory ethnography conducted at the New Delhi branch of the International Centre for Genetics and Biotechnology (ICGEB) research laboratory, we found that all scientists whose work we observed were opposed to the moratorium and constructed and perceived the position of anti-GM actors as "ignorant" or aimed at "publicity" seeking. Scientists' critique of the moratorium was often framed in terms of postcolonial discourse, as they argued that India could not afford the risk of "falling behind" in the development of biotechnology.

Next, we undertook research with lay publics on Indian public responses to GM crops and foods. Using a series of focus group discussions, we found the majority of our research participants developed negative views on GM crops and foods. Urban publics pointed out that they did not trust the government and the local authorities to provide a reliable regulatory system for the production of GM crops and therefore that they would prefer to avoid the consumption of GM food. Our research participants from the urban groups also suggested that the information about GM that was available in the Indian mass media was confusing and that they would welcome better structured TV debates that would provide a forum for both pro-GM and anti-GM actors to present their positions. Rural consumers also expressed negative views of GM crops and argued that using GM seeds was interfering with the preservation of indigenous seeds.

Finally, we organized a deliberative workshop, conducted with a range of national stakeholders, to explore research findings and how to advance the public debate on GM crops. We found that participants from different categories of stakeholders were eager to engage in a dialogue and emphasized the importance of considering diverse points of views in deciding the future of GM crops in India.

## Comparing GM Crops in Mexico, Brazil, and India

Comparing the three case studies, we found strong parallels and convergence on local experiences of GM crops and their relationship to agricultural sustainability. Reviewing the national controversies over GM crops, we found that in all three cases, the technical regulatory bodies charged with approvals for the release of agricultural GMOs had not provided what public policy scholar Maartin Hajer terms "authoritative governance," that is, they did not lead to decisions, developed through reasoned, open, and transparent deliberation, that were seen as worthy of acceptance by the broader community (Hajer 2009). Across all three jurisdictions, decisions made by approval committees had commonly been rejected by institutional and other stakeholders. To explain why the controversy surrounding GM crops had taken different forms in different national settings, we constructed a typology that transcends questions of technical risk. Factors that were seen to be relevant in structuring the controversy included: the perceived authority of the regulatory agencies, the cultural resonance of the crops in question, the level of intensity of protest movements, the extent to which GM can become represented as the symbol of wider struggle, and the degree of sustained effort by institutional actors to engage the public. All these are critical elements in a socially robust model of agricultural sustainability. These are presented in Table 11.1.

Comparing the field ethnographies, we found that GM maize in Mexico represents a unique case. The cultural resonances surrounding maize clearly indicate that any introduction of GM maize, especially in the southern states, would be perceived as a threat both to local traditional practices and to a historically situated sense of identity. In Brazil, on the other hand, soya has little cultural resonance (much of it is produced for export and for animal feed in any case) and there was little sense of GM soya as violating local senses of identity. India represents perhaps a more ambiguous case. Even though cotton is represented as a potent symbol of Indian national identity and self-sufficiency, in our local agricultural communities, *Bt* cotton was rarely represented as embodying a threat to traditional ways of life. We also found on-the-ground evidence, both in India and Brazil, of the prevalence of glyphosate-resistant weed species. In both cases, we found that seed companies and farmers were tending to blame each other, with farmers sensing that they had become increasingly "locked-in" to GM agriculture, finding it hard to revert to previously available indigenous seeds, and having to resort to using increasing quantities of pesticides.

Comparing the stakeholder interviews and questionnaires across all the national settings we found a clear divide between those who represented GM crops as part of a gradual and continuous path of science working towards agricultural improvement (mostly representatives from seed companies and some natural scientists), who utilized an Enlightenment-inspired GM as progress narrative (see also Chapter 2), and those who saw GM crops as a rupture or break with conventional breeding practices (mostly smallholder farmers, environmental and consumer NGOs, women's associations, and indigenous groups), who utilized a GM as "messing with nature" narrative (Macnaghten, Davies, and Kearnes 2015). For the latter, GM crops were seen as posing unknown risks, as engendering further dependency on seed companies, and as threatening traditional farming practices and lifestyles. There was also near consensus from the latter constituency that decision making thus far had lacked transparency and participation, with minimal levels of informed public debate.

Comparing the research laboratory ethnographies, notwithstanding the high quality and at times pioneering research being undertaken within and across all three laboratories, the

**TABLE 11.1** Factors Shaping the Controversy on GM Crops: A Comparison of Mexico, Brazil, and India

| *Country* | *Perceived authority of the regulatory agencies* | *Cultural resonance of the crop* | *Intensity of protest movements* | *GM as symbol of wider struggle* | *Degree of public engagement* |
|---|---|---|---|---|---|
| **Mexico** GM maize | **Low** Decisions by regulatory bodies seen as lacking in authority and transparency and judged at times to be illegal. | **High** Maize is an integral part of Mexican identity, history. and culture | **High** The anti-GM campaign has sustained its presence since 2002. | **High** GM maize is constituted as a symbol of the protest against neoliberalism and NAFTA. | **Low** There has been little sustained effort by institutional actors to engage the public. |
| **Brazil** GM soya | **Low/ Medium** Approvals have been successfully authorised by CTNBio since 2005 leading to widespread planting. But decisions remain contested. | **Low** Soya has little cultural significance in Brazil. | **High** (until 2003) **Low** (from 2005): Following the passing of the Biosafety Law the protests peter out. | **High** (until 2003): GM crops are situated within an anti-globalization discourse. **Low** (from 2005) | **Low** There has been little sustained effort by institutional actors to engage the public. |
| **India** GM cotton | **Low** Regulatory bodies seen as lacking in transparency and capacity. Perceived gaps in the regulatory system led to 2013 moratorium. | **High** Following Mahatma Gandhi, the "fragile thread" of cotton became a national symbol of Indian self-sufficiency. | **High** The anti-GM campaign has sustained high-profile protests. | **High** *Bt* cotton is a symbol of a struggle against multinationals and neoliberalism. | **Low** There has been little sustained effort by institutional actors to engage the public. |

Source: Compiled by the author.

research culture of the laboratories in each case was described as lacking in "reflexivity" and "inclusion." With the partial exception of some at LANGEBIO (Mexico), the researchers interviewed for the study appeared to lack the motivation or encouragement to work in interdisciplinary contexts with the human and social sciences, to develop informed understanding as to why GM crops had become controversial in their country, or to be afforded the opportunity to deliberate on the strategic context of GM crops with other actors in public forums. The research culture of the laboratories was similarly lacking in inclusiveness. Researchers did little to listen to, understand, or engage with social actors or views—and their often divergent framings of the issues—and had not developed a collective sense of responsibility for the outcomes of their research as they would be used in practice.

**Deliberative Processes**

"Deliberation" is defined as the critical examination of an issue involving the weighing of reasons for and against a course of action. A "deliberative process" is one where a group of actors receive and exchange information to critically examine an issue, and to come to an agreement that will inform decision making.

Next, we compared the public focus group research. We found that laypeople, mostly living in urban settings, were able to engage in meaningful debate across all three national settings. We found that national emerging food cultures were important in structuring public responses to GM foods: in Mexico the key factor was the appeal for maize cooking and products as an element of national self-identity; while in Brazil, public responses were shaped by growing concerns with the industrialization of food and the drive towards organics and local produce. Across the board, we found little public enthusiasm for GM crops and foods and a hardening of response as the discussions matured. People adopted negative views partly because they felt they had not been consulted and were inadequately informed, partly because GM foods were commonly perceived to be unnecessary and potentially harmful, and partly out of distrust for regulatory agencies and the seed companies. In Mexico GM maize was represented as a symbol for the apparent collusion between the political class and large corporations at the expense of the wider public interest. For some of the Indian participants, particularly from rural areas, the actual gene used in the genetic modification of plants was seen as relevant. Insertion of animal genes into plants was seen generally as less acceptable as this would transgress religious taboos.

Finally, we compared the stakeholder deliberative workshops. We found, generally, that the attempt to develop a new kind of conversation on GM crops and their governance was welcomed across all three settings, with a surprising degree of consensus among participants, and as an alternative to the polemic style of interaction that had tended to characterize the debate. We found that across all three settings there was consensus that improvements needed to be made in four domains: public dialogues, information, education, and governance. Across the three settings, stakeholders tended to prioritize the call for novel forms of public engagement, for the production of high quality and reliable information, for educational establishments to foster the development of critical citizens, and for governments to govern in the public interest. There were further calls for regulatory agencies to understand how regulations were being implemented (or not) in farming practices and, in the Indian workshop, for research institutions to develop greater socio-cultural sensitivity: to develop research that better responds to

India's strategic needs; to both understand and accommodate different framings of the issue, including religious sensitivities; and to do so in a way that is accountable, transparent, and responsive.

## A Responsible Innovation Framework for Agricultural Sustainability

So far we have described empirical research on the views of farmers, scientists, and lay publics on GM crops in Mexico, Brazil, and India. We have found a clear absence of "authoritative governance" across all three jurisdictions insofar as existing forms of regulation and oversight have been ill-equipped to anticipate, recognize, and respond to public controversy. More generally, the governance of GM crops falls into what Maartin Hajer calls an "institutional void," where there exist few agreed structures or rules as to how we should govern them in their "beyond risk" dimensions (Hajer 2003). Responding to this "institutional void," we now set out a novel way to govern GM crops informed by recent debates on responsible innovation: that if we are to innovate responsibly and robustly, we need new institutional capacities to better *anticipate* the wider driving forces as well as impacts of emerging technologies, to open up an *inclusive* debate with stakeholders and wider publics, to develop more *reflexive* scientific cultures, and to design new governance architectures that are *responsive* to these processes. The anticipation–inclusion–reflexivity–responsiveness (AIRR) responsible innovation framework has been pioneered in the UK-sponsored Engineering and Physical Science Research Council (EPSRC) and is being implemented by UK research councils and more widely across Europe. It offers new potential to reconfigure the debate on the governance of GM foods and crops in the UK, in Europe, and internationally, and hopefully to help move the debate away from its current polemic and impasse. We now explore what each of these dimensions might entail as a vision for the future governance of GM crops for agricultural sustainability. This final section makes use of a set of commentaries from participants in a policy workshop on the findings of the GMFuturos research, independently written up in the edited volume *Governing Agricultural Sustainability* (Macnaghten and Carro-Ripalda 2016).

### Anticipation

Anticipation is the first dimension of the responsible innovation framework. An anticipative approach requires the development of capacities to enable researchers and policymakers to understand the stakes of a techno-scientific issue, by systematically exploring possible and plausible futures and their associated societal and ethical dimensions. Given that GM crop technologies are by now relatively mature, and that at least first-generation GM crops have been adopted throughout much of the developed and developing world, an anticipative approach requires a systematic contextualization of GM crops' social and ethical impacts as a precondition for imagining how they could be otherwise configured. That is to say, we need a better understanding of the context out of which GM crops developed, of the kinds of social worlds toward which they have contributed, and thus, by implication, how such conditions need to be reconfigured to contribute to more humane, socially just, and environmentally sustainable futures.

In this regard, a number of our commentators highlighted neoliberalism as an important contextual factor to understanding the debate and associated controversies on GM crops. For science and technology studies scholar Les Levidow, the development of GM crops has its origins in neoliberal policy agendas and modes of thought, especially in Europe (Levidow 2016). Theologian Michael S. Northcott concurs. Using the classical Greek concept of *poïesis*, which he defines as "the unique capacity of humans to generate and sustain aesthetic and ethical ideals through their creative powers on earth," he views GM crops as deeply implicated in "the gradual neoliberal collectivization of agriculture by private corporations," which he sees as disrupting the intrinsically human potential to "enhance goodness, truth and beauty" through earth–human interaction (Northcott 2016, 202). Thus, any attempt to reconfigure governance debates will have to acknowledge this relationship and its transformative effects at the level of everyday practice.

Neoliberal policies, according to some academics, are at the heart of many of our most acute social, political, and economic problems. See this article in *The Guardian* for more on this viewpoint:

Monbiot, George. "Neoliberalism – The Ideology at the Root of All Our Problems." *The Guardian*. www.theguardian.com/books/2016/apr/15/neoliberalism-ideology-problem-george-monbiot.

In addition, we need methodologies to develop non-reductionist accounts of the relationship between GM crops and their impacts, methodologies that consider the problems GM crops cause in people's livelihoods alongside technical considerations of risk. Good anticipatory practice thus develops out of "local" cultural and country-specific historical inquiry. In the GMFuturos research, it was the variety of issues and the plurality of narratives that came together in the GM crop and food issue that consolidated the strength of its findings. On this point, we need to guard against the dangers of confining "the poor to categories not of their own making," and we need to ask the question frequently asked by policymakers: Can GM crops feed the world? While this may be an appropriate question in global spheres of elite policymaking, we need to consider *which world?* Following commentator and anthropologist Robert Simpson, we need to consider:

> Is it the precarious world of the third world farmer, struggling on the margins of global markets and for whom GM crops are simply one part of a complex mosaic of strategies aimed at surviving until the next harvest? Or is it about feeding the world as it is lived within advanced economies of the North and where appetites draw developing food producers into global food chains.
>
> (Simpson 2016, 223)

## Inclusion

Inclusion is the second dimension of the responsible innovation framework. To develop responsible governance requires inclusive and deliberative engagement with a broad range of stakeholders, including publics. In the GMFuturos research, we initiated broad-ranging engagement with publics and smallholder farmers across the three case study sites in Mexico, Brazil, and India. Dominic Glover found the conclusions of the GMFuturos research a necessary antidote to the narrative that presents the adoption of GM crops in "rising power" contexts as an unequivocal success story, "a story of radical and progressive technological change that has been embraced by literally millions of farmers, the great majority of them smallholder producers in the 'developing world'" (Glover 2016, 167). Glover highlights how the study counters this narrative: through the finding that both growers and consumers in each of the case settings felt ill-informed about the technology and excluded from decision-making processes; that in the ethnographic field research smallholder farmers evinced little trust in the technical advisors of the extension services; and in the general recognition that the issues at stake in GM crops reflect "conflicting interests and competing visions" that "are intrinsically difficult to reconcile." For Glover such findings point to "a pervasive breakdown in public trust in institutions of science, governance and regulation" (Glover 2016, 169).

To summarize, an inclusive approach requires the addition of new voices in the governance of science and innovation as part of a search for legitimacy. The further and significant challenges of promoting inclusive governance include that of developing ongoing social intelligence through longitudinal studies on public engagement, that of framing deliberative processes so that they open up the trajectory of GM crop technologies to questions of political economy, and that of involving citizens in different countries in decision-making processes at the international level. In addition, there is the difficulty of how to reconcile what may well be incommensurable narratives and positions. Such reconciliation may involve the art of using political judgment in the face of difficult and possibly incompatible choices. However, an inclusive approach to governance provides the grounds to enable plural narratives and framings to come to the surface (including those which tend to be overlooked), to enable deliberation on potentially competing narratives and framings, and to offer novel ways forward as they are developed.

## Reflexivity

Reflexivity is the third dimension of the responsible innovation framework, which Jack Stilgoe, Richard Owen, and I defined at the level of institutional and scientific practice, as "holding a mirror up to one's own activities, commitments and assumptions, being aware of the limits of knowledge and being mindful that a particular framing of an issue may not be universally held" (Stilgoe, Owen, and Macnaghten 2013, 1571). Our two commentators from the plant science community, Ian Crute and Keith Lindsey, were asked to reflect on the GMFuturos research and to consider "what kinds of scientific culture do we need for responsible agricultural innovation." In response, their views were roughly as follows: that previous advances in agricultural technologies in the twentieth century led to "spectacular increases in productivity" that "enabled a 250 percent increase in the global population (from 1.7 to 6.0 billion) to be fed from only a 40 percent increase in the area of cultivated land (from 1 to 1.8 billion hectares)"; that if we are to meet the basic needs of "a future world of at least nine billion people" and to do so "securely, sustainably and equitably," we will need further radical agricultural innovation; that the current system of regulation in Europe (where the technology associated with plant breeding is regulated as opposed to the novelty of any derived product) is unnecessary and costly, effectively risking that "the endeavor of crop improvement will become the preserve of large corporations;" and that the unifying principles which should underpin the "sustainable intensification" of agriculture should be "simultaneously to raise productivity, increase resource-use efficiency and reduce negative environmental impacts" (Crute 2016; Lindsey 2016).

Notwithstanding the virtues of such a public-interest–global-challenges model, this crop science perspective and viewpoint can be seen as limited in at least three respects, as elaborated by commentator Michael S. Northcott. First, the "relationality between people and land, culture and agriculture, is missed by crop scientists whose primary training is focused on maximizing production of an individual crop in a laboratory." Second, Northcott argued that the Enlightenment-derived "mechanistic and reductionist frame" that underpins much crop science laboratory practice "underwrites the belief that a laboratory made crop will not influence the environment, the farmers, the eaters and other species who interact with it." And third, arguing from political economy, Northcott observed that crop scientists underappreciate the risks associated with the "gradual neoliberal collectivization of agriculture by private corporations," and their potentially long-range impacts on questions of human freedom, dignity, and sovereignty (Northcott 2016, 201).

Physicist and science/theology writer Tom McLeish responds to the fundamental disconnect between science and society, as reflected in the GMFuturos findings, as partly a question of narrative failure. Given that the traditional "modernist–instrumentalist" narrative of science—which presumes that "science will inevitably lead to enlightenment and social progress," and "attempts to restrict the terms of debate to the evaluation and minimisation of technological risk and the maximisation of reward"—is increasingly discredited, McLeish asks how we might cultivate new narrative resources that can promote the "responsible care" of nature and society in sustainable futures. Using the "ancient wisdom" literature as a resource, and drawing on a detailed interpretation of the Book of Job, McLeish develops a theology of technology that speaks to an "ethics of human responsibility" in "husband[ing] the world," and a "teleology that centralizes and prioritizes the wellbeing of the world before the wealth of human beings" (McLeish 2016, 192).(This account is analogous to Michael S. Northcott's commentary that discovers the concept of *poïesis* across a range of ancient stories, including those of Jews and Christians, as a form of craft that "enhance[s] goodness, truth and beauty in the given order.") This "essential rebalancing," for McLeish, is positioned as a response to the tragic tone that informed current responses to GM crop technologies, as evidenced in the GMFuturos study, and as a counterweight to crop science's viewpoint that has tended to view genetic modification technologies as "allowing for the indefinite extension of human intervention in nature" (McLeish 2016, 195).

## Responsiveness

Responsiveness is the fourth and final dimension of the responsible innovation framework. To be effective, responsible innovation requires an institutional capacity to change shape or direction in response to improved anticipation, inclusion, and reflexivity. Across the three case sites of the GMFuturos study, we found little evidence of a responsive science policy and regulatory regime, a finding that arguably is replicated across most of the developed and developing world, including Europe. The key question is *why*. As commentator and risk governance scholar Judith Petts asks, while reflecting on her own experience, "why, given all the evidence on the need for new modes of risk governance and public engagement has this largely not happened?" (Petts 2016, 205).

There are perhaps four intersecting kinds of explanation. First, as argued earlier, there is the problem of the "institutional void." At least in relation to the non-risk dimensions of GM crops, there are few agreed structures or rules as to how we should govern them largely because of the conflicting interests and competing visions that are intrinsically difficult to reconcile. There is an obvious "interest" amongst certain communities (the global seed companies, for example, and perhaps certain government ministries) not to open up the debate on the governance of GM crops beyond their risk dimensions, not least because existing commitments are deeply embedded in extant styles of neoliberal policymaking.

Science policy analyst and commentator Rajeswari Raina adds a further perspective (Raina 2016). She identifies institutional rigidities in the Indian agricultural science and technology (S&T) system that militate against the possibility of a more responsive and deliberative system. Notwithstanding widespread criticisms, replicated in the GMFuturos study, the Indian S&T system continues to reproduce an industrial model and approach to agriculture that emphasizes increasing yields delivered through the supply of inputs in a centrally controlled linear mode. Such institutional rigidities help explain the historical lack of participation and transparency in decision making on GM crops and the inability of the established scientific and political leadership to engage meaningfully with alternative and more inclusive pathways to sustainability. No doubt her analysis can fruitfully be extended beyond India to other "rising power" settings as well as to the developed world.

## Recommendations for Governance

How might we develop a more anticipatory approach to the governance of GM crops? Our proposal is for an independent, publicly funded, and transnational interdisciplinary research program, with the social sciences and life sciences as equal partners (Macnaghten 2016). Such a program would seek a deeper appreciation of the context in which GM crops have developed in the laboratory and then been adopted in the field, from a range of disciplinary and interdisciplinary perspectives and with a particular focus on the relationship between the development and uptake of GM crops and neoliberal policymaking. It would also require deeper insight into how ordinary laypeople think about GM crops and foods, including in situations when they remain unfamiliar about the technology, and the differential factors that structure responses across different geographical and demographic areas. It would also seek to scrutinize the various and often contradictory claims that tend to preoccupy debates on GM crops and foods, such as, "Can GM crops feed the world?"; "Do GM crops transgress natural boundaries?"; "Do GM crops benefit large producers rather than smallholders or consumers?"; "Do GM crops have an impact on biodiversity?"; and "Do (or could) GM crops operate in the public interest?" In relation to such questions, research would seek to answer: "What is known?"; "What is not known?"; "What is possible?"; "What is plausible?"; "What if . . . ?"; and "Under what conditions?" And finally, from such "thick" understandings, it would explore through methods of foresight, technology assessment, horizon scanning, or scenario planning, the conditions under which, if any, GM crops could plausibly contribute to inclusive agricultural sustainability (alongside an evaluation of alternatives).

Responding to the need for improved inclusion in governance, we call for the provision of government-sponsored information, from a range of perspectives and encompassing a diversity of views. We call also for more rigorous implementation of labeling of GM foods, so that at least people are aware of whether they are eating GMOs. And we call again for publicly funded research aimed at fostering more inclusive and bottom-up public and stakeholder dialogue, at developing methodological innovation, and at improving evaluation and monitoring. Such research is fundamental to an inclusive model of agricultural sustainability and should be focused both at national and international levels, be continuous and longitudinal in scope, and have a particular focus on Global South contexts. It should aim to bring together a broad range of stakeholders, to use state-of-the-art deliberative methods to explore initial framings and narratives, to inject some new ones, and to move the discussion in ways that generate inclusivity and common purpose, avoiding logjams while permitting tolerance of remaining disagreements. Such "action" initiatives should aim to replicate "best practice" in dialogue practices, such as those specified by Michel Callon and colleagues, who place a particular emphasis on early and intense deliberation, on developing where possible shared definitions of the issues (including their political economy contexts), on ensuring diverse and broad participation (including those that do not represent established interests), on developing support structures that enable participants to develop mature and considered perspectives, and on the need for commitment to ongoing and longitudinal engagement (Callon, Lascoumes, and Barthe 2009).

Responding to the dimension of reflexivity, initiatives are needed to develop and promote more reflexive scientific cultures, particularly in the crop science/life science community and in policymaking arenas. Building on existing scientific practices of self-referential critique, initiatives are required to help make reflexivity a public matter where scientists and policymakers are encouraged to develop capacities to reflect on their own commitments, to be mindful of their framing of issues, to be aware of the limits of current knowledge, and to understand better the legitimacy and complexity of public and stakeholder views on GM crops. Second-order reflexivity is also needed where scientists are provided with skills to comprehend the assumptions that underpin their own scientific epistemologies and tacit models of practice (including the ways in which ideas of mechanism and reductionism may be deeply embedded in scientific practice and styles of thought). The development of new curricula and multidisciplinary collaboration and training are also required, alongside the involvement of social scientists and ethicists in scientific laboratories using approaches aimed at helping natural scientists to develop systematic reflection on the socio-ethical context of their work.

Finally, we need to respond to the dimension of responsiveness, crafting new policy architectures that are designed explicitly to be as responsive as possible. It is preemptive to delineate the specifics of a more responsive system and the hurdles that would need to be crossed to realize it. But, most importantly, this would require the development of new institutional structures and norms that are equipped to complement existing regulatory structures which hitherto have relied upon the assessment of risk to human health and the environment as the sole criteria for assessment and decision making. This may involve the establishment of new social advisory bodies alongside existing scientific advisory bodies, and in addition the use of stakeholder and other deliberative forums, to consider the broader cultural, societal, and ethical dimensions of new agricultural technologies (including but not exclusive to GM) and to provide social intelligence to inform good governance. Such bodies would need both to embrace a full range of disciplinary competences, from the social sciences and humanities alongside the natural sciences, and to be equipped to respond to each of the dimensions listed above. Albeit untested (the closest analogy was the UK Agriculture and Environment Biotechnology Commission), such institutional redesign would be a vital and necessary condition of good governance, and a prerequisite for GM crops and associated agricultural technologies to be developed and governed such that they respond to inclusive and socially just patterns of agricultural sustainability.

## References and Further Research

Callon, Michel, Pierre Lascoumes, and Yannick Barthe. 2009. *On Acting in an Uncertain World: An Essay on Technological Democracy.* Boston: MIT Press.

Carro-Ripalda, Susana, Marta Astier, and Patricia Artía. 2016. "An Analysis of the GM Crop Debate in Mexico." In *Governing Agricultural Sustainability: Global Lessons from GM Crops*. Edited by Phil Macnaghten and Susana Carro-Ripalda, 33–73. London: Routledge.

Crute, Ian 2016. "The Route to Food Security Is Not Through Glorification of the Rural Idyll and Demonization of Global Trade (or Vice Versa)." In *Governing Agricultural Sustainability: Global Lessons from GM Crops*. Edited by Phil Macnaghten and Susana Carro-Ripalda, 155–160. London: Routledge.

Egorova, Yulia, Rajeswari Raina, and Kamminthang Mantuong. 2016. "An Analysis of the GM Crop Debate in India." In *Governing Agricultural Sustainability: Global Lessons from GM Crops*. Edited by Phil Macnaghten and Susana Carro-Ripalda, 105–135. London: Routledge.

Glover, Dominic. 2016. "Public Participation, Accountability, and the Stewardship of Transgenic Crops." In *Governing Agricultural Sustainability: Global Lessons from GM Crops*. Edited by Phil Macnaghten and Susana Carro-Ripalda, 167–173. London: Routledge.

Grove-White, Robin, Phil Macnaghten, Sue Mayer, and Brian Wynne. 1997. *Uncertain World: Genetically Modified Organisms, Food and Public Attitudes in Britain.* Lancaster, UK: Lancaster University Centre for the Study of Environmental Change.

Guivant, Julia S., and Phil Macnaghten. 2016. "An Analysis of the GM Crop Debate in Brazil." In *Governing Agricultural Sustainability: Global Lessons from GM Crops*. Edited by Phil Macnaghten and Susana Carro-Ripalda, 74–104. London: Routledge.

Hajer, Maarten. 2003. "Policy Without Polity? Policy Analysis and the Institutional Void." *Policy Sciences* 36: 175–195. www.maartenhajer.nl/upload/HAJER%20Policy%20without%20Polity.pdf.
———. 2009. *Authoritative Governance: Policy-Making in the Age of Mediatization.* Oxford: Oxford University Press.
James, Clive. 2013. *Global Status of Commercialized Biotech/GM Crops: 2013*. ISAA Brief 46–2013: Executive Summary. Ithaca, NY: International Service for the Acquisition of Agri-Biotech Applications. www.isaaa.org/resources/publications/briefs/46/executivesummary/.
Levidow, Les. 2016. "Neoliberal Origins of Anti-GM Protest in Europe." In *Governing Agricultural Sustainability: Global Lessons from GM Crops*. Edited by Phil Macnaghten and Susana Carro-Ripalda, 179–185. London: Routledge.
Lindsey, Keith. 2016. "GM Futures: Perspectives from a Plant Molecular Biologist." In *Governing Agricultural Sustainability: Global Lessons from GM Crops*. Edited by Phil Macnaghten and Susana Carro-Ripalda, 186–191. London: Routledge.
Macnaghten, Phil. 2016 "A Responsible Innovation Framework for GM Crops." In *Governing Agricultural Sustainability: Global Lessons from GM Crops*. Edited by Phil Macnaghten and Susana Carro-Ripalda, 225–239. London: Routledge.
Macnaghten, Phil, and Susana Carro-Ripalda, eds. 2016. *Governing Agricultural Sustainability: Global Lessons from GM Crops.* London: Routledge.
Macnaghten, Phil, Sarah R. Davies, and Matthew Kearnes. 2015. "Understanding Public Responses to Emerging Technologies: A Narrative Approach." *Journal of Environmental Planning and Policy*: 1–19. Accessed November 10, 2016. doi:https://doi.org/10.1080/1523908X.2015.1053110.
McLeish, Tom. 2016. "The Search for Affirming Narratives for the Future Governance of Technology: Reflections from a Science–Theology Perspective on GMFuturos." In *Governing Agricultural Sustainability: Global Lessons from GM Crops*. Edited by Phil Macnaghten and Susana Carro-Ripalda, 192–197. London: Routledge.
*Nature.* 2013. "GM Crops: A Story in Numbers." *Nature* 497: 22–23. Accessed November 10, 2016. doi:https://doi.org/10.1038/497022a.
Northcott, Michael S. 2016. "Crop Science, the Heisenberg Principle and Resistance to Genetically Modified Crops." In *Governing Agricultural Sustainability: Global Lessons from GM Crops*. Edited by Phil Macnaghten and Susana Carro-Ripalda, 198–204. London: Routledge.
Petts, Judith. 2016. "Innovating Governance?." In *Governing Agricultural Sustainability: Global Lessons from GM Crops*. Edited by Phil Macnaghten and Susana Carro-Ripalda, 205–211. London: Routledge.
Quist, David, and Ignacio H. Chapela. 2001. "Transgenic DNA Introgressed into Traditional Maize Landraces in Oaxaca, Mexico." *Nature* 414, no. 6863: 541–543.
Raina, Rajeswari S. 2016. "Institutional Rigidities and Impediments: Agricultural Research and GM Crops in India." In *Governing Agricultural Sustainability: Global Lessons from GM Crops*. Edited by Phil Macnaghten and Susana Carro-Ripalda, 212–219. London: Routledge.
Royal Society. 2009. *Reaping the Benefits: Science and the Sustainable Intensification of Global Agriculture.* London: The Royal Society.
Simpson, Robert. 2016. "Focusing on GM crops." In *Governing Agricultural Sustainability: Global Lessons from GM Crops*. Edited by Phil Macnaghten and Susana Carro-Ripalda, 220–224. London: Routledge.
Stilgoe, Jack, Richard Owen, and Phil Macnaghten. 2013. "Developing a Framework of Responsible Innovation." *Research Policy* 42, no. 9: 1568–1580.

12

# Rethinking Sustainable Community Through Permaculture Initiatives in Zimbabwe

Pamela Richardson-Ngwenya

One approach to sustainable living that has gained popularity in many parts of the world over the past twenty years is permaculture. Through permaculture, defined here as a set of agricultural and social design principles that aim to simulate or directly utilize the patterns and features observed in natural ecosystems, a particular sustainability narrative is enlivened by ideas and practices that have the potential to reinvent communities and transform socio-natures in various ways. In this chapter, I look at three dimensions of how permaculture (and thereby, a particular "sustainability" narrative) has helped to reshape community in three different projects in Zimbabwe. Moreover, I travel through these case studies using a somewhat unconventional research approach, which focuses on serving the communities through creating public documentation materials (videos, newspaper articles, a blog) that could potentially foster mutual exchange on sustainability practices.

**Notes on Methodology**

Between October 2011 and August 2012, the author visited eighteen sustainability projects or organizations across Zimbabwe, conducting semi-structured interviews with thirty-two people involved in the initiatives. Of these projects, eight used permaculture as their guiding approach. Some of the projects were visited repeatedly or for several days. The project sites were toured and documented with video, photography, and/or in textual form. This material was used to create a blog, Sustainable Agri-Food Initiatives, Zimbabwe (www.sustainableagriculturezimbabwe.wordpress.com/), and several related newspaper articles were published in Zimbabwe. Framed by the motivations of public geographies (Fuller 2008) and hopeful geographies (Lawson 2005), this multimedia methodology was intended to create an accessible, informative, and inspiring resource for the organizations involved and for interested publics. The video production methodology was semi-collaborative, in that interviews and conversations were first carried out with participants, followed by collaborative planning of what the video was to portray. The research approach aimed to serve the efforts of these communities in a small way, by celebrating and documenting their activities and through the media outputs, potentially fostering exchange between them. Images and links are embedded in the following text, enabling the reader to interact with these resources and gain a richer insight into the case studies. It is hoped that this information and discussion will "invigorate alternative imaginations and strategic action" (Weis 2007, 8) for sustainable futures in Zimbabwe and will contribute to the significant body of ideas already circulating at these diverse project sites.

In Zimbabwe, arguments for economic growth, employment creation, and development, rather than sustainability per se, tend to dominate the popular media. Consumer awareness of or demand for "sustainably" produced goods is negligible. Drought and food scarcity are perennial issues; and in 2013, the United Nations estimated that 1.6 million of the almost 13 million people in Zimbabwe need food aid. Those in need of aid began to rise sharply as a result of a drought, caused by a El Niño weather pattern in 2015–2016, but began to drop again in 2017 as the drought's effects subsided. The state has set out its official "sustainable development" policy, framed by the document "Zimbabwe Agenda for Sustainable Socio-Economic Transformation (Zim Asset): Towards an Empowered Society and a Growing Economy, 2013–18." The national policy places a strong emphasis on social justice and redistributive policies, embedded in the current context of ongoing (racialized) processes of land reform and a revolutionary rhetoric. While calling for indigenization, equity, and empowerment, the document simultaneously mobilizes a much more conventional and familiar sustainable development discourse, based on "the judicious exploitation of the country's abundant human and natural resources" (Government of Zimbabwe 2013, 6). Here, sustainability and development are based on an imagined separation between society and environment, where even a revolutionary society's relationship to environment is structured through relations of control, exploitation, domestication, and guardianship of "Nature," available for the taking. In relation to agriculture, policy in Zimbabwe gives a nod to conservation agriculture (see Government of the Republic of Zimbabwe and Food and Agriculture Organization 2012) but tends to be more focused on high output and market-based agriculture: "The government will re-establish financial support for agriculture so that farmers will increase production, productivity and product quality" (Government of Zimbabwe 2013, 36). Since the fast-track land reform program commenced in 2000, there has been an exponential multiplication of smallholder farms. The land redistribution program is still underway and it is thus a critical time to consider broad-based sustainability issues. However, in relation to sustainable development of new farms, the state-owned media more often report on a lack of donor funds available to purchase agricultural inputs such as fertilizer and pesticides. It is of huge concern that the dominant discourse seems geared toward the (re)building of a highly modernized, corporate, and commercialized agro-food sector geared for export. Moreover, the moment is ripe for consideration of how to build "sustainable community," given the recent resettlement of more than 150,000 families across Zimbabwe.

Beyond the state discourse on sustainability, however, other approaches to sustainable development and food production can be found. These varied approaches articulate visions that both conceptually and practically move beyond economic growth, social equity, and poverty alleviation, and are not simply positioned as alternative ways to address hunger.

**A Brief Portrait of Zimbabwe**

The Republic of Zimbabwe (formerly called Rhodesia) is a landlocked country in southern Africa, which achieved independence in 1980. The territory had been a British colony since 1923 but was unilaterally declared independent from the United Kingdom in 1965, by a white minority government. Until 1979, Rhodesia as such was officially unrecognized and a guerrilla war ensued. Independence then came with the election of a black majority government ruled by the ZANU party, under the premiership of Robert Mugabe. An organized land reform process, in accordance with the Lancaster House Agreement, then unfolded. During the 1980s, the government resettled more than 50,000 families on around 3 million hectares (7 million acres). During the 1990s, less than 20,000 families were resettled and 11 million hectares (27 million acres) remained property of about 4,500 commercial farmers, the majority of them white (Human Rights Watch 2002). Following unofficial land invasions led largely by liberation war veterans, the Zimbabwean government announced the "fast track" resettlement program in July 2000. The program "allocated to new farmers over 4,500 farms making up 7.6m hectares, 20% of the total land area of the country, according to (admittedly rough) official figures" (Scoones 2010). Over 145,000 households gained access to "A1" (smallholder) schemes and around 16,500 households occupy A2 (medium–large scale) plots (Scoones et al. 2010).

In this chapter, three particular case studies are discussed. Through different permaculture practices, the three projects have reinvented "community" in interesting ways, including, first, an expanded, socio-natural notion of community. That is, the projects articulate a discourse of sustainability that shapes an understanding of environment–society relations as intertwined or co-produced, as socionatures. Second, the projects—two of them in particular—demonstrate an internal reordering of community to address unfair power relations. This resonates with the community and activist approaches to sustainability discussed in preceding chapters. Third, the three projects seek to connect with other communities across the world for mutual support and learning. As such, this chapter offers an insight into how a narrative around sustainability, promoted through a globalized literature and network around permaculture, has become actualized in the Zimbabwean context. Moreover, I show how such a narrative can have radical implications for how communities organize themselves, which are at once locally specific and contextualized, while continuously engaged and resonating with more global discourses, media, and networks.

Three guiding ethics underlie permaculture design: Earth Care; People Care; and Fair Share. For more information on the principles and practices of permaculture, see the PermaculturePrinciples.com website: www.permacultureprinciples.com.

The three projects outlined here are: Kufunda Village in Ruwa, Chikukwa Ecological Land Use Community Trust (CELUCT), and the Participatory Organic Research Extension and Training (PORET) Trust (the latter two in Manicaland). Kufunda Village is a living, learning village and hosting center located approximately 40 km (25 miles) from Harare. The village was founded in 2003 as an experiment in community building and sustainable living.

The project initiator, Maaianne Knuth, hails from Zimbabwe, but with Danish family on her father's side, and had spent several years undertaking her formal education in Denmark. As a student, she was deeply involved with establishing Pioneers for Change. Shaped by living "between worlds" and inspired by collective gatherings of people motivated to create a more sustainable world, Maaianne shared her vision of Kufunda with her family in Zimbabwe and negotiated to utilize part of the family farm in Ruwa to begin materializing this vision. She sent a call out by email to friends and contacts, communicating her idea for co-creating what has become a "communiversity" and inviting kindred spirits to join her. The call was answered and an initial team of more than twenty volunteers was composed of both local and international "Kufundees." Their learning and hosting activities cover two primary domains: (1) the practical physical domain of a sustainable healthy community, and (2) the less tangible but equally important domain of culture, learning, and collective leadership, which enables a community to thrive. In 2013, there were twenty-seven people living and working at Kufunda Village with their families. They work with permaculture and organic farming, health and nutrition, ethnobotany, eco-building, renewable energy, and other sustainable practices.

The Chikukwa Ecological Land Use Community Trust (CELUCT) in Chikukwa, Chimanimani district, is a community permaculture project that began in 1991. Chikukwa is a mountainous area that encompasses six villages and is home to about 1,000 people. At the time of the project's inception, the area was in a state of degradation, characterized by frequent landslides, soil erosion, food insecurity, droughts, and flooding. In 1991, a group of residents organized to try to change the situation. Two of the project initiators—the Westermanns—were teachers in the community who originally hailed from Germany and thus brought in "new" ideas and connections. To begin with, thirty farmers received some initial training in permaculture from a teaching friend of the Westermanns, John Wilson, a Zimbabwean who later cofounded the first permaculture organization in Zimbabwe, Fambidzanai Training Centre, located in Harare. In 1992, a refresher course took place with fifty-two people taking part. In the same year, Chikukwa experienced serious floods. As a result of this confluence of trainings, concern, and the crisis point precipitated by the flood damage, six community volunteers came forward to act as agents of change. They became the catalysts of the incipient community program that was to transform Chikukwa. After an initial experimental stage during 1991–1995, the organization CELUCT was officially founded and the community came together to build a hall where a farmer "permaculture school" began.

The PORET Trust in Chaseyama is closely linked to CELUCT through its founder, Julious Piti, who was an original member of the CELUCT team. The initial permaculture training at Chikukwa had taken place in his home. In Chaseyama, Piti has co-developed a 14-hectare (35-acre) permaculture site that acts as a living demonstration project. Piti registered the Trust with three other Board members in 2006, having spent the previous ten years using permaculture techniques to slowly transform the land from a rocky desert to a verdant green haven. It is now the central hub of the Chaseyama Permaculture Club, which has thirty-one members and developed around the site as the community perceived the benefits of permaculture and wanted to participate in the project themselves.

These three permaculture projects were selected for the ways in which they specifically and explicitly demonstrate how permaculture can cultivate transformations in communities in ways that relate to expanding, reordering, and connecting communities in a globalizing world. All three projects can be considered as sites of inventive and radical experiments in "what it means to live in community" (Author interview, Knuth, March 5, 2012). Through considerable reflexivity on the issue of "sustainable community," the three case studies all depart considerably from the state discourse on sustainability that endorses "the judicious exploitation of the country's abundant human and natural resources." I now present a brief discussion of permaculture, framing it as a sustainability narrative that endorses a set of practices and principles for transforming socionatures. I then elaborate on three ways in which permaculture can be said to foster a rethinking of sustainable "community" in the selected projects in Zimbabwe.

## Permaculture as a Sustainability Narrative

Permaculture is one approach to sustainability that is deployed by some sustainable agriculture projects and can be characterized as "a global movement with many local actualizations." Its motto is "care of earth, care of people, return of the surplus" (Puig de la Bellacasa 2010, 151). It is primarily a set of design principles and applied techniques, all deeply attentive to location and context. Deploying practices such as mulching, composting, recycling, and inter-cropping, permaculture cultivates a notion of sustainability that includes ecological living, renewable energy, green technologies, local food sovereignty, and radical forms of community governance. Burnett's *Beginners Guide* states that "[p]ermaculture is about creating sustainable human habitats by following nature's patterns" (Burnett 2008, 8). In permaculture, Maria Puig de la Bellacasa suggests that "the issue is not so much for humans to act upon the 'environment' but that 'we are nature working'. . . . The natural and the cultural are not considered separately" (2010, 152).

Through its practical engagements, we find that permaculture fosters a structure of mutually dependent and integrated socio-ecological relations, whereby humans are positioned as participants in a system where other parts of that system (animals, plants, waterways, etc.) also have significant rights and roles to play. Humans are seen less as managers and more as coworkers, with a vision of generating a system of health and abundance from which *all* (humans and nonhumans) benefit. Permaculture therefore requires a shift in sustainability thinking away from that commonly associated with the modernizing tropes of sustainable development agendas. Such narratives are generally guided by an imagined separation between society and environment, adopting a utilitarian, technocratic, managerial, and/or stewardship orientation towards the environment. Society's relationship to environment is then structured through relations of control, domestication, or guardianship of nature. Permaculturalists, on the other hand, are likely to understand humans and nonhumans as part of an interconnected web of life, with nonhumans considered as active participants. This different understanding affects the kinds of obligations and commitments that can arise. Such narratives are thus pivotal to the attitudes and practices that people adopt as they work "with" rather than "upon" their habitat. Essentially, these practices and attitudes affect the transformations that can take place.

Puig de la Bellacasa offers a fascinating discussion of the "ethical doings" of permaculture in dialogue with feminist and post-humanist ethics. She highlights a decentering of human subjectivity in the way that permaculture considers humans as "part of earth's living beings" (2010, 152). As discussed, this is a core dimension of permaculture's sustainability narrative and its socio-ecological understanding, and is the basis upon which commitments to life-sustaining practices are cultivated. In theory, humans are positioned as living in community with other socio-natural beings.

## Rethinking Community Through Permaculture

In this section, I will highlight how this socionature narrative translates into grounded practices and changes in Zimbabwe in the specific permaculture practices and transformations evidenced at Kufunda Village, Chikukwa, and Chaseyama. I frame this discussion of transformation by considering a rethinking of sustainable "community" at these sites.

## Expanding Community Life: Socionatures

As suggested above, all three projects can be considered as sites of inventive and radical experiment in "what it means to live in community." And, I suggest, a look into their practices reveals the emergence of an expansive and inclusive rethinking of community that goes beyond the reconfiguration of human-centered relations. At Kufunda, CELUCT, and PORET, a key technique of permaculture—that of encouraging organic ground cover—is practiced. Instead of sweeping the yards bare and burning the grasses in surrounding woodlands, the ground is covered and mulch is used around crops.

This facilitates the infiltration of rainwater and reduces surface runoff, evaporation, and soil erosion. It is therefore of benefit to the trees and plants growing here, which in turn, benefits households in the form of food and shade. The explanations received for this practice (which is uncommon in Zimbabwe), were not narrowly utilitarian and human-centered. Julious Piti, founder of PORET Trust, referred to enabling a healthy habitat for wildlife, including birds, bees, butterflies, worms, and such, and to helping restore the underground water catchment (Author interview, Piti, October 21, 2011). In turn, these aspects benefit human inhabitants. The ideal, then, is for healthy relationships to lead to a socio-ecological community where diverse actors can thrive and have life in abundance.

> Julious Piti, founder of PORET Trust, presents aspects of his permaculture project in Chaseyama in the video, "Introducing PORET Trust": www.vimeo.com/72088415.

From a similarly inclusive notion of community, permaculture encourages "returning the surplus," aiming for a zero-waste socio-ecological community with no need for external or nonrenewable inputs. One of the specialist cluster groups at Kufunda Village deals with waste management. As an example of returning the surplus, they have adopted a way of managing human waste to return nutrients to the soil. As a rural project, there is no sewerage or mainline-based water system here. Unlike most other rural communities in Zimbabwe where unproductive long-drop toilets and, in some places, open defecation, are the norm, at Kufunda Village, they recycle human waste using composting toilets. In this way, human energy outflow is returned to the agro-ecological system as a valuable input. The villagers act "in community" with their habitat, ensuring that they do not take more from the soil than they invest back in.

> For more general info, see the video "Introducing Kufunda Village: A Community of Learning": https://vimeo.com/69458759.
>
> Also, see this video, which youth trainees on a Kufunda youth leadership program made about the composting toilets: https://vimeo.com/72088414.

Another key aspect of these projects, particularly CELUCT, has been the recovery and holistic management of the local water sources and catchment flows. At the time of CELUCT's inception, the community faced a troubling situation with extreme weather events in combination with overgrazing of the slopes leading to landslides and erosion, and, therefore, the depletion of the local resource base. This in turn led to more frequent loss of property and extended periods of hunger, or food insecurity. Deploying permaculture practices, community volunteers began to transform their degraded habitat, repairing the damage that had been wrought by past overgrazing, monocropping, and poor environmental conservation practices. The volunteers transformed gullies and rehabilitated springs, restoring their water supply and managing surface water runoff. They built a community-wide water storage and distribution system, as a result of which nearly all homesteads now have piped water, fed by a rehabilitated spring.

In tandem with restoring water catchments, these projects have improved food security and nutritional diversity. At Kufunda Village in the summer, an uncanny array of wild mushrooms sprouts up from the carpet of leaves beneath the trees. This abundance of mushrooms is commented upon by nearby communities, as much of the local area is suffering from deforestation and overgrazing, diminishing the growth of mushrooms, which need moisture, shade, and humus to thrive. To reduce food insecurity, permaculture also promotes the use of open-pollinated indigenous crop varieties, nutrition gardens, home orchards, seed-sharing, hedging, and terracing. At Chaseyama in particular, which is in a low-lying, dry, rocky valley, the main focus of PORET Trust is sustainable food production: "People must see that permaculture is not just about planting trees and conserving the environment; we must also grow food and get rid of hunger," said Piti (Author interview, Piti, October 21, 2011). However, they choose not to focus solely on increasing the immediate productivity of crops, but rather practice zero tillage, meaning that no plowing is done and the soil is minimally disturbed during planting. Nitrogen-fixing crops such as sunhemp, pigeon pea, and comfrey are also grown to organically enrich the soil, rather than applying chemical fertilizers. In such ways, these projects adopt a set of practices that avoid fast-track productivity benefits and instead aim to nurture and care for the long-term health of *multiple* life-forms in their communities.

Permaculture, as practiced at these three sites, thereby promotes an *expansive rethinking* of community. Through the principles of caring for earth and returning the surplus, and through the promotion of a holistic socio-ecological imagination, participants in these projects have been compelled to rethink and expand their arena of concern to incorporate more-than-human actors. This contrasts with a more conventional view of sustainable development that so often rests on the "judicious exploitation" of conceptually externalized (nonhuman) resources.

I now turn to a discussion of how these permaculturalists have also made changes to the ways in which they structure and govern relations among themselves through locally incorporating the permaculture principles of "caring for people" and "Fair Share."

## Reordering Community: Negotiating Subjectivities Through Community Governance

Conflicts within community are often the key cause of a project collapsing. CELUCT and Kufunda are two projects that demonstrate resilience, having run for over twenty-two and ten years respectively. Significantly, these projects have designed ways of deliberately countering inter-human conflict to reorder their communities. Indeed, building a vibrant and sustainable community has become their key focus. Participatory practices have been made integral to achieving this. At Kufunda Village, I was fortunate to join one of their regular Monday morning community meetings. We sat in a circle and a "talking stick" was passed around the circle, with all respecting the right of the holder to speak (or not to speak) at that time. As the object was passed around the circle, each person shared how they were feeling and what had been going on with them in the past week. This process is designed to encourage democratic participation and the honoring of each person's viewpoint. After this process of "tuning in," specific village tasks and issues were discussed with items then delegated to specialist cluster groups. There was no individual leader; instead all were invited to participate in making collective decisions.

At Chikukwa, their learning practices follow a similarly participatory model, having been peer-to-peer since the project began. They started their own "permaculture school" in 1995, where people in the community would take turns to address each other on topics with which they were deemed to have experience. The community would then be encouraged to put into practice what they were learning, to "look and learn" by visiting good examples, and eventually clusters of specialist groups formed. Drawing on postcolonial models of pedagogy that redistribute agency and value indigenous knowledge systems, this is not unlike the "communiversity" concept adopted at Kufunda Village. Through their Communiversity platform, Kufunda runs ongoing, long-term learning programs, and also offers several short-term workshops.

The Kufunda Village website (www.kufunda.org) states, "At their essence, our programmes are all about nurturing and releasing each person's unique and creative gift and talent in community. We believe in community and in what becomes possible when whole, or free, people come together for learning and co-creation." For this passage, and further information about its Communiversity program, see "Communiversity Courses": www.kufunda.org/courses.

These radical practices of communication and learning have implications for the subjectivities of those constituting such communities. While respecting and valuing many "traditional" (ancestral and gender-based) roles through, for example, gender-specific meetings, consultations, and celebrations, both communities have adopted the above practices to adjust or reform power relations and integrate a spirit of democracy. In contrast to many other rural Zimbabwean communities, these villages have elevated the status of women and young people through an emphasis on participation and encouraging the individual spirit, as well as the greater community, to flourish. Maaianne Knuth, founding villager at Kufunda, reflected:

> it feels like now, as we're coming into our tenth year, we're going more into some of the 'internal' [community relations]. So issues that are being worked through at the village now are about the relationship between men and women . . . as they are showing up at Kufunda and [we may say], 'oh, but the tradition' . . . but [we must ask], how do *we* want it? And how do we honour the tradition, but still live together in a way where we also honour that we're each strong and unique individuals?
>
> (Author interview, Knuth, March 5, 2012).

At CELUCT, many of the initial and current volunteers are women (partly as this is a community with high levels of male-outmigration). Patience Sithole, a founding member, described how the project has acted to enable local women to build their social and economic capital. Women in the community were encouraged to make their own decisions, manage and solve their problems together, and create their own livelihoods through activities such as beekeeping, jam-making, and selling garden produce (Author interview, October 25, 2011). In both projects, youth are also a central focus and are involved through leadership training programs. Youth are always invited to participate in village meetings on a (more or less) equal footing with elders. They are found to be active leaders in various fields at both sites.

This is rather different from more traditional community dynamics where women and younger people are often expected to be quiet and defer to their elders. The communal practices and commitment to the "fair share" principle promoted by permaculture have acted to bring people together in a more democratic way, and in the act of coming together to take collective responsibility for problems such as deforestation, soil erosion, and flooding, people have begun to address other issues such as HIV/AIDS, gender relations, and social justice. In these ways, participatory communal practices have allowed for intra-community power relations and entrenched rural subjectivities to be renegotiated.

Despite participatory communication and learning tools being utilized, various forms of conflict have arisen in both projects. As Phineas Chikoshana from Chikukwa put it, "conflicts are everywhere. As an individual, you always have conflicts within yourself . . . and when one is not at peace with him or herself, then that person is bound to have conflicts with others as well. So it's very important to cultivate a spirit of peace within the community" (Author interview, Chikoshana, October 25, 2011). In 2006, in response to emerging community conflicts, CELUCT initiated their "Building Constructive Community Relations" (BCCR) program. Through the BCCR program, CELUCT members worked to sensitize each other to conflict resolution methods, in particular the "Three Circles of Knowledge" approach. As Phineas shared with me, "development needs peace [in order] to thrive" and so "we must learn how to transform conflict situations" (Author interview, Chikoshana, October 25, 2011). In other words, as an organization, CELUCT identified that one cannot avoid conflict and so were compelled to integrate conflict management trainings into their core practice. These inventive forms of community governance are key in enabling the project to expand and endure (Author interview, Sithole, October 25, 2011), thereby sustaining community cohesion. So the "sustainability" of these projects, one can argue, is deeply connected to their innovative approaches to expanding, reordering, and harmonizing their communities.

Patience Sithole discusses aspects of CELUCT in the video, "Introducing Chikukwa Ecological Land Use Community Trust": https://vimeo.com/74118407.

## Connecting Communities in a Globalizing World

Active and long-term connections with other communities, connections often based on common practices such as permaculture, is a third vital characteristic factoring into the experiences of these projects. Permaculture's motto of "care of people, care of earth, and fair share" is not unlike older, traditional southern African notions of *ubuntu.* For example, social giving to those in need, community sharing, and respect for nature are often explained as indigenous practices that the ancestors favored. Whether this is true or imagined, it does seem that, in this rural southern African context, people easily adopted (some) permaculture practices as "traditional" rather than viewing them as foreign ideas. So while permaculture itself is a globalized approach that fosters global interconnection through knowledge exchange, its emphasis on contextuality and local knowledge renders its practices locally relevant.

### The Nature of Being Human

*Ubuntu* is a Bantu Nguni (southern African) term referring to the nature of being human, usually denoting the positive aspects of "human kindness." It describes a humanist ethic that highlights the qualities of compassion, empathy, humility, sharing, caring, and gratitude. This meaning is often expressed through the phrase "I am what I am because of who we all are." In post-Apartheid South Africa, the term took on more political dimensions with frequent reference made to the term by leaders such as Nelson Mandela: "In Africa there is a concept known as ubuntu—the profound sense that we are human only through the humanity of others; that if we are to accomplish anything in this world, it will in equal measure be due to the work and achievements of others" (in Stengel 2010, ix). At Mandela's funeral in 2013, U.S. president Barack Obama referred to *ubuntu* in his speech; "There is a word in South Africa—*ubuntu*—a word that captures Mandela's greatest gift: his recognition that we are all bound together in ways that are invisible to the eye; that there is a oneness to humanity; that we achieve ourselves by sharing ourselves with others, and caring for those around us" (Obama speech, December 10, 2013). Such popular references have given the term a more global reach.

Although permaculture fosters minimal usage of external physical inputs (and creative use of local resources), the most successful and enduring of projects in Zimbabwe tend to have a high level of external input in the form of donations, visitors, and trainers. So while all three projects discussed here are attentive to local conditions and traditions, they remain open to cosmopolitan influences. As such, I found that at these sites, people conveyed a broad outlook and sense of global interconnectedness, were fluent in macroscale environmental concerns, and also had a relatively radical sense of social justice. Such concerns are the hallmarks of sustainability thinking, often articulated in terms of fostering "sustainable" market-based linkages such as fair trade. Fair trade operates by providing a certification of standards to ensure a fair price and secure working conditions for producers in the Global South, for products largely marketed to sustainability-conscious consumers in the Global North. However, at these sites, the approach to local–global connections means that these projects go beyond a consideration of market-based exchange. While attending to the specific local contexts in which they are geographically embedded, these projects promote globalized connections that are based on a notion of kindred and connected communities.

Kufunda Village, for example, is remarkable in Zimbabwe in the way it is shaped by contemporary discourses and communication technologies circulating at the international level, such as the Internet, social media, the global village concept, and the transition movement. This is evident in, for example, the inception of the project through an international email invitation. Since inception, international volunteers and trainers, researchers and visitors like myself, as well as youth trainees, regularly participate, becoming an ephemeral part of the local Kufunda Village community. Moreover, through social networking sites such as Facebook, such participants are encouraged to maintain an emotional and discursive connection to and involvement with the village. In fact, most of the material donations that fund the youth program and other developments are generated through these online and translocal connections. The village is also featured in a published book on intentional communities (Wheatley and Frieze 2011). Kufunda is thus actively participating in global networks and mobilizations, while simultaneously being consciously embedded in a unique context and constellation of villagers. In communion and connection with other communities in disparate geographical locations, Kufunda Village receives not only donations and in-kind training from "outsiders," but is also exposed to alternative configurations of community power, and to wide networks of support. By both connecting "inwards" and connecting "outwards," Kufunda is able to sustain itself in material and ideational ways. By doing so, it participates in the articulation of a narrative about sustainable communities that engages with globalization through making intercommunal connections.

CELUCT is also funded by various overseas donors and has received international attention in permaculture circles, as well as through a documentary film made by Gillian Leahy. At least one of the project initiators came from outside Zimbabwe and mobilized "new" ideas that perhaps reframed locally available concepts and approaches. CELUCT now acts as a host organization for training workshops and also has national outreach programs. They network and collaborate with other regional organizations such as Kufunda, and they work with international donor organizations such as Find Your Feet (based in the UK). PORET Trust also has influential international relationships, such as founder Julious Piti's networking and advisory work in the United States, Germany, Austria, and East Africa, as well as through their blog (http://poret-zimbabwe.org/). Through such outreach and communications, there is an explicit effort to exchange ideas and learning with other sustainable communities.

Gillian Leahy's film about CELUCT, *The Chikukwa Project*, can be accessed here: www.thechikukwaproject.com/.

The PORET Trust blog, called "Poret," can be accessed here: http://poret-zimbabwe.org/.

Finally, these projects communicate a sense of responsibility to share their experiences globally. This, I suggest, enhances the resilience of the projects, which are highly conscious of their own significance and value. All three projects regularly use social media to share their learning experiences. This promotes further connections with and inputs from kindred communities, international organizations, donors, and other interested persons. In short, the shape, resource pool, and longevity of these particular projects—and thereby their own sustainability—is related to their simultaneous attentiveness to local conditions while fostering connections with other communities and globalized discourses.

## Conclusion

This chapter has focused on permaculture as a sustainability narrative that endorses a set of practices and principles engendering an expansive, more-than-human sense of community, a reordering of power relations, and connections with kindred communities at the global scale. The three Zimbabwean projects presented have fostered socio-ecological transformations and reinventions of "sustainable community" that reflect this. The unique constellation of interacting forces, in specific places, shapes an articulation of sustainability that can offer hope to others living in this context. Given the political–economic situation in Zimbabwe and the fragmenting tendencies of recent land reform, alternative approaches to agriculture and community-building have great potential to take hold. Furthermore, it is acknowledged that crises often precipitate moments of opportunity, or "fault lines for transformative possibilities" (Weis 2007, 8). Considering the upheavals of recent history, it is an apt time to explore more caring agro-cultures and holistic ways of "living in community" in Zimbabwe, and to create and share knowledge of these alternatives with a wider public.

I have outlined three ways in which permaculture effected a rethinking of sustainable community. The case studies discussed here demonstrate, among other things, the rehabilitation of degraded soils and water catchments, the improvement of human nutrition, and the importance of conflict management. They also offer insights into how groups of people have collectively reflected and acted upon *themselves* in order to make positive transformations in their intentionally expansive, reordered, and interconnected communities. They therefore hold important potential for communities facing various sustainability problems.

However, agricultural development policy in Zimbabwe has not yet taken up this potential and, as noted, industrialized, market-based agriculture, framed as a judicious exploitation of resources, is definitely favored. For alternative approaches such as permaculture to succeed, policies need to recognize and actively support such initiatives. The current dependence of such projects on international sources of funding is also a huge risk, common to many projects in Zimbabwe. But despite this current lack of policy support and local investment, there is hope; for where projects like CELUCT, PORET Trust, and Kufunda Village generate an abundance of community life, they transform socio-ecologies in a deeply holistic sense and they inspire others to emulate their practice of sharing. Through maintenance of these expansive, inclusive, and connective community practices, many forms of life can benefit and so become incentivized to participate, endure, and ultimately to thrive.

## References and Further Research


Braun, Bruce, and Noel Castree, eds. 2001. *Social Nature: Theory, Practice, and Politics.* Oxford: Blackwell Publishing.

Burnett, Graham. 2008. *Permaculture. A Beginner's Guide.* Westcliff on Sea: Spiralseed.

"The Chikukwa Project." *The Chikukwa Project*. www.thechikukwaproject.com.

Fuller, Duncan. 2008. "Public Geographies: Taking Stock." *Progress in Human Geography* 32: 834–844.

Glanzberg, Joel. 1999. "Permaculture as a Way of Seeing and Acting." In *A People's Ecology: Explorations in Sustainable Living.* Edited by Gregory Cajete, 225–242. Santa Fe: Clear Light Publishers.

Goforth, Monika, and Terry Leahy. 2011. *Permaculture Successes in a Zimbabwean Community*. http://permaculturenews.org/2011/10/01/permaculture-successes-in-a-zimbabwean-community.

Government of the Republic of Zimbabwe and Food and Agriculture Organization of the United Nations. 2012. Zimbabwe Country Programme Framework 2012–2015. ftp://ftp.fao.org/tc/cpf/countries/zimbabwe/cpf_zim_2012-2015.pdf.

Government of Zimbabwe. 2013. *Zimbabwe Agenda for Sustainable Socio-Economic Transformation (Zim Asset): Towards an Empowered Society and a Growing Economy, 2013–18.* www.herald.co.zw/wp-content/uploads/2014/01/Zim-Asset.pdf.

Human Rights Watch. 2002 *Zimbabwe: Fast Track Land Reform in Zimbabwe.* Vol. 14 No. 1 A. www.hrw.org/reports/2002/zimbabwe/ZimLand0302.pdf.

Kufunda Village. www.kufunda.org.

Lawson, Victoria. 2005. "Hopeful Geographies: Imagining Ethical Alternatives. A Commentary on J. K. Gibson-Graham's 'Surplus Possibilities: Postdevelopment and Community Economies.'" *Singapore Journal of Tropical Geography* 26, no. 1: 36–38.

Lindhagen, Elin. 2010. "Chikukwa: A Lesson in Self-Reliance." *Permaculture Magazine,* Summer. https://barefooteducation.wordpress.com/permaculture/applied-permaculture-training-limpopo-africa/the-magic-of-chikukwa-a-lesson-in-self-reliance/.

Ngwenya, Pamela. "Introducing Kufunda—a Community of Learning." https://vimeo.com/69458759.

———."Introducing the PORET Trust." https://vimeo.com/72088415.

———. Sustainable Sgri-food Initiatives, Zimbabwe. Wordpress. www.sustainableagriculturezimbabwe.wordpress.com.

———. 2007."Introducing Chikukwa Ecological Land Use Community Trust." Pemba Productions. Video. https://vimeo.com/74118407.

Obama, Barack. 2013. "Remarks by President Obama at Memorial Service for Former South African President Nelson Mandela." First National Bank Stadium, Johannesburg, South Africa. December 10. www.whitehouse.gov/the-press-office/2013/12/10/remarks-president-obama-memorial-service-former-south-african-president-.

PORET. PORET Trust. www.poret-zimbabwe.org.

Puig de la Bellacasa, Maria. 2010. "Ethical Doings in Naturecultures." *Ethics, Place and Environment* 13, no. 2: 151–169.

Scoones, Ian. 2010 "Zimbabwe's Land Reform: Challenging the Myths." *The Zimbabwean*, October 20. www.ids.ac.uk/files/dmfile/zimbabwean1pdf.pdf.

Scoones, Ian, Nelson Marongwe, Blasio Mavedzenge, Felix Murimbarimba, Jacob Mahenehene, and Chrispen Sukume. 2010. *Zimbabwe's Land Reform: Myths and Realities*. Rochester, NY: James Currey/Boydell & Brewer.

Stengel, Richard. 2010. *Mandela's Way: Lessons on Life, Love, and Courage.* New York: Crown Archetype.

Telford, Richard. Permacultureprinciples.com. Permaculture Principles. www.permacultureprinciples.com/.

USAID. 2017. "Food Assistance Fact Sheet, Zimbabwe: November 27, 2017." www.usaid.gov/zimbabwe/food-assistance.

Weis, Terry. 2007. *The Global Food Economy: The Battle for the Future of Farming.* London: Zed Books.

Westermann, Eli. 2008. *The Three Circles of Knowledge: How to Build Constructive Community Relations by Understanding Conflicts in Rural African Communities.* Harare: Tien Wah Press, CELUCT.

Wheatley, Margaret, and Deborah Frieze. 2011. *Walk Out Walk On: A Learning Journey into Communities Daring to Live the Future Now*. San Francisco: Berrett-Koehler Publishers.

13

# The Entanglements of Oil Extraction and Sustainability in the Ecuadorian Amazon

Angus Lyall and Gabriela Valdivia

Oil extraction is a useful optic for thinking and writing about the future of sustainable resource use. Whereas concerns over the burdens of oil extraction tend to be of a planetary scale (e.g., discussions around fossil-fuel addiction, energy security, and climate change), in this chapter we zoom in to the case of the Ecuadorian Amazon, where indigenous peoples have raised profound questions about oil extraction practices and outcomes. Amazonian peoples' resistance to oil extraction, in particular, has received significant global attention and is considered emblematic of indigenous peoples' sustainability thinking. At the same time, this common narrative about indigenous political action hides the complex ways that Amazonian peoples relate to oil extraction. We focus on the case of Playas del Cuyabeno (Playas hereafter), a Kichwa community located in the northern Ecuadorian Amazon to illustrate how "sustainability" is not an abstract concept that can be applied worldwide seamlessly, but that thinking and acting sustainably emerges from locally rooted visions of the past, present, and future.

In this chapter, we introduce the dominant ways in which sustainability and oil extraction are currently discussed in Ecuador, and how debates around oil extraction tend to reproduce a particular way of thinking about sustainability. Next, we lay out our conceptual framework for examining sustainability in Playas. Then, we briefly trace the experiences that shaped how Playas residents see themselves in relation to oil, first resisting and then acquiescing to oil extraction within their territory. In the subsequent section, we examine the conditions and subjectivities through which people of Playas came to position themselves not only as supporters of oil extraction, but as potential oil producers themselves, despite popular narratives that associate indigenous peoples with anti-oil politics. We interrogate concepts of sustainability by tracing the conditions that made it possible for Playas' population to imagine an indigenous oil company as a vehicle of sustainability, noting that sustainable development planning is not the exclusive practice of elite state, nongovernmental, and multilateral institutions. We highlight the intersectional dimensions of the decisions of indigenous leaders to embrace oil extraction in the name of social *and* environmental sustainability. No single relation can explain positioning vis-à-vis oil in local contexts; rather multiple relations and complex histories construct ways of seeing and acting.

## Sustainability in the Ecuadorian Amazon

The popularity of sustainability discourse around oil extraction in Ecuador heightened during the administration of President Rafael Correa (2007–2017). One of the most emblematic events bringing together oil and sustainability is the case of drilling for oil in the Yasuní

National Reserve, one of the most biodiverse areas in the Amazon Basin. This high-profile case exemplifies competing visions of sustainability at the national level and highlights the "for" or "against" camps that can develop around oil extraction. On August 15, 2013, Correa appeared before television cameras to announce that he would authorize oil drilling in the Yasuní National Reserve. He argued that this decision was based on two moral propositions: on the one hand, the state was responsible for eliminating poverty, for which it needed greater revenues from oil extraction; on the other, it was responsible for enhancing well-being, and thus would employ first-class technologies and transportation mechanisms to minimize impacts of oil extraction on local populations. Correa's position parallels a trend in "progressive extractivism" developing in Latin America, where states support the expansion and intensification of resource extraction (e.g., oil drilling, mining, and agro-industry) in the name of sustaining future national well-being. These same practices, however, continue to sacrifice the environments and lives of people in zones of extraction (see Lang and Mokrani 2013).

The dispute that ensued between government officials and environmental activists brought to the fore a debate that has been ongoing since the establishment of the Ecuadorian oil industry in the 1970s: indigenous organizations and environmental activists have openly opposed the practices of oil extraction, despite the benefits accrued to the broader nation, on the grounds that it puts the environments and social values closest to oil operations at risk (see Kimerling 2006; Sawyer 2004; Valdivia 2008). The debates about and protests over opening the Yasuní to oil exploitation renewed and intensified this opposition, and pushed some indigenous peoples and environmentalists to doubt the very possibility of "responsible" oil operations. Moreover, according to detractors, the proposal to expand operations into Yasuní dissociates social "sustainability" from ecological "sustainability" and enrolls "sustainability" discourse in practices that entangle the state and indigenous groups in vastly uneven power relations.

President Correa maintains that to sustain the social and ecological well-being of the Ecuadorian society, oil and mineral revenues need to be invested in the development of "ecologically friendly" industries, such as ecotourism, renewable energy, and a "knowledge economy" based on research and services. From this position, capitalist development is seen as a universally desired response to the problem of how to generate long-term, collective well-being. Intensifying extractive practices, proponents argue, is a necessary step in the transition to an economic model that is less destructive, or *post*-extractivist, where increased resource rents will spur alternative economies that will rectify Ecuador's overreliance on primary exports.

Detractors of this approach respond from a position of ecological universals, where well-being can only be sustained through the preservation of ecological systems, not their destruction. They characterize the proposition of sustainability-via-more-extraction as "predatory": the government's framing of sustainability rationalizes the territorial dispossession and sacrifice of certain rural and historically marginalized groups in the name of the well-being of the greater collective. From this position, the preservation of places and support for groups resisting the expansion of extractive frontiers is key to preserving the values that underpin collective well-being. Although both positions emphasize sustainability as a normative project, entangled with ecological, social, and economic factors, the Correa administration lends primary importance to socioeconomic achievements, whereas environmental activists emphasize holistic ecological functioning. As a result, we see an increasing divide between the two groups over the question of how to act upon and create sustainable futures.

The narrowing down of the politics of oil extraction exemplified in the Yasuní case facilitates debate, working through established camps that support and oppose extractive practices. But it also obscures the many pathways and creative formulations that exist in-between and become entangled with these camps—some of which offer novel and alternative understandings of sustainability. Here, we examine the case study of the Kichwa community in Playas—separate from the Yasuní case, but also in the northern Ecuadorian Amazon—to explore the entanglements, potentials, and limitations of sustainability in extractive economies.

Our approach is community-centered, that is, we begin from the position of people who live in areas directly affected by oil extraction, rather than from a national or global perspective. The community of Playas offers a long and rich history of involvement with oil exploitation and a case through which to reflect on the current expansion of the oil frontier in Ecuador. As other Amazonian peoples living in protected areas of the Ecuadorian Amazon, Playas residents often are associated with biodiversity conservation discourses and resistance to oil development. Yet, 2007 marked a shift in the community's position on oil and conservation. At first, community leadership decided to form a community-managed oil company, one that would be respectful of socio-cultural norms and ecosystems while securing an income for the community. In their view, responsible and sustainable management could only happen if the community owned and ran the company. In 2008, for reasons that will be explored in detail in this chapter, the idea of an "indigenous company" was shelved and the leadership signed an agreement to allow the national petroleum company, Petroamazonas, to operate within their territory, as a way to achieve local well-being. We examine the conditions, experiences, and relations shaping how Playas residents responded to and negotiated oil extraction as a sustainability project and practice, in an effort to encourage in-depth analysis of how sustainability is conceptualized and put to work in the Ecuadorian Amazon.

We draw on our combined ethnographic fieldwork to look beyond preconceptions of a deterministic, "natural" relation between indigenous groups and environmental conservation. In July and August of 2014, the first author documented in-depth interviews with seventy-four respondents in Playas and neighboring communities, in addition to engaging in participant observation, focus groups, and archival research. We also draw on the second author's long-term research on the effects of the oil economy among Native Amazonians, including Kichwa from Playas, and from six months of field experiences with Ecuadorian Amazon Kichwa between 1999 and 2002. All quotes are drawn from our field notes, unless otherwise indicated.

## Indigenous Positioning

"Indigenous positioning" is an analytic that considers subjectivities (or sense of self and others) as a relational process, one in which people identify themselves in relation to their histories, landscapes, and social relations and through which they engage in and make sense of everyday life (see Li 2000). Amazonian indigenous peoples are often associated with discourses about the preservation of natural environments. This indigenous subject position is closely aligned with the idea of the "noble savage," which associates indigenous peoples with moral simplicity and innocence—thus, closer to nature—and civilization (e.g., urban lifestyles) and market economies with corruption and loss. Sustainability, in turn, is often associated with this positioning—assumed as dependent on the *preservation* of native ecosystems and peoples. Through the latter part of the twentieth century, indigenous social movements and their allies have partially debunked this narrative about the Amazonian "noble savage" and its natural links to sustainability. They argue that positioning Amazonian peoples as "noble savages" not only simplifies their agency but also falsely designates their beliefs and experiences as pre-modern and pre-political. Focusing on the simplistic assumptions of indigenous sustainability, these movements often challenge government leaders and conservation practitioners to recognize indigenous peoples as citizens with rights, and that ecosystem preservation is not a natural or inherent property of indigenous and rural practices, but is often entangled with struggles over social, economic, and national inclusion, struggles shaped by historical ethnic, gender, and class exclusions. While the association between indigenous peoples and ecological preservation is still dominant in the Ecuadorian Amazon, acknowledging the intersectional nature of sustainability allows a more holistic understanding of the challenges and opportunities that Amazonian peoples face concerning the extraction of natural resources.

**"Subjectivity," Intersectional Relations, and Sustainability**

"Subjectivity" refers to how individuals understand themselves and are understood by others as social subjects, through both imposed and self-determined ways. Subjectivities form the basis of agency: it is through the internalization and resistance of external framings of the self that individuals are able to act. Subjectivity is a relational concept that recognizes that the outlook of each human being forms in relation to their particular social, political, cultural, and economic contexts and histories. Subject identities are intersectional, meaning that they are not determined by any single relationship but instead emerge from among multiple, overlapping relations that vary with historical and spatial contexts. These identities are constantly practiced by individuals, in relation to other individuals and social structures, and are thus open to reinterpretation and different enactments. In the words of Judith Butler, subjectivities are not only "performed," i.e., that individuals take on a role or act a particular identity—being a man or a woman, for example. Rather, our being *something* (man, woman, indigenous, white) is "performative": our acts produce effects, they consolidate how we see ourselves and are seen by others.

Identities are subject to norms established by institutions, cultural patterns, and socioeconomic structures that police appropriate and deviant behaviors and affect how individuals and groups of people identify themselves and are identified by others. For example, categories such as "Ecuadorian," "female," "poor," and "white" are often associated with specific social roles, highlighting differences among populations and assigning a value to these differences through which power can be exercised. For example, a governmental agency may identify a heterogeneous group as a specific sort of subject (e.g., undocumented immigrant), which may affect how people are perceived as subjects and even how they perceive themselves. The subject position "indigenous person" calls on an individual to occupy this position or "slot" as the dominant way through which they are identified in society, and the kinds of claims they can articulate, despite the complexity and diversity of the social experiences of indigenous or "ethnicized" people. Faced with identification, a person may either accept a subject position entirely or partially, or try to refuse it, consciously or unconsciously.

For an overview of social theory of the subject and subjectivity, see Nick Mansfield's (2000) *Subjectivity: Theories of the Self from Freud to Haraway*. For more advanced theory on subjectivity, see the influential work of Michel Foucault and Judith Butler.

See how Judith Butler explains the links between behavior, society, and identity in this video titled "Judith Butler: Your Behavior Creates Your Gender" published on the Big Think YouTube channel:

www.youtube.com/watch?v=Bo7o2LYATDc.

Next we examine how indigenous positionality and subjectivities play out in Playas. We explore how Amazonian peoples intervene in the projects and practices of sustainability in Ecuador as groups with claims to indigenous territories. Second, we turn to how the specific interventions of indigenous groups in sustainability projects and practices draw on complex geographically and historically entangled relations that may or may not coincide with the priorities and practices often associated with the indigenous subject.

## Playas del Cuyabeno

Formed in 1970 with the help of regional Kichwa organizations, Playas is located in the Cuyabeno Wildlife Reserve, Sucumbíos Province, at the meeting of the blackwater Cuyabeno River and the whitewater Aguarico River. Sandbars or beaches (*playas*, in Spanish) frequently form at the intersection of these two rivers, lending their name to the settlement. Playas is a zone of frequent flooding, meandering rivers, intermittent lagoons, and swamps. The first families that settled Playas came from diverse places, including peri-urban Quito, the province of Azuay in the western Amazon, the San Miguel River (border between Ecuador and Colombia), and the Napo Province. Approximately four hundred people live in Playas.

Like other Kichwa peoples in the Amazon, many residents of Playas have used a dual-home system that stems from their long relationship with the Ecuadorian state. In the 1960s, the state promised Amazonian Kichwa families communal land titles if they settled forests and organized as *comunas*: centrally organized and directed indigenous communities that actively transform environments and engage in market activities. The small group of families that founded Playas gained collective land title through this program. Over the last four decades or so, many residents of Playas have maintained two homes, one located in the community center at the meeting of the Aguarico and Cuyabeno Rivers to facilitate participation in state-sanctioned "organized life" (where a meeting area, local government office, elementary school, and sports fields have been located for much of Playas' existence) and another home in more remote areas—often several kilometers away—where they cultivate gardens and hunt for everyday subsistence. The dual-home system allows individuals to enjoy community-owned goods and benefits, gained through their decision to engage the formal channels of recognition by the state, and to maintain traditional practices of reciprocity and solidarity with their extended kin in more remote areas, where they could escape the frictions of state-determined, collectivized life.

The geographic location of Playas is not optimal for market-based activities. When first settled, the community did not have sufficient resources to cover transport costs to the nearest markets, which were accessible only by a four to eight hour ride on the river, depending on the canoe and conditions of the river. Until recent years, Playas residents were considered poorer than other Amazonian peoples; they sought wage labor from the oil industry or neighbors and often were unable to pay for gas for their motorized canoes to service a small ecotourism enterprise. Playas residents volunteer their time to maintain communal areas and make small monetary contributions to fund the costs of their community leaders to represent their claims to state, nongovernmental organizations (NGOs), and other institutions, though not all members contribute equally.

With the discovery of highly commercial crude oil in Sucumbíos Province in 1967, many Kichwa men became wage laborers for the oil industry, where they witnessed firsthand the profound ecological and social changes associated with the expanding oil frontier. With the noise, contamination, and deforestation, animals fled, fish died, families lost their everyday sustenance, and men and women left communities in search of wage labor. This experience shaped the political and ethical position of Playas residents with respect to oil extraction: no company would be allowed to operate in their territory. Environmental activists enthusiastically supported communities such as Playas and underscored the importance of their anti-oil politics in protecting the ecosystems of the Cuyabeno Reserve.

## The Uprising and the Aftermath

Residents of Playas proudly recall what they refer to as the "uprising," which they initiated on September 8, 2008, against the Ecuadorian state-owned oil company, Petroecuador. On that morning, eighty community members made the river journey from their farms to congregate for a monthly meeting in the swampy town center, where a schoolhouse and a few wooden houses with zinc roofs were located. Three barges carrying heavy machinery and military officers appeared on the Aguarico River. The men of the community, having worked for oil companies, recognized the machinery as oil drilling equipment. The president of the community, Alberto Noteno, recalled to us rallying the community to detain the barges. Men and women rushed to canoes, motored up to the barges, and boarded them. Andrea Chávez and other women describe snatching guns from military officers and tossing them into the river.

This was the beginning of a standoff between the families of Playas and military personnel that lasted for forty-five days, during which the barges of Petroecuador did not advance beyond Playas' shores. Community members tell us heroic stories of responding to tear gas grenades by hurling homemade balloons of hot pepper and tobacco juice that would burn just as strong as teargas in the eyes of the military officers. According to newspaper reports, community members were injured by rubber bullets and some military officers were burned by Molotov cocktails.

Playas' leadership activated its contacts through affiliations with a regional indigenous network and relationships with conservationist NGOs in the capital city of Quito to increase the visibility of the event. Regional and national media arrived on the scene, while government representatives and lawyers scrambled from Quito to negotiate with community leaders. They explained that a portion of the community located further upriver had given the company permission to drill. This, however, was no innocent misunderstanding. Petroecuador negotiators had visited Playas several months earlier to share their plans and, according to multiple testimonies, had explained that possible environmental impacts would be negligible. Yet, the men of Playas recalled their experience working in oil drilling on the neighboring Napo River since the 1960s and understood that the risks were substantial. They refused to negotiate based on the incomplete information that had been presented and turned Petroecuador away. In turn, company representatives had negotiated separately with a more receptive minority of the upriver community. According to accounts from Playas' leadership, the revelation of this tactic angered people in Playas even more. Eventually, military reinforcements arrived and the barges continued to their destination under the cover of overwhelming force. Government representatives continued to visit Playas in the following weeks and months, offering infrastructure development projects to ease tensions. The community, outnumbered and overpowered, agreed to negotiate. Community leaders requested housing, basic services, and better access to education, although they were not confident they would receive any significant concessions from the government.

## The First Millennium City

The outcome of this negotiation produced a new government program poised to transform the Ecuadorian Amazon in the coming years, much to local people's surprise. In October 2013, President Correa "inaugurated" Playas anew, as the first ecological city or Ciudad del Milenio (Millennium City) in Ecuador. Correa publicly stated that investing oil rents in technology and modernization in Playas would be the first step in countering the historical racism of environmentalists who wished to keep Amazonian peoples poor and marginalized for the sake of protecting nature, and he framed it as a transformational step in how we think about sustainability and progress. Economic, social, and environmental sustainability in the Amazon, he argued, depends on eliminating poverty, which stems from material scarcity and, he added, leads to environmental degradation (e.g., deforestation, overexploitation of flora and fauna) and social disintegration, as adult family members migrate.

Government contractors bulldozed the school and houses in the swampy town center of Playas and covered the terrain with sand on which they paved eight square blocks and built up an urban center. The new city included 17 hectares (42 acres) of roads, sidewalks, and street lamps; a high school with language and computer laboratories, equipped with the Internet; a medical clinic; and a police station. Sixty-eight families received a house of steel, cement, and synthetic materials, equipped with plumbing and electricity, appliances, furniture, computer, telephone, and cookware. As a last minute surprise each family was given two bicycles for everyday transport, as cars are not allowed in Playas. The new city came to life overnight with the ignition of a towering generator that consumes 6,000 gallons of diesel per month and an industrial water treatment system, both maintained by a full-time team of technicians. Millennium Cities like Playas are urbanization projects financed by oil rents, which aim to achieve a marriage of modernity and social and environmental responsibility: access to the comforts of fossil-fueled modern life, but with no vehicle emissions; access to sewage systems, but with ecological sewage processing to introduce waste into the river safely; a medical clinic with doctors assigned each year, as part of their required rural service; integrated rural development programs, including the provision of cacao, coffee, and maize seeds as potential cash crops.

Explore these sites for more information, including images, on Playas and other Millennium Cities in Ecuador:

- "Ciudad del Milenio Is Already a Reality in Playas de Cuyabeno" from the Instituto para el Ecodesarrollo Región Amazónico: www.desarrolloamazonico.gob.ec/ciudad-del-milenio-ya-es-una-realidad-en-playas-de-cuyabeno/.
- "Ciudades del Milenio: Similar la Civilización," a recent critical news report, published on PlanV.com: www.planv.com.ec/investigacion/investigacion/ciudades-del-milenio-simular-la-civilizacion.
- "Ciudad del Milenio 1," a recent, critical television report, published by EcuaVisa.com: www.ecuavisa.com/video/video-vision-360/programas-anteriores-360/ciudades-del-milenio-1.
- "La Comunidad del Milenio Playas de Cuyabeno, un Ejemplo de la Nueva Amazonía," an official government page and report on the inauguration of the first Millennium City: www.hidrocarburos.gob.ec/la-comunidad-del-milenio-playas-de-cuyabeno-un-ejemplo-de-la-nueva-amazonia/.

Community members were enthusiastic about the new settlement. Young people no longer had to leave the community to go to high school. "It was elegant," remarked 13-year-old Marcelo, and immediately transformed daily life. Rebecca Noteno explained that families could now store meat in refrigerators rather than having to share it out or spend days preparing it for storage. She added that they now had clean water for drinking, cooking, and washing instead of the dirty water they had had to carry from the river. They no longer had to hoard matches. They could learn to use computers and the Internet to communicate with family members living elsewhere.

Interviews and observation indicate that within a few months this transition to modern life had its difficulties. In order to receive a house, families had relocated from farms dispersed along dozens of kilometers of river into the urban village. Money for gasoline and gasoline itself were scarce, but had become necessary to power canoes that would take them back to their farms to work and obtain food. In addition, families grew anxious about how they would pay for electricity and water once the municipal government began to charge them. The only sources of regular income were a few positions as laborers and boat operators with the oil company. Money trickled through the economy from the sale of beer, rice, and canned foods, but not enough to ease anxieties. All phone lines had been disconnected for lack of payment within the first three months of service. During the local elections in early 2014, 90 percent of adult community members ran for office, hoping to win one of the salaried positions. In early July 2014, a major oil spill nearby generated excitement over the jobs created in the clean-up—ironic and surprising, given the anti-oil stance community members had once supported so enthusiastically.

Whereas the deal Playas cut with the government became problematic due to the lack of money and market commodities in circulation in this new urban space, hope remains due to the 2010 reforms of Ecuador's Hydrocarbons Law, which ensure that 12 percent of all profits from oil extraction are directed to local development projects (see Acosta 2011). For the community of Playas, this translates into approximately 10 million USD annually that they must invest in infrastructure. Community members hope to develop a luxury hotel in the city to boost commerce and bring in tourist dollars. Then, they plan to build a mall.

## Oil and Sustainability

Why did Playas residents initially oppose the state company, and why did they later acquiesce to its presence? Initially, Playas residents drew on previous experiences with oil companies to oppose oil exploitation within their territory. Residents knew firsthand the impact of oil on their use of the river for drinking water and fishing, and its chilling effect on their twenty-year ecotourism project. They had seen the detrimental effects that oil spills had on tourism over the years as thousands of barrels of spilled crude periodically flowed the length of Aguarico and through their territory on its way to Peru. They reasoned that the impacts would be worse if drilling opened in their own parish.

Yet, interviews with Playas residents also suggest that in addition to protecting natural resources, the immediate motivation for resistance was to protect a business plan. Over the previous year, community members had held dozens of assemblies to discuss and plan their own oil company that would be called Alain Petrol. The name refers to *alianza indígena* or "indigenous alliance." Originally, the community wanted to use "Sacha Petrol," but, Alberto Noteno explains, this name was already in use by an indigenous Kichwa organization in Napo that provides services to oil companies. Playas residents had met with indigenous leaders from regional and national networks. Members of the Cree nation from Ontario, Canada, visited Playas' leadership to share their own experiences of planning and forming an indigenous oil company.

Alain Petrol was the centerpiece of a plan among the leaders of Playas to sustain social, economic, and even environmental conditions. Their business plan emerged from their historical experience of oil development. During the 1990s, the Ecuadorian government had carried out eight separate calls for bids from foreign companies to explore and drill in the Amazon, an effort to open up the oil sector and attract investors. The effort also involved reduced environmental and social oversight, which translated into an absence of state presence in rural areas and the conspicuous absence of local material benefits (see Sawyer 2004). Increasing social mobilization in the 1990s and early 2000s brought attention to this perceived state-abandonment—often referred to as the "rolling back of the state"—and successfully secured tighter controls to avoid and mitigate social and environmental impacts, including reforms to the 1998 Constitution (see Narváez 2004), the Law of Environmental Management (1999), the Forest Law of Conservation of Protected Areas and Forest Life (2002), and the Law of the Prevention and Control of Contamination (1999). New regulations included a requirement for the prior consultation of local communities beginning in 2002. Such an increasing concern with regulating impacts informed Ecuador's 2008 Constitution, the first constitution in the world to recognize nature's rights.

**From Neoliberalism to the Rights of Nature**

Neoliberalism, broadly speaking, refers to the belief that human well-being is maximized through unfettered market competition and that the role of the state is to create the conditions and institutional frameworks that support markets. It includes political–economic practices that emphasize private property, free markets, and free trade as necessary to enhance the entrepreneurial freedoms of individuals and economic growth. Deregulation, privatization, and withdrawal of the state from many areas of social provision, health, and education are common practices of neoliberalization. In Latin America, during the 1980s and 1990s, governments struggled to respond to a sovereign debt crisis and adopted a set of neoliberal economic reform policies—in many cases imposed by states and lending institutions from the Global North—known as the "Washington Consensus." In reality, neoliberal practices led to increased inequalities and economic stagnation in most cases, as countries reduced their state budget and institutional apparatus. In Ecuador, the GDP actually decreased in what is now remembered as "the lost decade" of the 1980s.

One consequence of neoliberal policies was increasing contamination and environmental degradation produced by unregulated industries in the agricultural, fishing, mining, petroleum, and other key sectors in the Ecuadorian economy. In the 1990s, social movements, led by organized indigenous groups and organized on an unprecedented scale, overthrew neoliberal governments and powerfully threw into question neoliberal dogma. This popular rejection of neoliberal governance opened a path for an alliance of academics, NGO professionals, and progressive leaders to win the presidency in 2007 and profoundly transform Ecuadorian institutions, including the National Constitution, rewritten in 2008 by a popularly elected assembly. Among the changes introduced in the Constitution is the recognition of the "Rights of Nature," including the right of ecosystems to renew their natural cycles and be restored from states of degradation. In addition, people and communities enjoy the right to demand that the Rights of Nature be respected. By the same token, this Constitution carves out the right of the state to suspend the Rights of Nature in the interest of exploiting natural resources deemed "strategic" for national development.

For more information on neoliberalism in Latin America, see Mark Goodale and Nancy Postero's *Neoliberalism, Interrupted: Social Change and Contested Governance in Contemporary Latin America.*

For information on modern Ecuadorian social and political history, see Marc Becker's *Pachakutik: Indigenous Movements and Electoral Politics in Ecuador.*

See also this video, "Naomi Klein on Global Neoliberalism," where Klein explains her view of neoliberalism in Latin America, and the relationships among private corporations, governments, and social movements; published on YouTube by Big Think: www.youtube.com/watch?v=sKTmwu3ynOY.

For more information, see the writings of environmental activist, economist and president of the Constitutional Assembly Alberto Acosta. Also, see articles 71–74 of the National Constitution of the Republic of Ecuador, available at: https://therightsofnature.org/wp-content/uploads/pdfs/Rights-for-Nature-Articles-in-Ecuadors-Constitution.pdf.

Amidst this social agitation and regulatory transition, leaders from the Amazonian indigenous organization CONFENAIE met with indigenous entrepreneurs in Calgary, Canada, in 2000 (see Fontaine and Le Calvez 2010). Their aim was to learn about the experiences of members of the Cree nation in forming and administrating a company that provided a variety of services in Canada's oil and gas sectors and that extracted and commercialized natural gas. Four Ecuadorian indigenous leaders, including a representative of the northeastern Amazonian territories where Playas (and parts of Yasuní) is located, formed a committee and formalized a consortium with Canadian Cree to extract and commercialize gas in the Amazon. The consortium obtained the support of their respective governments and 250,000 USD from the Inter-American Development Bank. However, by 2003 the project had dissolved due to conflicts within CONFENAIE and instability within the Ecuadorian government and the state oil company.

In 2007 and 2008, several of the same leaders involved in the gas initiative held meetings with the president of Playas and fourteen other communities along the Aguarico River. They understood that Petroecuador was planning to expand drilling along the Aguarico River in wells that had been identified in previous decades. In these meetings, participants resolved to form Alain Petrol and gain the support of their respective community assemblies. When they presented the project proposal to representatives in the Correa government, it was disregarded as fantastical. They would need at least 50 million US dollars of capital investment, they were told. The group responded with documentation indicating that they had interested investors in Houston, Texas, and advisors from Talisman Energy, a Canadian oil and gas company with operations in North America and Colombia (see El Ciudadano 2010). This startled the government and high-ranking officials of Petroecuador, and the Ministry of Mines and Petroleum denounced Alain Petrol in the national media and among communities along the Aguarico River as a scheme to extort communities (see La Hora 2008; Ecuadorinmediato 2008).

Playas' president Alberto Noteno explains that when government representatives arrived during the uprising of 2008, community leaders presented them with two proposals. Either the government authorize Alain Petrol, they demanded, or provide the families of Playas with housing, basic services, and education. Although Alberto Noteno and other community members had their hopes set on the success of the indigenous oil company, they had no choice other than to compromise in order to receive the latter, given the government's overwhelming show of military force. Ultimately, every community but two dropped out of the project due to the public defamation campaigns and militarized response. The remaining two communities eventually became the first Millennium Cities: Playas and the nearby community of Pañacocha.

## Intersectionality and the Quest for Sustainable Oil Futures

Prominent narratives on a harmonious relationship between indigenous peoples and nature (e.g., the "noble savage"), in addition to Playas' own history of involvement with conservation, might make it surprising that the community tried to form an oil company. However, attention to the various subject positions and experiences of residents of Playas helps better explain how and why they positioned themselves as potential oil producers. Before the clear and present threat of a foreign, or mestizo, company coming into their territory, the leadership in Playas reacted in relation to their previous experiences with oil companies in other parts of the Amazon, which are marked by ethnic discrimination and environmental degradation. In the following, we describe examples of these relations and highlight historical and geographic factors that shaped the decision to create an indigenous company. These factors do not necessarily coincide with popular conceptions of environmental conservation and sustainability.

### Caring for Lived Environments

One of the ways that Playas residents reconciled their commitments to place and environment with their oil exploitation ambitions was through narratives of care. Alberto Noteno explains to us, "you don't contaminate your own home." He argues that part of the motivation behind Alain Petrol was that if people from Playas were drilling the oil, they would be more careful than an external oil company would be to avoid contaminating or otherwise affecting flora and fauna in the territory. They would take care to maintain pipes and other infrastructure to prevent oil and other contaminants from entering the river, and they would not produce excessive noises that startle wildlife, which would reduce habitats and negatively impact hunting. Given these concerns, many people in Playas decided that the safest strategy to confront oil in their territory was to drill it themselves. Playas has a long-running relation with oil production in the province and with contamination along the Aguarico River or, in other words, with an already contaminated nature—as opposed to a nonexistent "pristine nature"—that helps us to explain the position inhabitants took with respect to oil production in Playas itself.

For more than forty years members of Playas have migrated throughout the Napo and Aguarico rivers to participate in one of the few sources of steady income they could count on. From the 1970s through the 1990s, most young men from Playas worked for oil companies for some amount of time to have income for tools, clothing, salt, and gas, among other commodities, and for sending their children to study in other parts of the province. In this period, most men studied until their early teenage years and then worked as guides and temporary laborers in the seismic activities integral to oil exploration. At age eighteen, they completed a year of military service to obtain a certificate that, in turn, permitted them to access steady work as laborers—again, for the oil companies. Thus, many recall observing firsthand the contamination of rivers due to the negligence of their mestizo coworkers and supervisors, who did not maintain pipes and infrastructure or take care to limit the noise pollution and deforestation that thinned wildlife.

Both young and old in Playas recall oil spills that have affected fishing and killed livestock from the 1970s to the present. Several older inhabitants recall a spill in the 1970s that occurred just prior to annual flooding in mid-July, thus killing most of the pigs in town after they were immersed in "diesel." Carlos Machoa, a farmer and fisherman in his fifties, recalls multiple periods stretching back to his teenage years during which "the fish smelled like diesel." In response to these spills, only the current government has been credited by residents with bothering to stage cleanups. In other words, "nobody worried about us," Juan Chávez, a former community leader, explained to us.

Awareness about the health consequences of this negligence has emerged in recent decades through interactions with NGOs. A variety of NGOs funded by the Italian aid agency and the European Union in the 1990s and 2000s have inculcated an awareness of the impacts of oil spills and of forms of daily contamination, helping people to contextualize their experiences of oil contamination within the framework of health and well-being—for example, the connections between the incidence of illness and having to collect drinking and washing water from rivers covered with oil. The memory of Texaco's activities in the province from 1974 to 1986 are infamous throughout the province and country largely due to a two-decade-long series of environmental justice trials and an ongoing campaign by the government to shame Texaco's current owner, Chevron, into paying for damages. Texaco spilled millions of gallons of oil and contaminants into the rivers around Shushufindi, just a few kilometers west of the Cuyabeno Natural Reserve. This experience has also shaped fears of people in Playas about the possible impacts of oil drilling by foreign, mestizo, or other non-native groups along their segment of the Aguarico River.

Therefore, despite popular narratives about an intransigent anti-oil politics among indigenous peoples affected by oil contamination in the Amazon, many residents of Playas came to view forming their own oil company as the best defense against careless foreigners and mestizos in their territory. Environmental sustainability, in this sense, was dependent on responsible oil extraction. Yet, this was not the only, or the primary, motivation of Playas residents. As explored in the next and final section, the potential social impacts of external companies largely shaped the plans of Playas' leadership.

## Valuing Indigenous Difference

Oil is good business, very good business. Alberto Noteno recognized this and so did much of the rest of the community. But what was the drive to enter into *very good business* in a community located at the margins of monetized, market society? Residents of Playas sought to capture oil revenues for two main reasons that responded to an ethnically defined social hierarchy, which powerfully shaped concerns and planning in Playas. First, given previous experiences in which oil companies had not valued their labor equally to that of mestizo workers, many affirmed that they could prevent discrimination by mestizos if they had their own company. Second, although money has not historically been a resource for sustaining livelihoods in Playas, the consumption of market goods is conceived of by many as an effective strategy of ethnic whitening, or *mestizaje*, a goal that has emerged in the wake of generations of ethnic exclusion and violence.

### Ethnicity and *Mestizaje* in Latin America

There are numerous ethnic identities in Latin America. In Ecuador alone, fourteen different ethnic "nationalities" are recognized by the state. Ethnic categories in Latin America are often associated with common, non-European ancestry, and/or cultural practices. "Indigenous" as a general category applied to all non-European subjects emerged in and through colonization, beginning in the fifteenth century. During the colonization of the Americas—the largest genocide in human history—indigenous peoples were subject to brutal military and socioeconomic oppression. Today, indigenous groups continue to suffer dispossession and discrimination throughout the Americas. Étienne Balibar (1991) argues that because no nation can accurately trace the origins of its existence (and must do so through foundational myths), one might conclude that modern nation-states only came into existence as such through the cultural production of non-European subjects who would function as the "Other" or the opposite against which to define European nation-states and their proper citizens.

*Mestizaje* broadly describes the social practice of biological and cultural mixing between different ethnic groups with the supposed purpose of producing "whiter" generations that will be closer to the social ideal of whiteness imposed by force and by political and cultural means since colonization. Ethnic "whitening" or "*mestizaje*" also emerged as a policy of nation-building in the late nineteenth and early twentieth centuries in some Latin American countries. This was a directed process of producing and valorizing a "mixed" indigenous and European national identity. *Mestizaje* during this period did not valorize indigenous and European identities equally; European identities (expressed in biology, language, culture, history, capitalism) were seen as more valuable and necessary to complete the assimilation of indigenous peoples into nation-building projects.

For more information on racial democracy, *mestizaje*, and genetics research, see Peter Wade's 2010 *Race and Ethnicity in Latin America,* and his video, "Race, Nation and Genomics: Biology and Society," published on YouTube by the British Academy: www.youtube.com/watch?v=gKhY-_pVZqA.

In effect, the promotion of an indigenous oil company responded to a long history of ethnic exploitation. The parents of Playas' founders fled the Ecuadorian–Peruvian border in the war of 1941 and ended up along the Colombian border in Putumayo. Playas came together in the late 1960s, as a handful of families then fled guerrilla and paramilitary violence along the Colombian border. Many of the founders had already known of Playas because either they or their parents had followed Colombian traders to the area to extract rubber and to hunt for animal skins. This work had been compensated in kind: rifles, machetes, clothing, salt, matches, and other market goods. However, community members today recall that many members of previous generations were pushed into indebted servitude by traders. "My grandparents were slaves," recalls Alberto Noteno.

An ethnic hierarchy in the region has since been reinforced by diffuse social interactions. Many people in Playas recall going to school downriver throughout much of the second half of the twentieth century, where Italian missionaries taught indigenous youth to speak, read, and write in Spanish. Speaking Kichwa was seen as backward by missionaries and prohibited under the threat of violence. Similarly, in the military barracks, young men of Playas in the 1970s and 1980s were routinely beaten with whips or tortured with electric cables by mestizo superiors. Such relations, defined by an ethnic difference that was punished violently, were reproduced in work for oil companies, where the young men of Playas received lower pay due to their Kichwa identities.

Ethnic hierarchy is reproduced within Playas today. The new high school, for example, has attracted students from neighboring Cofan, Secoya, and Shuar ethnic communities and many of the majority Kichwas, proud of their city, have taken to claiming ethnic superiority over their classmates from other communities. Young men in particular joke about the inferior Spanish abilities of the Cofan students, pejoratively referring to them as "tunics" in reference to traditional dress. Another particularly salient manifestation and recreation of ethnic hierarchy emerges in an ongoing feud between the two largest families in Playas: the Chávez and the Noteno. The Chávez family often reminds people that their grandfather and patriarch, Franklin Chávez, was originally from the capital city of Quito. However, this claim is rebuffed by members of the Noteno family, who offer an alternative genealogy, placing Franklin's origins deep in the Peruvian Amazon. They argue that the Chávez family changed their indigenous last name a generation ago. Felix Noteno explains to us, "they just want to seem as if they were more from the outside." This feud is representative of other attempts among individuals in Playas to position or associate themselves with people, histories, and things from the urban and decidedly mestizo "outside".

The inhabitants of Playas have given up on Alain Petrol, but insofar as the school represents their children's opportunity to leave Playas, most families have been sufficiently satisfied with the Millennium City. Many parents and children would like to see the youth move to Quito, Guayaquil, and other major cities in Ecuador in order to become "professionals" or, as it is often said in Playas, to become "someone." Parents like to remind their children that they had to travel up to a month, leaving Playas on military barges or the mailman's boat, to get to missionary schools, often floating on tree branches for hours on the return trip. Having a school in Playas makes an eventual and definitive journey out of Playas and into the center of national mestizo culture that much easier. Alberto Noteno has similar plans for his children. Alberto was one of our respondents who had been tortured with electric cables during his military service. His mestizo superiors told him that "torture was practice" for when he might be captured by the Peruvian army and forced to give up information about Ecuadorian positions. Thus, although Alberto remembers his shaman grandparents with pride and identifies with his ancestry as a Kichwa, he explains that he would not teach his children or grandchildren to be shaman or even to speak Kichwa. He wants them to be urban "professionals" and to have an "easier" life, one less marked by discrimination and violence than he, his parents, and his grandparents had experienced.

Today, conspicuous consumption of specific goods is widely conceived in Playas as strategic for positioning oneself on an ethnic hierarchy. When men receive payment from short-term work for the oil company or from tourism, most spend the larger share of this income on beer—a much-lamented situation among many women of the community. Beer has become construed as the mestizo alternative to the traditional alcoholic drink made out of yucca, *chicha*. Elena Nepas, who moved to Playas from a peri-urban area near Quito, ridicules community meetings as spaces for "Indians to drink *chicha*," but this drink is being relegated only to such meetings, whereas the consumption of large quantities of beer has become common daily practice. Another form of conspicuous consumption is the purchase of sound systems and urban music. Throughout the city, speakers are positioned in the windows of houses to project urban hip-hop and rap music out towards the sidewalk. Several young men explain that they had proudly positioned speakers out of their windows for at least several weeks to ensure that their neighbors could hear that they had been able to purchase them. In this context, having one's own oil company represented the possibility of accessing not just money but status within a steep ethnic hierarchy that located indigenous people at the bottom, but permitted a sort of whitening-by-association with consumer goods from broader market networks. Ethnic discrimination, therefore, needs to be considered in order to understand how oil drilling became a component of planning among Playas' leaders and residents.

## Policing Unruly Subjects

In addition, the "protection" of women was one of the most thoroughly considered and clearly articulated reasons for forming Alain Petrol. As a locally run indigenous oil company, male residents of Playas hoped that it would keep non-indigenous men out of the parish and away from the women of Playas. According to Alberto Noteno, "[w]e had the experience that here in [the nearby town of] Tarapoa, in the county of Cuyabeno, companies were coming in and abusing the women." To counter this potential threat, they developed their own plans, regulations, and proposals for a local company that would grow out of the labor and direction of local residents—one that would enable the surveillance of women. A teacher from Playas concurs with this rendering of the brief history of Alain Petrol, recalling to us: "[t]here were anecdotes that the oil companies generally come and, you know, a lot of people come, only men . . . and here there are many girls just growing up . . . and anything could happen, at least that's what the people thought." Likewise, he observed that many people were on edge about the presence of workers brought in to build the Millennium City. To everyone's relief, he said, "there wasn't a single rape." The male-dominated assembly that made the final decision to form Alain Petrol drew on this perceived and routinely practiced need to protect the women from *other men* in their rationalizations.

According to women in the community, the men of Playas were not concerned about violence per se. Gender violence is quite common in Playas. Case in point, Playas' Millennium City includes a police station that was abandoned after police were threatened by armed men from the community for having tried to intervene in domestic abuse. "We have our own laws here," the police were warned. It was not violence towards women that the men wished to evade by forming their own oil company and preventing the entrance of foreign and mestizo men, but rather, as several women explained to us, they were concerned about the "taking" of women by other men—that is, that women might fall in love with foreign or mestizo oil workers more easily than with indigenous men and become part of their world.

Thus, Kichwa men in Playas respond to exclusion not only through processes of ethnic whitening, but also through the preservation of gender domination. This case reflects longstanding gendered relations entangled with oil. In the 1960s, when the oil frontier was expanding, indigenous women searching for food would sometimes visit oil camps, a predominately mestizo space, where accounts suggest that violence (sexual and otherwise) was not uncommon. In turn, the indigenous woman's body becomes imagined as a space to control, since it is through its sexuality that *mestización* takes place and where the possibilities of reproducing—as well as interrupting—established norms on cultural and biological difference are located (see Valdivia 2009). The proposal for an indigenous oil company in Playas illustrates how indigenous, female bodies may be seen as objects of control in hierarchies of difference in the Amazon and, more broadly speaking, how particular ways of seeing social hierarchy can influence planning. The male indigenous government planned to form their own indigenous oil company to extract the oil themselves in response to environmental and social concerns, among which ethnic and gender relations powerfully shaped the decision to embrace indigenous oil extraction.

## Conclusion: Towards Intersectional Sustainability

While at first sight the expansion of resource frontiers and sustainability might seem at odds, the community of Playas demonstrates that sustainability is not a "thing" imagined and reproduced the same everywhere, defined by universal principles naturally aligned with indigenous peoples' interests and philosophies or in stark opposition to Western or market society's modes of production. In this chapter, we described how legacies of exploitation, exclusion, and patriarchy shape the possibilities of extractivism and resistance along resource frontiers in the Ecuadorian Amazon.

The analytic framework of indigenous positionality helped us to move beyond simplistic narratives to examine how indigenous subjectivities and struggles form and shift according to changing conditions and ongoing histories of exclusion and violence. How people in Playas self-identify, and are identified by others, is fundamental to their shifting position on sustainability and oil extraction. When they see themselves as indigenous peoples facing an expanding oil frontier, they view themselves in relation to nation, government, ethnic, and gender hierarchies; their own experiences with the oil industry; and their own ways of living in place. At first, long-running encounters with oil positioned the people of Playas to decide that drilling for oil themselves might be their best option to avoid environmental degradation at the hands of negligent, non-indigenous companies backed by a state that had ignored their contamination for decades. Residents also reflected on their historical experiences of ethnic violence, exploitation, and exclusion from the schoolhouse to the military barracks to the workplace, and responded to the ethnic hierarchies that emerged from these relations to inform plans for an indigenous oil company in Playas. Responding to these forms of discrimination, in turn, influenced masculinities in Playas and positioned men to view oil production as a strategy to renew control over women's bodies and protect themselves from perceived threats from the "superior" racialized Other.

The case of an indigenous oil company in Playas undermines a host of narratives that yoke together indigenous agency with environmental protection. Although indigenous groups often share similar histories of colonial exploitation and ethnic exclusion, their ideas about how best to employ nature do not necessarily coincide with one another, or with prevailing ways of knowing or practicing sustainability. Oftentimes indigenous groups do defend environmental sustainability, particularly when capitalist interests threaten to expropriate their means of production, yet indigenous political agency is shaped and reshaped by multiple relations that are not limited to relations with the environment. Here, we have highlighted the ethnic and gender relations that shape a community's decision to embrace oil extraction.

Positionalities of resistance and self-determination, such as those evidenced in Playas, did not go uncontested. The Ecuadorian state overwhelmed Playas with military force to ensure state claims to oil in their community. The Correa administration promised to mobilize oil revenues to provide affected families in the area with access to multiple services and development programs and thus strengthen sustainable urban living practices (e.g., energy efficiency, recycling, bike paths, etc.), and improve standards of living. These promises of development, complete with urban services, infrastructure, and amenities, echoed those that Playas residents had envisioned as part of their future in the Amazon and with their own indigenous oil company.

The experiences of Playas complicate abstract conceptions of sustainable development by signaling the shortfalls of social engineering from above, in the case of the Millennium City, and the social, economic, and environmental entanglements involved in planning from below in the case of Alain Petrol. In this case, the need to ask "sustainability of what and for whom" is starkly evident. The residents of Playas are not the first indigenous peoples to have proposed the extraction of natural resources within their territory as a strategy for environmental and social sustainability on their own terms. Prominent examples include the aboriginal companies of Canada's oil, gas, and tar sands sectors and those of Australia's mining industry. When indigenous peoples negotiate with oil and mining companies or form extractive companies themselves, they are often described as falling prey either to confusion or to corruption. Both views are simplistic and incomplete. It is important to recognize and support indigenous movements that prioritize environmental sustainability. It is equally important to recognize indigenous projects that facilitate extractivism and environmental degradation, in order to understand how power relations shape conditions for imagining what sustainability might be. This should not pigeon-hole us into blaming "confused" and/or "corrupt" peoples for cooperating in extractive activities, or to simplify and "environmentalize" indigenous political agency as visible only when it involves ecological conservation. Rather, the point is to establish accounts that force activists, students, and researchers to go beyond common narratives to understand the complex positions of marginalized peoples, which do not naturally coincide with principles and conditions of sustainability that are assumed to be universal. Ultimately, indigenous peoples' positions with respect to extraction are *not* determined by romantic notions of an indigenous group's uncomplicated relations with nature, but rather are shaped by a given group's specific experiences of overlapping and competing resource governance regimes, entangled with specific histories, ways of life, and identities.

## Postscript

Millennium Cities like Playas are being replicated in other areas in Ecuador. In August 2014, a community on the coast that had mobilized in resistance to a hydroelectric dam became the nation's third Millennium City. President Correa originally proposed the construction of two hundred Millennium Cities in the Amazon alone, although plummeting oil prices in 2014 and 2015 significantly hampered these plans. As of 2016, four Millennium Cities were completed and eleven more were being developed near sites of oil, mineral, gas, and hydroelectric production in the Amazon, Andes, and coastal regions of Ecuador. On the one hand, these cities threaten environmental sustainability insofar as they require rapid integration into labor and commodity markets, encouraging the construction of roads and placing emphasis on monoculture cash crops; they concentrate populations into small areas, often leading to pollution and the overexploitation of resources near those spaces; and, in particular, they give way to the state-led expansion of destructive forms of natural resource extraction and even lend legitimacy to these activities in national politics. On the other hand, they place in doubt the social and economic sustainability of communities struggling to adjust rapidly to such radical transformations. Some families have already abandoned the two Millennium Cities that were inaugurated in 2013, mostly due to the lack of employment and food.

The Millennium Cities program emerged as a development practice in response to community resistance to expropriation, and it emerged along with a particular manifestation of sustainable development discourse meant to justify displacement in the name of environmentally friendly socioeconomic inclusion. Millennium Cities may serve to promote some, limited forms of social or ecological well-being, but they threaten others. What is increasingly certain, however, is that they appear to be more efficient at demobilizing communities in resistance to territorial expropriation than they are at shaping economic, environmental, and social relations that might be sustained into the foreseeable future.

## Acknowledgments

Angus Lyall wishes to acknowledge the contributions of Nancy Carrión Sarzosa, with whom he conducted portions of this research in Playas. Lyall and Sarzosa worked together as co-investigators in a study of social transformations along the Ecuadorian oil frontier, co-designed by Lyall with Cristina Cielo, Ivette Vallejo, Fernando García, and Lisset Coba of the Latin American Faculty of Social Sciences (FLACSO) in Ecuador. For further publications based on this research in Playas, see Cielo and Sarzosa 2018; Cielo, Coba, and Vallejo 2016.

## References and Further Reading

Acosta, Alberto. 2011. "Ecuador: Unas Reformas con Muy Poca Reforma." *Ecuador Debate* 82: 45–60.
Balibar, Étienne. 1991. *Race, Nation, Class: Ambiguous Identities*. New York: Verso.
Becker, Marc. 2013. "Ecuador's Buen Vivir Socialism." In *Latin America's Turbulent Transitions: The Future of Twenty-First Century Socialism*. Edited by Roger Burbach, Michael Fox, and Federico Fuentes, 98–113. London: Zed.
Cielo, Cristina, and Sarzosa, Nancy Carrión. 2018. "Transformed Territories of Gendered Care Work in Ecuador's Petroleum Circuit." *Conservation & Society* 16: 8–20.
Cielo, Cristina, Lisset Coba, and Ivette Vallejo. 2016. "Women, Nature, and Development in Sites of Ecuador's Petroleum Circuit." *Economic Anthropology* 3, no. 1: 119–132.
Conklin, Beth A., and Laura R. Graham. 1995. "The Shifting Middle Ground: Amazonian Indians and Eco-Politics." *American Anthropologist* 97, no. 4: 695–710.
Ecuadorinmediato. 2008. "*Gobierno Desconoce Cualquier Relación con 'Supuesta Empresa AlainPetrol.'*" April 9. Accessed July 15, 2014. http//:www.ecuadorinmediato.com.
El Ciudadano. 2010. "*Dirigentes Waoranis y Kichwas Destacaron Ayuda del Gobierno en Sus Comunidades.*" June 14. Accessed August 12, 2014. www.presidencia.informatica.gob.ec.
Fontaine, Guillaume, and Marc Le Calvez. 2010. "Dos Perspectivas Indígenas En Torno a la Extracción De Gas en la Amazonía Andina: el Proyecto Camisea (Perú) y la Empresa Comunitaria Amazonía Gas." In *Desarrollo con Identidad? Gobernanza política y económica en zonas indígenas*, coordinated by C. Gros, and J. Foyer, 83–119. Lima, Quito: FLACSO-IFEA.
Gerlach, Allen. 2003. *Indians, Oil, and Politics: A Recent History of Ecuador*. Lanham, MD: Rowman and Littlefield.
Kimerling, Judith. 2006. "Indigenous Peoples and the Oil Frontier in Amazonia: The Case of Ecuador, ChevronTexaco, and Aguinda v. Texaco." *New York University Journal of International Law and Politics* 38, no. 3: 413–664.
La Hora. 2008. "*Falsas Promesas de Petrolera.*" April 12. Accessed July 15, 2014. www.lahora.com.ec.
Lang, Miriam, and Dunia Mokrani, eds. 2013. *Beyond Development: Alternative Visions from Latin America.* Amsterdam: Transnational Institute.
Li, Tania Murray. 2000. "Articulating Indigenous Identity in Indonesia: Resource Politics and the Tribal Slot." *Comparative Studies in Society and History* 42, no. 1: 149–179.
Mansfield, Nick. 2000. *Subjectivity: Theories of the Self from Freud to Haraway*. New York: New York University Press.
Martin, Pamela. 2011. *Oil in the Soil: The Politics of Paying to Preserve the Amazon*. Lanham, MD: Rowman and Littlefield.
Narváez, Iván (2004). *Derecho Ambiental y Temas de Sociología Ambiental*. Quito: Editora Jurídica Cevallos.
Nightingale, Andrea J. 2011. "Bounding Difference: Intersectionality and the Material Production of Gender, Caste, Class and Environment in Nepal." *Geoforum* 42, no. 2: 153–162.
Sawyer, Suzana. 2004. *Crude Chronicles: Indigenous Politics, Multinational Oil, and Neoliberalism in Ecuador.* Durham, NC: Duke University Press.
Sundberg, Juanita. 2006. "Conservation Encounters: Transculturation in the 'Contact Zones' of Empire." *Cultural Geographies* 13, no. 2: 239–265.
Valdivia, Gabriela. 2007. "The 'Amazonian Trial of the Century': Indigenous Identities, Transnational Networks, and Petroleum in Ecuador." *Alternatives: Global, Local, Political* 32, no. 1: 41–72.
———. 2008. "Governing Relations Between People and Things: Citizenship, Territory, and the Political Economy of Petroleum in Ecuador." *Political Geography* 27, no. 4: 456–477.
———. 2009. "Indigenous Bodies, Indigenous Minds? Towards an Understanding of Indigeneity in the Ecuadorian Amazon." *Gender, Place and Culture* 16, no. 5: 535–551.
Vallejo, Ivette. 2014. "Petróleo, Desarrollo y Naturaleza: Aproximaciones a un Escenario de Ampliación de las Fronteras Extractivas Hacia la Amazonía Suroriente en el Ecuador." *Anthropologica* 32, no. 32: 115–137.
Wade, Peter. 2010. *Race and Ethnicity in Latin America*. 2nd edition. London: Pluto.

14

# Sustainability and Gender Equality

## Exploring the 2030 Agenda for Sustainable Development

Sarah Bradshaw

This chapter will consider how the notion of sustainability has manifested itself in the international policy agenda in the so-called "post-2015" context. It provides a case study not of a particular place, but of a space—the global policy arena—and explores the role of sustainability within the development of a set of universal Sustainable Development Goals (SDGs). It pays particular attention to the intersection of sustainability and gender within the post-2015 processes to define a new global framework for sustainable development—the 2030 Agenda for Sustainable Development.

But why gender? As the sustainability discourse is interested in balancing social, economic, and environmental considerations, it should then have a focus on equality issues, including gender equality. While gender equality has been an issue for development since the 1970s, it has been most visible within discussions of social policy and has been less visible in the economic and environmental policy arenas. When gender, or rather women, have entered the economic and environmental policy discourses, this inclusion has often been more efficiency than equality driven, with women seen as a means to better deliver wider economic growth or sustainability goals.

There is also a competing sustainability discourse that focuses on notions of the "carrying capacity" of the earth or "planetary boundaries," which has a more scientific than social rhetoric. How we understand the notion of sustainability relates to the construction of knowledge, and which and whose knowledge becomes the dominant discourse. Contemporary ideas of sustainability are linked to climate change concerns, and while it is a scientific and male-dominated knowledge that informs policy debates, this fact is not recognized; the debates are constructed as gender neutral. When they have been included, women have been positioned not as subjects but as objects of those debates. Consideration of the sustainability discourse then highlights that it is not just *if* issues are included that matters, but it is the *how* and the *why* that are key.

Advancing women's rights has been the aim of feminist and gender activists, women's groups and social movements, national governments and regional associations, and also international policy regimes. This chapter focuses upon the latter. It will begin by exploring how gender has been addressed within international initiatives for bringing development to so-called "less developed" or "Third World" countries, and how gender has been included within environmental debates. Bringing the two together, gendered critiques of the sustainable development agenda will be presented. The chapter then considers the contemporary context and the processes to set a sustainable development agenda post-2015, using a gender lens to problematize the notion of "sustainability" within this global policy space. The chapter argues that the international policy discourse around sustainability is based on a scientific and masculinized knowledge that constructs what "the problem" is to be solved. When women are included in policy solutions it is a means to more efficiently deliver wider policy objectives. As the dominant sustainability discourse is an efficiency rather than an equality discourse, it does little to improve the situation and position of women.

The case study for exploring how gender equality has been included within the wider sustainability discourse is the post-2015 context. The year 2015 became a milestone year in that it saw the culmination of the process to establish a set of Sustainable Development Goals (SDGs) initiated by both the United Nations Conference on Sustainable Development (UNCSD) in 2012 to review the advances made in the twenty years since the first Earth Summit (also known as Rio+20), and the 2010 Millennium Development Goals (MDGs) Review Summit. It is also an important year for the wider sustainability agenda as the United Nations Framework Convention on Climate Change (UNFCCC) came of age, celebrating its twenty-first international meeting, the Conference of the Parties (COP). The outcomes of extreme climatic events were also addressed in 2015 with the adoption of the Sendai Framework for Disaster Risk Reduction for 2015–2030. Furthermore, in gender terms, 2015 saw celebration of the twenty years since the Beijing Declaration and Platform for Action was agreed upon and offered a chance to take stock of advances made in promoting gendered rights at the local, national, and international levels promoted by the Platform for Action. In short, 2015 was a year when many global efforts that link gender and sustainability came together.

## Gender, Development, and Environment

Gender has emerged as an alternative way of understanding how men and women are positioned in society, underscoring that the roles men and women are expected to fulfill are socially constructed. In contrast to the concept "sex," which highlights biological difference, gender challenges the static nature of the differences between men and women (see Oakley 1972). It focuses on the social and cultural meanings that govern what it means to be a man or a woman, and suggests the possibility for change. While women are part of a gender analysis, gender should not be limited or reduced to women alone, but should focus on the unequal power relations between men and women and boys and girls (see Baden and Goetz 1997).

Following an influential study by Ester Boserup in 1970, which highlighted how "development" may affect gender roles and relations in ways that may not benefit women, gender and development emerged as a field of academic enquiry and as a policy practice. It began with the Women in Development (WID) approach, which sought to better integrate women into what was constructed as a benign development process. The approach was critiqued for its focus on women only and its rather narrow understanding of empowerment as being achieved via women's education and employment. The World Bank, for example, highlighted a number of actions that could be taken to improve girls' attendance at school, such as providing a source of clean water for girls to carry home after school. This focus on "practical gender needs" helps to improve the situation for all people through improving the efficiency of women and girls to fulfill their roles, such as collecting water for the household.

**Strategic Gender Interests**

The concept of women's practical and strategic gender needs was first put forward by Maxine Molyneux (1998) and was further developed and operationalized by Caroline Moser (1993) as a planning tool. It highlights that not only including women in projects and policies is important, but how they are included matters also.

While the inclusion of women's needs in development projects is fundamental, many of the needs defined as the basic or practical needs of women, such as the provision of water, basic services, and food, are also needs of all male and female members of the family. They are identified as the needs of women because they generally assume the responsibility for them (in their reproductive role). Addressing these needs then responds to and reinforces existing gendered divisions of labor rather than questions them.

Gender activists suggest that to promote gender equality a focus on "strategic gender interests" is needed. These are perhaps best understood as a process of questioning why women are the ones responsible for activities such as collecting water, and the unequal power relations that inform unequal gender roles and relations. Projects then would focus on the elimination of institutionalized forms of discrimination, such as the right to landownership, measures against intrafamily violence, and alleviation of women's unpaid domestic work, for example.

Yet importantly, such projects do not fundamentally target gender equality, for example by challenging underlying unequal power relations and changing social norms so that boys share the burden of collecting water with girls. Rather than equality, they are focused on efficiency. Ensuring that girls can continue to collect water and still have time to attend school may improve school attendance but may also reinforce gender stereotypes. WID also focused on increasing women's involvement in productive work, such as agriculture and paid employment, as a means to economic empowerment, which in turn, it suggested, would lead to wider empowerment. Taking a WID approach, the World Bank has similarly privileged the productive over the reproductive, seeing women as an untapped resource for economic development. This thinking does not recognize that women's entry into income-generating activities may leave them time-poor as they juggle a double-day of work outside and inside the home.

By contrast, the Women and Development (WAD) approach specifically focused on the nature of productive work and the relationship between patriarchy and capitalism, considering unequal gender relations through a neo-Marxist lens and adopting a dependency theory perspective. It was the Gender and Development (GAD) tradition of the 1980s that emerged as the strongest critique of WID. It focuses on gender roles and relations that are at the basis of women's exclusion from development, and raises issues with the nature of development. GAD projects are more holistic, not just about income and education, but also giving special attention to the oppression of women in the family. GAD projects seek to address women's "strategic gender interests." So instead of seeing educating women in modern farming techniques as a solution to their low participation in agriculture, it seeks to eliminate the institutionalized forms of discrimination around land rights that deny women access to landownership. While attractive in theory, GAD presents much more of a challenge to put into practice as it questions existing unequal gender power relations. With its origins in academic thinking, it may be seen as providing a more theoretical understanding of the problem than the "solutions" offered by the practically oriented focus of WID.

Much has been written over the years about WID, WAD, and GAD (see Moser 1993) and a good source of information is BRIDGE, which is a specialized gender and development research and information service based at the Institute of Development Studies (IDS), in the UK. You can access its website at this URL: www.bridge.ids.ac.uk/.

In the 1980s a number of key works on women, development, and the environment were also published, beginning with those espousing the Women, Environment, and Development (WED) approach. WED drew on the WID tradition and was critiqued in the same way. In particular "ecofeminist" constructions of women as closer to nature, based on the sexual division of labor, were critiqued. A major critique was that they were essentializing—that they prioritized biology as an overarching explanatory variable (see Leach 2007). Some even went as far as constructing women *as* nature through giving birth. This led to a number of subsequent approaches that recognized the material basis of the gender division of labor and differences between women and included the Gender, Environment, and Development (GED) approach. GED emerged in the mid-1990s and paralleled the move in development toward a GAD approach by applying gender analysis tools to the environment. This and other approaches such as "feminist environmentalism" (see Agarwal 1992) and the "feminist political ecology" approach all share a common idea of gender–environment relations as embedded in dynamic social and political relations (see Rocheleau et al. 1996).

Within the development discourse in the 1990s, research to explore the impact of harsh neoliberal economic policies that forced cutbacks in social welfare programs saw wide-scale acceptance that poverty has a female face. As such, in policy terms, claims were then made that poverty reduction policies automatically addressed gender inequality. However, if gender equality is the goal then gender equality programs and projects are the only means to achieve this goal. While gender activists talked of a "feminization of poverty," highlighting that poverty was a gendered experience, policymakers translated this into a "feminization of poverty alleviation" whereby women are constructed not as the most poor, but as the most efficient means by which to reduce the number of poor. Expected to take on a new role as the beneficiary of projects that provide economic resources for the household and in particular children, this new role is conceptualized as part of women's existing gendered roles as mother and carer. This feminization of responsibility places women at the service of development, rather than being served by it and it does little to change the situation and position of women (see Chant 2008). The feminization of policy responses has also entered other areas such as the environmental arena, where parallel developments could be seen in the environmental policy discourse.

Policymakers could draw on WED ideas that women's responses to environmental issues are different from men's. Actors such as the World Bank used the presumed closeness of women to nature that is core to WED thinking, and more importantly the "naturalness" of what are ultimately socially constructed gender roles, to appropriate women's unpaid labor in activities to protect the environment. In the 1990s it was highlighted how global development actors were able to use this discourse to suggest a win-win approach to environment and gender, constructing women as both the chief victims and the chief caretakers of the environment (see Jackson 1993). For this reason, Sherilyn MacGregor, in her article "A Stranger Silence Still" (2010), suggested that to celebrate women's responses to environmental problems, particularly when connected to their gender-specific responsibilities for social reproduction, might be "somewhat dangerous." It seems then this dual identity of women as virtuous-victims continues to inform policy response today.

In the contemporary context, much of the environmental discourse is either explicitly or implicitly focused on climate change and climate change adaptation. Despite the rich literature on gender and environmental concerns, after promising beginnings, it has been suggested that gender was a latecomer to the climate change debate, at least in policy terms. While recent years have seen the production of gendered literature, overall many agree with MacGregor's suggestion that there has been a "strange silence" on the gender dimensions of climate change, particularly in global policy discourse. This could be due to the current emphasis on climate science and technological solutions, meaning the agenda has become "scientized" and "securitized" and as such dominated by a masculinized knowledge. Before exploring this masculinized construction of knowledge further and how it may influence the policy agenda, first a brief summary of the evolution of the international policy agenda as it relates to gender, environment, and sustainability will be helpful.

## The International Gender and Sustainability Policy Context

As the opening chapter of this book notes, discussions around who is included and who is excluded from any given sustainable development program are often absent from academic literature, and there has been little discussion of how processes of inclusion and exclusion are generated, particularly at a global level. Consideration of how gender has entered into, or intersected with, the sustainable development agenda compensates for these ommissions.

In 1979, during the first UN Decade for Women (1976–1985), the Convention on the Elimination of All Forms of Discrimination against Women (CEDAW) was adopted by the UN General Assembly. CEDAW is often described as an international bill of rights for women, and is important for two key reasons: first, it affirms the reproductive rights of women; and second, countries that have ratified or acceded to the Convention are legally bound to put its provisions into practice. While far-ranging in the issues covered, environmental concerns are not explicitly covered within CEDAW. However, the decade for women also saw the initiation of a series of United Nations World Conferences on Women starting with a meeting in Mexico in 1975, followed by meetings in Copenhagen (1980) and Nairobi (1985), where women's actions and roles in environmental management were raised. The fourth, and last full meeting, was in 1995 in Beijing and the resultant Beijing Platform for Action (BPfA) now acts as a framework for annual sessions to review progress across a range of issues, including the environment. In terms of environmental concerns, the BPfA highlights the need to involve women actively in environmental decision making at all levels and to integrate gender concerns and perspectives in policies and programs for sustainable development. This might suggest that gender should have been an element of any emergent sustainability agenda.

The Convention on the Elimination of All Forms of Discrimination against Women (CEDAW), adopted in 1979 by the UN General Assembly, requires states to commit themselves to undertake a series of measures to end discrimination against women in all forms. For more information, see these UN Women web pages:

- "Fourth Conference on Women": www.un.org/womenwatch/daw/beijing/platform/.
- "Convention on the Elimination of All Forms of Discrimination against Women": www.un.org/womenwatch/daw/cedaw/.

Sustainable development had been the central concern at the Earth Summit meeting in Rio in 1992. Before the Summit, the Global Women's Lobby sought to influence its outcomes, including actions by groups such as the Women's Environment and Development Organization (WEDO), and Development Alternatives with Women for a New Era (DAWN). Despite this, one key outcome of Rio—the UNFCCC—is seen by many to be a gender-blind document. Although equity is a major issue in climate change debates, gender equality/equity has been neglected within these debates and gender has been included almost as an afterthought. The first Conference of Parties (COP) took place in 1995, and during the succession of annual COPs, a group of concerned delegates formed over a period of time into the GenderCC. It was not until COP13 in 2007, however, that the group was internationally recognized for its inputs into this global process. After lobbying by women activists under the banner "No climate justice without gender justice" in 2008, the Secretariat finally called for gender-sensitive measures within UNFCC processes. Since that time, the efforts of various individuals and groups to promote gender within the climate change negotiations have made some progress—for example, the inclusion of gender in the fourth (2014) International Panel on Climate Change report. However, gender inclusion in this context is often limited to describing the particular ways in which women are affected by the impacts of climate change, and many official climate change documents merely rehearse gender stereotypes that serve to construct women as victims.

In contrast to these oversights within the UNFCCC, another key product of the Rio Earth Summit, Agenda 21, did recognize women as important actors in environmental protection and poverty alleviation, particularly in relation to its emphasis on initiatives by local governments and community groups. However, the focus on the community, while welcome, often meant gender and other inequalities were not considered or that women were equated to the community, through their gender-specific social roles. The transformative agenda promoted by the women's lobby, which sought to change how the economy and livelihoods were constructed, was not reflected in the outcomes of the summit, which took a more instrumentalist approach to gender. This led DAWN and other groups to suggest in the 1990s that "sustainable development" was a huge contradiction.

There are a number of women's groups and movements that have been important in trying to influence UN processes around gender and the environment.

Development Alternatives with Women for a New Era (DAWN) grew from a coming together of like-minded women from across the world to analyze the (negative) impact of development on women and present alternative development strategies ahead of the UN Conference on Women in Nairobi in 1985. Their analysis, presented at the Conference, was later published as an influential book, *Development, Crises and Alternative Visions: Third World Women's Perspectives*, written by Gita Sen and Caren Grown (Monthly Review Press, 1987; it is available at this page on DAWN's website: http://dawnnet.org/publication/development-crises-and-alternative-visions-third-world-perspectives/). Over thirty years later DAWN continues to exist today as a network of gender activists. Access their website at this URL: www.dawnnet.org/feminist-resources/.

The Women's Environment and Development Organization (WEDO) began life in the early 1990s and aims to ensure that women's rights; social, economic, and environmental justice; and sustainable development principles—as well as the linkages among them—are at the heart of global and national policies, programs, and practices. To this end it acts as an important agent for global advocacy organizing women for international conferences and actions. Access their website at this URL: http://wedo.org/.

One outcome of the Earth Summit in 1992 and Agenda 21 adopted at the Summit was the recognition of the need for civil society participation in UN activities related to sustainable development. Nine sectors of society were identified; these are officially called "Major Groups" and include a Major Group for Women (WMG), for Children and Youth, and also for Indigenous Peoples. Major Groups are able to engage actively with intergovernmental processes at the UN. However, member states ultimately decide on the modalities of participation of the Major Groups and their access to the negotiations that influence international policy outcomes is then dependent on and varies across intergovernmental processes. For more information on the Major Group for Women, see this UN Sustainable Development Knowledge Platform web page: https://sustainabledevelopment.un.org/majorgroups/women.

Why might gender analysts and activists see sustainable development as a contradictory notion? One issue relates to how we know or understand the notion of sustainability—and thus relates to the construction of knowledge. The other relates to which and whose understandings, or knowledges, become the dominant discourse, and how knowledge is selectively used to inform policy dialogue.

## Exclusion of Women/Gender from Sustainability: The Construction of Knowledge

Donna Haraway (1991) has suggested that traditional models of scientific objectivity divide subject from object, knower from known, assuming a view from nowhere, while claiming to be everywhere equally. This suggestion of the possibility of objective knowledge produced from an outside-of-nature and perspective-free viewpoint is something that feminists and others have contested. In 1983, the first significant collection of articles on feminist epistemology was published, edited by Sandra Harding and Merrill B. Hintikka, suggesting gender to be a variable in conceptions of rationality leading to a critique of masculinist constructions of knowledge and the notion of "objectivity." Out of this developed the idea of "embodied objectivity" arising from an understanding that knowledge is always situated, always local and limited. Despite the importance of these critiques and the widespread influence they have had across the social sciences, the current emphasis on climate change within the environmental discourse and the focus on carrying capacities within sustainability discussions once again prioritizes scientific knowledge and technological solutions and suggests a return to an objective understanding of social reality. In contrast to the scientific approach, there have been calls to "relocalize" the negative effects of global warming, to develop an "environmentalism of everyday life," and to incorporate the concept of climate justice into a rights-based approach to climate change. Those who see environmental justice through a gender lens also make clear the links between calls for environmental justice and reproductive justice, linking ecology and social reproduction (see Bradshaw and Linneker 2014).

While this might seem a rather abstract ontological and epistemological discussion, how the world is known and how social reality is constructed have implications for policy and practice. For example, "risk," a key concept in social, economic, and environmental literatures, while often presented as an objective fact, is a subjective construction, and whoever controls the definition of risk controls the rational solution to the problem. In climatic terms the 2°C target, that is, the long-range increase in temperature over the pre-industrial era, currently expresses acceptable climatic risk and is presented as having been arrived at via scientific reasoning. However, as Joni Seager (2009) argued, it is in fact underpinned by a framework of values based on power. Moreover, such targets suggest we can identify levels of acceptable danger, and that we have the means to hold global warming to these limits. Both are doubtful assumptions. The focus on the objective and rational also produces what has been described as a "global managerialism," which presupposes that all people can adapt to climate change with the right resources and knowledge. This technical solution ignores structural inequalities such as gender and class that limit the ability for some to control resources and use knowledge to put in place adaptation strategies, while constructing these "other" people as to blame for not adapting.

While the issue of responsibility is key within debates around sustainability, it is often discussed only at the macro or national level, focused on inequalities of power between countries, not at the micro or individual level, with a focus on inequalities such as those based on gender. Gender equality, then, has been largely excluded from discussions of environmental sustainability and economic development through the construction of these as macro-level, gender-neutral concerns.

As Julie Nelson's 2008 work has highlighted, "scientized" knowledge is not limited to scientists; economists also continue to present their advice as based on science, not nature, as scientifically sound and economically rational. This ignores the fact that many economic models include normative assumptions about how the world works or what should be valued and how. For example, policies promoted by the World Bank and International Monetary Fund in the 1980s and 1990s under so-called "structural adjustment" programs saw cuts to services, such as healthcare, and increases in the costs of basic food as part of a project to promote economic growth. While bringing some gains in economic growth, these measures were only socially sustainable because the costs of the policies were borne by families, and in particular women. Roles shifted from the state to the household, from the public to the private, and as such these roles became invisible. Even when the impact of the policies became known, cost/benefit analyses favoring a macro focus on national growth meant the associated micro impact on women's time and well-being were constructed more as just part of women's social responsibilities rather than economic costs.

The construction of economic and environmental issues as scientific issues has positioned them as gender neutral and has meant women have been largely invisible within policy debates. However, they have not been absent from policy in practice. Instead women have been constructed as those with the responsibility to deliver policy goals. For example, more recently Conditional Cash Transfers have become a popular policy means to mitigate the unequal and sometimes negative outcomes of economic growth policies. These target cash at women on condition that they ensure children attend health checks and learn how to improve child health outcomes through attending workshops, for example. Participation in such activities is costly for women in terms of time spent walking and waiting. While they are premised on the idea that women do not work and have free time, the opportunity cost of participation may be high in terms of loss of earnings. Studies to date have suggested some positive impact on child health and education, but few studies show positive outcomes for women. Rather, they suggest an increase in women's "time poverty" with little change to their income poverty (see Bradshaw 2008). So what of sustainable development? Is it more than just "business as usual" in terms of gender equality?

## Gender and Environment in the Pre-2015 Development Context

While Rio launched sustainable development as a key concept, it might be argued it did not actually translate into mainstream development, which by the end of the 1990s became dominated by the Millennium Development Goals (MDGs). The MDGs widened the previous narrow economic-growth focus to include health and education, for example, yet issues of sustainability were not at the core of the goals, but instead confined to one goal.

MDG7 focuses on "Ensuring Environmental Sustainability"; it contains a number of elements and has a range of aims, including to reduce biodiversity loss. It also includes targets relating to preserving forests and fish stocks, as well as $CO_2$ emissions and consumption of ozone-depleting substances. It has a specific focus on sustainable access to safe drinking water and basic sanitation and also on the urban environment in terms of improving the lives of "slum" dwellers. Critics of MDG7 suggest even if achieved it will not necessarily have made advances in sustainability. It is said to tackle symptoms, not causes, failing to address access to assets and the institutions that govern them. Of particular concern is the fact the MDGs fail to deal with structural inequalities, which are at the root of environmental degradation—including gender inequality.

Rather than being mainstreamed throughout the MDGs, again there is a stand-alone gender goal: MDG3 aims to "Promote Gender Equality and Empower Women." Examination of the related targets highlights that equality and empowerment are measured in terms of education, seats in parliament, and women's non-agricultural employment. This focus on education, employment, and formal politics suggests more a WID than a GAD approach, and, as noted previously, the limitations of WID have been well documented. It also suggests a focus on building women's productive capacity, something in line with the World Bank's construction of women as an "untapped resource" and promoting gender equality as "smart economics," something critiqued by gender academics. The other goal specifically focused on women is MDG5 with an aim to improve maternal health. While a worthy intention, rather than ensuring all women's reproductive health rights, the focus is on women as mothers. Limiting services to maternal health does not provide access to reproductive health services, and does not actively promote women's reproductive health rights as per the BPfA.

Some of the strongest critiques of the MDGs came from the women's movements, and for some gender writers the notion of sustainability has become merely empty rhetoric. Yet as the MDGs program ended in 2015 and a set of Sustainable Development Goals were designed to replace them, there appeared to be an opportunity to place gender as core to the development agenda.

## International Processes for a Post-2015 Sustainable Development Agenda

In June 2012, the United Nations Conference for Sustainable Development was held to consider the Rio declaration twenty years on. It noted that "eradicating poverty is the greatest global challenge facing the world today and an indispensable requirement for sustainable development." The conference proposed a need to recognize the links between the economic, social, and environmental, and the need to integrate these aspects. While this is laudable, a gender vision was missing from the outcome document and the proposed "Future We Want," despite the work of gender scholars and activists.

Resources related to the Rio+20 Summit including the document "The Future We Want" are available at the summit's website: http://rio20.net/en/.

The main outcome of the Rio+20 conference was the establishment of a new process to further the pursuit of a sustainable development agenda via an "inclusive and transparent intergovernmental process" with a view to developing a set of global Sustainable Development Goals (SDGs). To this end an Open Working Group (OWG) was established in January 2013 and included seventy member states in thirty multi-constituency or "shared" seats. The process established by Rio+20 to develop a set of new goals paralleled that initiated after the UN General Assembly's High-Level Plenary Meeting on the MDGs in 2010 (the MDG Summit). The MDG Summit proposed to advance the development agenda beyond 2015, and to this end a High-Level Panel (HLP) of Eminent Persons was established in July 2012. The twenty-seven-member HLP brought together leading representatives from civil society, the private sector, academia, and local and national governments.

In its deliberations to develop a vision of, and recommendations on, a global development agenda to report to the Secretary-General, the HLP called on many and various agencies and actors, including the newly formed UN Sustainable Development Solutions Network (SDSN). Launched in 2012, the SDSN seeks to "mobilize scientific and technological knowledge" to focus on the "challenges of sustainable development." The SDSN argue that planetary boundaries should not be presented as an excuse to avoid promoting the development of all countries. They also suggest, however, that a "business as usual" approach post-2015 will result in many regions and peoples failing to improve their well-being, growing inequalities, and little "development." The SDSN is interesting as it explicitly seeks to find a balance between demands for development and concerns around sustainability, to balance respect for the planet with the promotion of rights, and in so doing brings together the scientific and social science communities, those working on environmental and those on equality issues.

Of the two processes set up to define the agenda for post-2015, the HLP was the first to produce its report including a set of twelve illustrative goals, which were then considered by the Open Working Group as it began its discussions. The OWG expanded the number of goals to seventeen and these were supported by the Secretary-General in his Synthesis Report "The Road to Dignity," published in December 2014, and as such provided the basis for intergovernmental negotiations to agree the final set of SDGs launched in September 2015. So, what is new in sustainability terms?

Information on the High-Level Panel (HLP) of Eminent Persons, including reports the panel has issued, is available at this URL: www.un.org/sg/en/management/hlp post2015.shtml.

Access the website of the UN Sustainable Development Solutions Network (SDSN) at this URL: http://unsdsn.org/.

The Secretary-General's "The Road to Dignity" report is available at this URL: www.un.org/disabilities/documents/reports/SG_Synthesis_Report_Road_to_Dignity_by_2030.pdf.

## The SDGs: What Do They Tell Us About Understandings of Sustainability

One issue highlighted in the post-2015 process is that within the sustainability discourse the urban has become more visible, with a stand-alone urban goal proposed by the Open Working Group. The HLP suggested cities may be the "driving force of global sustainable development" not least since over 50 percent of the world's population now lives in urban areas, and approximately 60 percent of the area of cities expected by 2030 is yet to be built. Yet while, for example, a large proportion of those most at risk from the effects of climate change are concentrated in urban centers, scientists, governments, and international agencies provide less support for adaptation in urban compared to rural areas. This highlights once again that the construction of knowledge may not be as objective as assumed and that only selective knowledge is used to inform policy. Who constructs climate change knowledge, and how, is also important in practical terms as it means that climate change tends to be seen as the responsibility of ministries of the environment, while planning bodies are responsible for developing urban infrastructure. Thus, even as the concept of "sustainability" expands, the continued compartmentalization of responsibilities limits changes in how it is understood in practice.

The UN's Sustainable Development Goals are listed on this UN website: https://sustainabledevelopment.un.org/?menu=1300. The icon for each goal links to a page describing the goal in more detail.

Applying a gender lens highlights how this continued compartmentalization of issues is seen across the SDGs. Within the proposed goals there is a stand-alone gender goal and also a general call for indicators across the goals to have data disaggregated by sex. The gender goal includes targets such as "end all forms of discrimination against all women and girls everywhere," which may seem a more long-term ambition than realistic aim. Review of the goals focused on climate change and environmental issues (SDGs12–15) highlights the lack of a gendered discourse. This is because the focus of these goals is on the planet, not people, and the issues are once again then constructed as scientific, gender-neutral concerns. Thus in gender terms the new sustainability goals see "business as usual": equality, environmental, and economic issues are all present, but the links between them are not drawn.

One concept that could provide the missing link is the notion of "greening," which has seen governance and economic growth integrated into the sustainability discourse. While "green economics" might allow consideration of the blind spots of traditional economics, gender analysts have suggested caution is needed. Care needs to be taken with what "green" means. While it should imply internalizing and valuing (not commodifying) the use of environmental resources, the Women's Major Group has highlighted how it could in fact result in the financialization of the earth's natural processes, rebranded as "ecosystem services." Rather than dismiss the idea completely, however, some suggest that enlightened gender policies could steer the green economy away from narrow male priorities and limited market-based perspectives toward a more sustainable pattern, and gender budgeting might be a tool to be used within this. For this to occur, some have suggested that a change of terminology is needed, with a focus on achieving a "sustainable and equitable economy" rather than a "green" economy (see Dankelman 2010). They are suggesting then that in its current construction sustainability does not necessarily imply a focus on equality.

The UN gender discourse of 2015 might suggest an advancement on that present at the turn of the millennium. For example, the gender goal does recognize women's unpaid social reproduction and care burden with a target that calls for "shared responsibility" of housework within the home. However, the inclusion of the term "as nationally appropriate" allows countries to justify not acting on this on cultural grounds. The idea of "cultural relativism" is often used to justify inaction in promoting gender equality. There are also important gender silences within this sustainable development discourse—in particular, around women's sexual and reproductive health and rights. That is, the right to decide if and when to have sex and with whom, and the right to decide the number, timing, and spacing of children. Such gendered rights reflect power imbalances at the global and intimate levels. It is low-income countries that have continued high fertility rates. It is low-income women who have little choice over their own fertility. These gendered rights then are key for ensuring women's autonomy and mobility, and necessary for them to enjoy economic gains at an individual and collective level.

In the SDGS only reproductive rights are recognized as rights, and sexual rights are reduced to access to sexual and reproductive health, such as family planning. While access to contraception should be a right for all women, implicit within a family planning focus is the desire to reduce fertility. This is related not to individual choice but overall population levels and the related economic growth gains. For example, the World Bank's "Reproductive Health Action Plan" (2010, 1) notes that improved reproductive health creates a "more productive work force" and one consequence of high fertility is high population growth, which can "constrain countries at low levels of socioeconomic development." Despite the SDGs capturing a wider range of issues related to women's reproductive rights, underlying assumptions remain that continue to draw on women as providers of wider development aims.

Sexual and reproductive health and rights have only been explicitly discussed within an environmental sustainability frame in the post-2015 discussions by the SDSN. Here there is explicit mention of fertility rates, and it is directly linked to notions of planetary boundaries. The planetary boundaries discourse is one based on science with carrying capacities constructed as objective fact by those working in the field, who still tend to be men. Yet while male knowledge is used to construct the problem, the solution is constructed as lying with women. For example (male) economists and scientists construct low economic growth as a problem, on the one hand, and high population growth, on the other. A solution is presented as investing in family planning services to meet (poor) women's unmet desire for contraception. It is highlighted how this will increase the number and productivity of the female workforce and reduce the overall population (in low-income countries) and thus impact positively on national per capita economic growth and also on the carrying capacity of the earth. This then is clearly an efficiency argument. Promoting women's reproductive rights, including the right to reproductive health services, is an important element for promoting gender equality, but here rights might be better read as responsibilities—the responsibility of individual, low-income women to limit their fertility for the good of the economy and the planet.

## Conclusion

The contemporary sustainable development discourse as reflected in the 2030 Agenda for Sustainable Development sees gendered rights present, if problematic, in the social and economic strands of the sustainability discourse. But what of the environmental discourse? Here there is silence, but it is perhaps a welcome silence.

When viewed through a gender lens, the current constructions of sustainability within the international policy context suggest a "business as usual" approach—one that sees women included in sustainable development initiatives, but in which their inclusion is not based on equality, but rather on efficiency. This does not necessarily promote or even protect their rights, while at the same time it widens their responsibilities. As noted at the start of this chapter, gender critics in the 1990s suggested that the proclaimed success of environmental projects had often been gained at the expense of women—adding unremunerated environmental protection roles to women's existing burdens, which bring positive collective benefits but few personal gains. In the contemporary context it seems women continue to be promoted as the efficient policy solution. Their inclusion in the 2030 Agenda for Sustainable Development may result not in greater equality, but greater responsibilities for women as the ones charged with ensuring that not only development, but sustainable development, is achieved. This raises the question of how best to include women within ongoing sustainable development programs in a way that ensures they are served by these programs, rather than being at the service of them.

## References and Further Research

Agarwal, Bina. 1992. "The Gender and Environment Debate: Lessons from India." *Feminist Studies* 18, no. 1: 119–158.

Alston, Margaret, and Kerri Whittenbury, eds. 2013. *Research, Action and Policy: Addressing the Gendered Impacts of Climate Change*. Amsterdam: Springer.

Baden, Sally, and Anne Marie Goetz. 1997. "Who Needs [Sex] When You Can Have [Gender]? Conflicting Discourses on Gender at Beijing." *Feminist Review* no. 56, Debating Discourses, Practising Feminisms (Summer): 3–25.

Boserup, Ester. 1970. *Woman's Role in Economic Development*. New York: St. Martin's.

Bradshaw, Sarah. 2008. "From Structural Adjustment to Social Adjustment: A Gendered Analysis of Conditional Cash Transfer Programmes in Mexico and Nicaragua." *Global Social Policy* 8, no. 1: 188–207.

———. 2013 *Gender, Development and Disasters*. Cheltenham, UK: Edward Elgar.

Bradshaw, Sarah, and Brian Linneker. 2014. "Gender and Environmental Change in the Developing World." *IIED Working Paper*. London: IIED. http://pubs.iied.org/10716.

Cela, Blerta, Irene Dankelman, and Jeffrey Stern, eds. 2013. *Powerful Synergies: Gender Equality, Economic Development and Environmental Sustainability*. New York: United Nations Development Programme (UNDP).

Chant, Sylvia. 2008. "The 'Feminisation of Poverty' and the 'Feminisation' of Anti-Poverty Programmes: Room for Revision?" *Journal of Development Studies* 44, no. 2: 165–197.

———, ed. 2010. *International Handbook on Gender and Poverty*. Cheltenham, UK: Edward Elgar.

Dankelman, Irene, ed. 2010. *Gender and Climate Change: An Introduction*. London: Earthscan.

Haraway, Donna. 1991. *Simians, Cyborgs, and Women: The Reinvention of Nature*. New York: Routledge.

Harding, Sandra, and Merrill B. Hintikka, eds. 1983. *Discovering Reality: Feminist Perspectives on Epistemology, Metaphysics, Methodology, and the Philosophy of Science*. Dordrecht: D. Reidel.

Jackson, Cecile. 1993. "Doing What Comes Naturally? Women and Environment in Development." *World Development* 21, no. 12: 1947–1963.

Leach, Melissa. 2007. "Earth Mother Myths and Other Ecofeminist Fables: How a Strategic Notion Rose and Fell." *Development and Change* 38, no. 1: 67–85.

MacGregor, Sherilyn. 2010. "A Stranger Silence Still: The Need for Feminist Social Research on Climate Change." *Sociological Review*, 57, Issue Supplement S2, Special Issue: *Sociological Review Monograph Series: Nature, Society and Environmental Crisis.* Edited by Bob Carter and Nickie Charles, 124–140.

Molyneux, Maxine. 1998. "Analysing Women's Movements." In *Feminist Visions of Development: Gender Analysis and Policy,* edited by Cecile Jackson and Ruth Pearson. London: Routledge.

Moser, Caroline. 1993. *Gender and Development Planning*. London: Routledge.

Nelson, Julie A. 2008. "Economists, Value Judgments, and Climate Change: A View from Feminist Economics." *Ecological Economics* 65, no. 3: 441–447.
Oakley, Ann. 1972. *Sex, Gender and Society.* London: Temple Smith. Reprinted with new Introduction. London: Gower.
Rocheleau, Dianne, Barbara Thomas-Slayter, and Esther Wangari, eds. 1996. *Feminist Political Ecology: Global Issues and Local Experiences*. London: Routledge.
Seager, Joni. 2009. "Death by Degrees: Taking a Feminist Hard Look at the 2°C Climate Policy." *Kvinder, Køn & Forskning* 3: 11–21.
Visvanathan, Nalini, Lynn Duggan, Laurie Nisonoff, and Nan Wiegersma, eds. 2011. *The Women, Gender and Development Reader*. Second edition. London: Zed Books.
World Bank. 2010. "The World Bank's Reproductive Health Action Plan 2010–2015." Washington, DC: World Bank.

# 15

# Sustaining What and By What Means?

## The Importance of Local Social Relations in the Sahel

Matthew D. Turner

The question of sustainability lies at the heart of debates about development and conservation practice in the Sahelian region of West Africa. The term "sustainability" necessarily incorporates normative concepts concerning proper forms of land use and desired environmental conditions. For regions of the world such as the Sahel—regions that are seen as socially and ecologically fragile—perceptions of the region's livelihoods and ecology are interwoven. In global environmental discourse, the Sahel invokes images not of pristine nature but of a highly humanized landscape with widespread land degradation. As such, it is often one of the first examples of "unsustainability" raised in global policy forums.

This chapter begins by reviewing common misunderstandings of human-and-environment relationships in the Sahel and how these in turn have shaped environmental assessment/management practices in the region. As will be described in more detail, there are four reasons that these simple narratives persist about Sahelian people: (1) the high mobility of people and land-use systems, which make it difficult for outsiders to know where and how people are using natural resources; (2) environmental change in the region does not, for the most part, show clear causal relationship between climate and human land uses; (3) scientific methods have relied on short-term, spatially circumscribed, visual observations of landscape with little input from rural people about their land uses or environmental change; and (4) a reliance on scientific models that have facilitated prescriptions for changing human livelihoods through an assessment of environmental conditions without adequately understanding the complexity of causes (climate and human land uses). In these ways, environmental scientific tradition in the Sahel has often been conducted in ignorance of actual land-use dynamics and the environmental understandings of rural people. Because scientists and resource managers fail to understand the causes of environmental change, environmental and development programs have been shaped by under-examined perceptions of Sahelian environments, peoples, and livelihoods.

Sustainability ideas and efforts have led to few positive effects on Sahelian landscapes and the people who inhabit them. This leads to a question of sustainability *of what* and *for whom*. In seeking to move beyond a history of widespread development and conservation failure, we must reject goals and concepts developed elsewhere and engage seriously with the on-the-ground realities of livelihoods and ecologies in the region. The Sahel's harsh environment has resisted easy technical fixes to the region's environmental and development problems. Solutions must be built from the knowledge and institutions of the region's people. We must look to sustain the livelihoods of Sahelian people that rely heavily on productive ecologies.

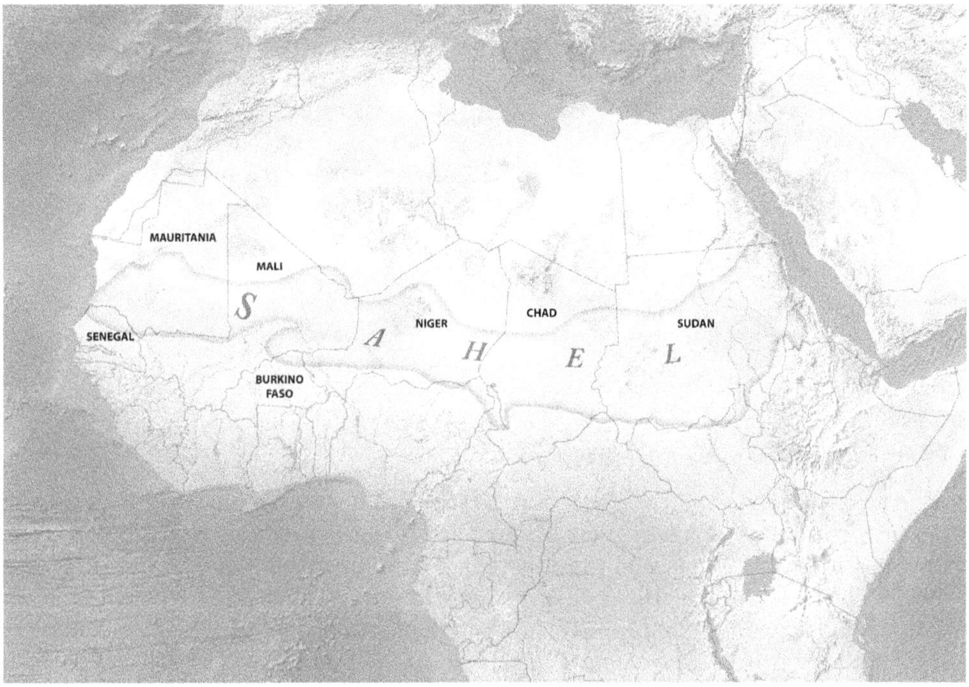


Map of the Sahel.

*Source:* Munion/Wikimedia; https://commons.wikimedia.org/wiki/File:Map_of_the_Sahel.png.

Rural livelihood strategies in the Sahel incorporate diversification, mobility, and wealth savings to cope with high levels of drought risk. As will be further developed in this chapter, these livelihood features depend upon the relationships among groups of people, facilitated by both customary and more formal institutions. In this way, questions of environmental governance lie at the center of concerns over sustainability of what and for whom. Using the Fakara region of western Niger as a case study, the chapter will explore the importance of changing social relations and their effect on the relationship between people and their environment. An understanding of these relationships reveals a range of institutional innovations that would improve this relationship—some seemingly not directed at the environment at all. For the analyst and activist, sustaining rural livelihoods necessarily moves us beyond the herder with his goat or the farmer with her millet crop to consider a wide range of factors that shape current dynamics and the prospects for future solutions.

The scientific methods used to study the Sahel described here are used in the UN Environment Program (UNEP)'s *Sahel Atlas of Changing Landscapes: Tracing Trends in Vegetation Cover and Soil Condition.* It is available for download at this UNEP Document Repository URL: http://wedocs.unep.org/handle/20.500.11822/8599.

## Sustainability Narratives of the Sahel

"Sahel" is Arabic for shore, which refers to the region in West Africa lying just south of the "sea" of desert called the Sahara. The Sahelian region refers to the semi-arid area lying between the true desert to the north and wetter savanna regions to the south.

It is a region that has produced some of the most important precolonial civilizations in West Africa, civilizations that took advantage of their location to control trade networks running parallel to one of the steepest ecological gradients in the world unassociated with drastic changes in topography. Across 1,200 km moving north of the equator, tropical rainforests grade into wet savanna, followed by dry savannas and grasslands of the Sahel before reaching the desert. The diversity of cultures and ecologies in the region provided many opportunities for learning and trade. During Europe's Middle Ages (AD 400–1400), the Sahelian region supported the precolonial empires of Ghana and Mali, along with the fabled cities of the Timbouctou and Djenné. Early European exploration and eventual conquest during the late nineteenth century were driven by nothing more than rumors of the region's wealth along with competition among European powers.

**Defining the Sahelian Region**

When climatologists and ecologists refer to the Sahelian region, they are most often referring to the strip of grass- and bush-land, lying south of the Sahara Desert, receiving on average 100–600 mm (4 to 24 inches) of rainfall annually over the long term. When social scientists refer to Sahelian countries they are referring to countries that contain this strip of land along with desert to the north and more humid areas to the south. In West Africa, these countries include: Senegal, Mali, Burkina Faso, and Niger. More broadly, the term can extend to Chad and Sudan to the east. In these ways, the Sahelian region often refers not only to the strip of land, but to the broader region encompassing Sahelian countries.

Today the region is no longer seen as a symbol of wealth and prosperity. By any measure, Sahelian countries (Senegal, Mali, Burkina Faso, Niger, Chad) are some of the very poorest countries in the world. Timbouctou is no longer seen as a center of science and learning but as a synonym for a far-away and forgotten place. The region has experienced a succession of droughts dating back to the late 1960s. The broader public, if it has even ever heard of the Sahel, would think of the famines of the 1970s and 1980s and the current crises of food insecurity and warfare.

To researchers, students, and policymakers, the Sahel has proven to be a useful symbol of something more. In academic papers, policy reports, and conferences, the Sahel is presented as the negative example of all that can go wrong in the relationship between environment and development. As such, its "ubiquitous" poverty and environmental deterioration has made it a convenient example for a succession of arguments related to sustainability. The Sahel serves as the socio-ecological endpoint of despair resulting from the "driver" of <u>un</u>sustainability being championed by particular institutions and people. Here are examples of how the region has been used as evidence for simple, inadequately examined ideas relating human society to the environment:

1. *Desertification.* The Sahel was a prominent early example of desertification during the 1930s and its droughts in the 1970s spurred international action in the form of the UN Conference on Desertification in 1977.
2. *Tragedy of the Commons.* Its common pastures served as popular examples of the environmental tragedy that was assumed to develop under conditions of common ownership during the 1960s and 1970s.

3. *Population*. High population growth rates in certain Sahelian countries were used as evidence for population-driven land degradation.
4. *Global Political Economy*. The boom and bust of its cotton economy was used as an early case illustrating the role of political economic forces in producing peasant vulnerability and resource overuse in the 1970s.
5. *(Un)sustainable Development*. The juxtaposition of high poverty and environmental decline in the Sahel served as a major stimulus for the sustainable development consensus of the 1980s and 1990s that linked poverty with environmental mismanagement and environmental decline.
6. *Conflict Induced by Resource Scarcity*. Since the 1990s, environmental security analysts commonly use the Sahel as an example of a region where social conflict is fueled by resource scarcity.
7. *Environment Driving Economic Backwardness*. Since the 1990s, the Sahel has been used as an example of how physical geography shapes the economic destiny of countries.

In short, the Sahel has been many things to many people. Certainly, any region has multiple characteristics and experiences over time. Still, the ease with which the Sahel has served as a prominent example of simplistic environmental narratives for outside analysts suggests other factors at work. Outsiders' ignorance of the region, despite years of development and conservation activity, has encouraged such generalizations. Clearly there is an issue of power—the Sahelian poor and their financially strapped governments have little ability or inclination to counter narratives of environmental and human despair. In fact, they may gain more development and conservation resources from outsiders by supporting the dire depictions.

These different sustainability narratives have had real impact on the people and landscapes of the Sahel. Since the colonial era (1890–1960), development initiatives to protect the environment and rationalize production have met with widespread failure. The forced removal of trees from cropped fields in parts of the southern Sahel during the colonial era to allow for plowing and "proper" soil management has led to reduced fertility and increased the possibility of wind erosion. Some of these trees, such as *Faidherbia albida* (scientific name), are now considered by scientists as the quintessential agroforestry tree since it drops its leaves during the rainy season when crops are growing while bringing nitrogen from the atmosphere to the soil (a legume). Improved livestock breeds and crop cultivars introduced to increase productivity have not led to a "green revolution" or the alleviation of the harsh conditions of the Sahel. Introduction of the "ranch" by fencing in pastures and sedentarizing livestock production in one place failed miserably in the 1970s. The adoption of the French model of the forest service (*Eaux et Forêts*), where funding for the service and individual agents is explicitly tied to revenue from fines, led to significant hostility between forest guards and local communities and resulted in weak protection of forests. Boreholes dug to open up new rangelands for livestock led to the settlement of people and livestock and local overgrazing. Land registration and privatization programs that were promoted to spur investment and improve land management have been strongly resisted by local people. In sum, false diagnoses of environmental problems and the importation of outside technologies, improved species of plants and animals, and management schemes have been failures.

Thus, the region has been in part reshaped by environmental protection programs that reflect the preconceptions and interests of outsiders more than Sahelian realities. This is not to say that the region has no environmental or development problems. In fact, some of the causal connections already mentioned hold some truth under certain circumstances. Moreover, there are real sustainability questions that need to be addressed. What are the environmental problems that are threatening the maintenance of a particular environmental

The tree *Faidherbia albida* is now considered by scientists as an excellent agroforestry tree due to its role in fertilizing the soil; this image shows it growing in fields at the edge of a village at the end of the dry season in Senegal.

*Source:* Matthew D. Turner, June 2013.

parameter? And what are the causes of these problems? The broad definition of sustainable development hammered out in Rio (and described in Chapters 2 and 3) provides little guidance for addressing these questions.

In some cases, tree planting has been suggested to have resulted in widespread "regreening" of the region. See these videos available on YouTube for examples. A note of caution, however, is needed in considering them. While it is important to recognize the importance of programs to support farmers' efforts to improve their environment, such broad claims should be treated with some skepticism since recent increases in tree cover within the Sahelian region have also occurred in areas without tree planting and correlates regionally with improvements in the rainfall since early 2000s. Moreover, the same techniques in tree planting that failed in the 1970s and 1980s are now successful (with improved rainfall). Just as it is dangerous to conclude that declines in vegetative cover (e.g., desertification) were due to human mismanagement during the 1970s and 1980s, we can't conclude that improvements in human management is the sole cause of increased tree cover since the early 2000s.

- World Agroforestry Centre. "Roots of Recovery: Trees, People and Regeneration of the Sahel." www.youtube.com/watch?v=YQfhKjjEQeM.

- EDA–DFAE–FDFA–Swiss Foreign Ministry. "Regreening Africa's Landscape: Trees as Natural Fertilizer." www.youtube.com/watch?v=hJuMX8LOpO0. This covers, among other topics, the value of *Faidherbia albida* in agroforestry strategies.

More complex understandings of sustainability based on grounded engagements with Sahelian realities are needed. This history is not explained solely by corruption or unthinking resistance to change by Sahelian governments and people. The Sahel's conservation-and-development history is one of failure on the part of the international community in terms of environmental diagnosis, local engagement, and imagination. Environmental mismanagement does not result from single causes, nor is the Sahelian region uniform. Solutions to environmental problems will therefore be different for different Sahelian locations. This chapter presents some of these grounded realities to show how more sustainable outcomes require a coordinated set of multiple interventions, with the fuller involvement of Sahelian people and governments. We begin with a description of the livelihood strategies of rural people in the Sahel, followed by an exploration of the Sahelian environment as it relates to these livelihood strategies.

## Rural Livelihoods in the Sahel

Sahelian people make a living in one of the harshest and most variable climates in the world. In this risky environment, people diversify their livelihood activities through cropping, livestock husbandry, and labor migration. All these activities involve significant mobility of people, livestock, and areas of cultivation. As will be discussed, mobility has both contributed to outsiders' ignorance of these livelihood systems and led to a persistent view of their primitiveness. Ignorance and misrepresentation of livelihoods have strongly shaped conservation and development in the region.

The Sahelian region has a semi-arid monsoonal climate, with rain falling during a single summer rainy season. Herbaceous natural vegetation is dominated by annual plants that are resilient to extreme dry periods. Mean rainfall and the length of the rainy season declines from south to north. Rainfall varies significantly from year to year and within a year from place to place. It is not uncommon for harvests to differ widely across the fields of a single village. The productivity of natural vegetation and crops is also constrained by soil fertility. Soils are sandy and acidic, with low levels of plant nutrients. The relative importance of soil fertility and soil moisture will fluctuate throughout a cropping season at any location, but in general, low soil fertility is more likely to limit production in areas where moisture is higher.

Reflecting the high variability of the Sahelian climate, rural people diversify their livelihoods, as mentioned, by combining livestock husbandry, crop agriculture, and labor emigration. Given the high risks involved, it makes sense not to invest all of one's time, labor, and wealth into a single pursuit. Farming is practiced only in the southern, moister portions of the Sahel (400–600 mm [16 to 24 inches] of annual rainfall) while the other two activities are practiced throughout the region. Families shift their emphasis depending on their opportunities and resources, but the majority of families in the southern Sahel, whether their ethnic/caste identities are tied to farming, herding, or artisanal pursuits, are involved in at least two of these activities at any one time. Their relative importance within rural households and communities varies seasonally and from year to year. For example, a family may devote most of its efforts toward crop agriculture during the rainy season, choosing to pay a different family to herd their livestock over the period. After harvest, the family could take the livestock back to herd themselves, with several family members leaving to look for work in the city until the next growing season.

Poor soil quality leads to farming and herding strategies that seek to redistribute nutrients and move production activities to more fertile areas. Beyond fallowing, farmers will seek to harvest nutrients from surrounding areas and concentrate them on their few permanently cultivated fields through livestock manuring. As a result of these practices, conducted over many decades, nutrients from uncultivated pasture areas within 5 kilometers (3 miles) of villages are transported to cultivated fields adjacent to villages. Nutrient gradients around older villages and livestock water points are created in this way. The process has been described as "soil mining" in the literature, but from a local perspective, farmers view this as environmental improvement. It allows for higher harvests (if there is sufficient rainfall) but it can cause declines, over the long term, in the productivity of outlying pastures. One person's degradation (soil nutrient mining) is another's improvement (soil nutrient enrichment). Herders' response to nutrient limitation is to move their livestock to drier northern pastures during the rainy season, where vegetation is more limited by soil moisture than soil nutrients and therefore is generally of higher nutrient content. The seasonal movement of the region's livestock from the southern Sahel (where cultivation is practiced) to the northern Sahel during the rainy season (and back at end of rainy season) is referred to as "transhumance." Long-distance seasonal movements are practiced in other parts of the world as well, though the factors and ways of organizing vary from place to place.

Due in large part to highly variable rainfall, the tripartite livelihood strategies (farming, herding, labor emigration) in the Sahel involve a significant level of mobility compared to most production systems in Europe and North America. In other words, they involve the seasonal movements of people and their productive activities across broad regions. Take labor migration as an example. Rural people whose own crops have failed will move to areas that enjoyed good harvests to work as laborers for others. Once the harvesting work is done, some may move to a city seeking employment in petty trade or as laborers. Likewise, herders will move livestock to areas with sufficient fodder. These movements display both a regular seasonal dimension (e.g., "transhumance") and less predictable movements that respond to the local availability of fodder. Farming shows less mobility than labor migration and livestock husbandry, but is still mobile. Farmers prefer, unless limited in terms of their access to land, to shift the locations of their fields and leave abandoned fields in fallow to restore nutrient levels. In many cases, farmers will seed more fields than they expect to bring to harvest in the hopes that a portion of these will receive sufficient rainfall.

Early European observers saw the mobility of livestock husbandry and farming in the region as a sign of primitiveness. This, along with multi-cropping and the retention of trees in fields, was seen as a symptom of an "undisciplined" African mind. Mobility was viewed as working against the development of private landownership and the security of tenure necessary for investment. With respect to the environment, the "rootlessness" of these activities was seen as preventing ties to particular parcels of land, and thus as working against the development of sustainable management practices. Moreover, colonial governments saw all forms of mobility as limiting their ability to collect taxes and to provide governmental services. In the following example, a French colonial administrator reflects on the difficulty of creating a census of herds within his district for taxation purposes:

> For more than six years I have attempted by all means to census the herds of the nomads whose administration I have been assigned. I have tried to counter their tricks, going even as far as engaging in night marches of 40 to 50 kilometers (25 to 31 miles) to

> surprise camps that I had visited two or three days before. I must confess that it is impossible for me to state that I have succeeded in my task. This is despite a force of more than 100 native infantry on camel at my disposal.
>
> (Mangeot 1930, 312, author's translation)

Outside perceptions like this persist in shaping development and conservation programs today. This is despite research that supports the ecological and economic rationality of their adaptations to a highly variable climate. Just as important, many tend to assume that all Sahelian landscapes are heavily shaped by human land uses despite the fact that population densities are low—particularly in the north. Therefore, unlike other regions of the world where pristine nature is invoked, damaged and humanized landscapes are invoked in the Sahel. As discussed elsewhere in this book, the reality is more nuanced.

The outside criticism presented so far in this chapter has primarily been put forth by scientists, government officials, and development experts: actors whose experience with the day-to-day realities with rural life is often limited. So, how do rural people in the Sahel view their environment and their relationship to it? Their views are diverse, but one can identify several commonalities. First, they view their livelihoods as very much shaped by the nature of rainfall during the rainy season. Year-to-year variability of rainfall is an important feature of their life that is beyond their control. Outsiders have tended to portray such views as fatalistic—suggestive of a passivity and lack of entrepreneurial spirit. Improved understanding of Sahelian realities shows that this view is best portrayed as realistic. Second, people plan for the lack of rainfall with a focus on storing surplus from good years to bad. Therefore, people are often as interested in development projects that help to increase production during good years as in drought-proofing their activities during unfavorable years. Third, with respect to the environmental impact of human land uses, rural people generally are more concerned with soil nutrient depletion and loss of trees and shrub cover. Changing variation in the cover of annual grasses and forbs are seen as more a function of variable rainfall and less the result of human land uses. In this way, while local people may point to heavy use as leading to temporary shortages of certain resources, they are more inclined to point to cropping and wood collection as the primary land use affecting the productive potential of landscapes on which they depend.

## Sahelian Environmental Dynamics

As described above, poor soils and one of the harshest climates in the world make the Sahel a difficult place to support human and nonhuman life. By the end of a typical dry season, Sahelian landscapes are often devoid of vegetation even when protected from human land use. Moreover, outsiders—from colonial or African government administrators, to environmental scientists on short-term consultancies and workers for environmental or development NGOs—remain ignorant of spatial patterns of human land use. Finally, signatures of the environmental effects of human land use are not clear. Unlike the deforestation of tropical rainforests or the degradation resulting from mining, the effects of grazing or farming in highly variable environments such as the Sahel are less obvious since vegetation varies significantly from year to year due to climatic variability. In short, connecting human land use to environmental health in the Sahel is often very difficult.

Relating human resource use to the environment is a basic requirement for environmental assessment and resource management. As described previously, outsiders' persistent ignorance of human livelihood patterns has proven to be a barrier for improvements in resource management. In this section, we turn to outsiders' understandings of environmental dynamics and how these are influenced by human activities. Just as high livelihood mobility prevents outside understandings of resource use in the region, high variability of rainfall has proven to be a barrier to understanding environmental dynamics. Fluctuating rainfall over space and time complicates our understanding of factors behind changes in vegetation. Still, just as with livelihoods, where ignorance has persisted due to misrepresentations of the systems as primitive (and therefore something to be replaced), ignorance about the Sahelian environment has persisted due to an overreliance on imported approaches to environmental assessment. In fact, our ignorance has arguably been produced (or at least sustained) through the application of these approaches. Over the past fifty years, international attention to the Sahelian environment has been episodic, with long periods of indifference punctuated by bouts of crisis-driven attention. Longer-term research programs that would develop new assessment models and management ideas have been rare due to this history and due to the poverty of the region's governments.

Tools of range management, agronomy, wildlife ecology, and forestry were generally developed in the temperate north and applied by outside experts and Africans trained in these techniques at European and North American universities. These applied fields depend on conceptual models relying on steady-state or equilibrium assumptions in ecology (see Botkin 1990) which draw boundaries around the resource-extracting activity (grazing, hunting, farming, fuelwood harvesting) and the resource (pasture, wildlife population, field, and forest) with outside influences (e.g., climate) that are considered constant. The utility of such approaches for short-term fieldwork is that no matter what the cause of the variation in the environmental parameter, the management prescription is to change the level of resource extraction. For places such as the Sahel, where climatic fluctuations are high, prescriptions to adjust resource extraction pressures during dry periods are (incorrectly) seen by policymakers as evidence of over use.

Let's take as an example the effects of human resource extraction (primarily grazing) on the natural vegetation in the Sahel. Much has been written about this, particularly as relates to the concept of "desertification"—a vague term that refers to the changing of a vegetated productive landscape to an unproductive desert-like landscape. The literature on desertification makes recurrent references to a "marching desert"—expanding southward due to human land uses. The evidence used to support widespread desertification in the Sahel is derived from the following features of environmental assessment:

1. A high reliance on one-time, visual descriptions of the landscape to evaluate environmental conditions. Very little experimental or quasi-experimental work (e.g., enclosures, grazing trials, transect analyses, fallow succession) was performed.
2. A heavy reliance on vegetative cover as a measure of environmental change and early notions of environmental degradation. Discontinuous cover of vegetation was incorrectly assumed to be a sign of environmental degradation.
3. Environmental observations biased toward locations near roads within a day's travel radius around major towns.
4. A proclivity to generalize from field observations at one locality and time (season and year) to the broader region and longer time frames.
5. Ecological observations made with little understanding or reference to how the observed patch relates spatially to rural production systems.
6. The reliance on spatial correlations between patterns of vegetative parameters and signs of human land use (farming or the presence of livestock) as a means for inferring human-caused environmental change. This produces a strong tendency for circular argument—human signs and signs of environmental change were often regarded as one and the same.

In sum, the environmental scientific tradition in the Sahel has relied on short-term, spatially circumscribed visual observation to detect environmental change. The seemingly straightforward approaches to environmental assessment produced by equilibrium assumptions, along with limited funding and interest, contributed to a dearth of research findings that could address important sustainability questions. More specifically, there has been little work distinguishing the effects of rainfall variation and human land use on vegetative productivity, as well as on more enduring environmental change beyond the admittedly ephemeral changes to vegetative cover. Today, definitions of desertification generally refer to land degradation that is tied to the longer-term productive potential of land. These definitions point us toward the chemical and physical properties of soil, but unfortunately, there is a dearth of information tying human activities to changes in soils. So, despite widespread recognition of the limitations of prior understandings of "desertification" and the adoption of new definitions, the legacy of old methods limits the information we have about the relative importance of climate versus human land uses in driving more persistent change.

**Desertification**

In English-speaking West Africa, the idea of desertification as a concept can be traced to E. P. Stebbing, a colonial forester working in India and then Nigeria. Stebbing's writings, beginning after the Dust Bowl in the United States, highlighted the specter of productive land becoming barren (see Stebbing 1935). Desertification became officially an international crisis with the International Conference on Desertification in 1977 following the prolonged 1968–1973 Sahelian drought and further with the UN Convention to Combat Desertification in 1994 (see Middleton and Thomas 1997). Current definitions have sought to focus on more persistent changes to the productive potential to the land caused by human land uses. As such they have shifted attention toward human-caused changes in the fertility and structure of dryland soils.

See the Conventions web page: www.unccd.int/en/Pages/default.aspx.

## Environmental Assessment and (In)justice

Given the history of environmental assessment in the Sahel, much of what had been described as land degradation caused by the activities of rural people was actually less persistent and associated with climate variation. For example, an analysis of satellite imagery has shown that the edge of the Sahara Desert has not expanded southward over time but oscillates along its length due to year-to-year variations in rainfall (Tucker et al. 1991). Still, prior assessments and ongoing negative views of the rationality of African production systems has led to attempts, in the name of conservation, to change the livelihood practices of rural people. In this section, we consider how misunderstandings of Sahelian livelihoods and overly simplistic views of Sahelian ecology converge to shape conservation and so-called sustainable development programs that, in the extreme, are attacks on rural people's ways of life.

This was most obvious in the case of livestock-rearing specialists (pastoralists) portrayed as environmental degraders who build up massive herds and overstock Sahelian rangelands. They were seen to over-accumulate livestock to the detriment of the environment due to a combination of their economically and ecologically irrational veneration for livestock, and their ability to graze livestock on common pastures, thus avoiding the full costs of overstocking (therefore aligning with the tragedy of the commons argument). Added to perceptions of widespread land degradation, these views led to efforts to encourage destocking and reduce livestock mobility so that the costs of overgrazing would be directly felt by the herder. Unlike in East Africa, these efforts were generally more indirect, and did not involve government coercion to sedentarize pastoralists. Best seen as "malign neglect," government and NGOs have chosen not to protect livestock movement pathways from expanding crop agriculture or to help pastoralists restock after drought. In these ways, the misreading of pastoralist livestock production as economically irrational and inherently destructive to the environment has resulted in added hardship to Sahelian pastoralists.

**African Livestock Husbandry and Overstocking**

Outsider assessments pointing to the economic irrationality of African livestock husbandry has a long history, starting with the "cattle complex" analysis by Herskovits (1926) of East African pastoralism as accumulating cattle for cultural prestige rather than for subsistence or market gain (see C. Kervan's 1992 *Customary Commerce* as a counter to this discussion). Later assessments pointed to the low rates of offtake (selling or butchering animals) from African herds. Such assessments assumed that the goal was the production of meat when in fact African herders are focused on milk production (dairy production actually requires larger herds). In addition, these assessments did not take into account the size of herds that need to recover after drought (see Scoones 1994 for more on this). More importantly, herds were assumed to be owned by the herder which is not necessarily the case; many herds comprise animals with multiple owners (see Turner 1993).

Rural families in the Sahel suffered through recurrent drought during the 1968–2008 period. Since that time, rainfall variability within and between years has increased with a return of good rainfall years interspersed among bad years. The few studies in the region point to the remarkable resilience of Sahelian vegetation with fairly rapid post-drought recoveries. Still, conventional assessments of desertification have continued to equate low vegetation during drier periods to human-induced land degradation. Unbeknownst to rural Sahelians, they are often blamed for the barren landscapes on which they are struggling to make a living. In other words, for much of what has been described as land degradation in the Sahel, the victims of drought have been blamed for their own suffering.

These social justice concerns are only heightened when one considers that the climate shifts experienced in the Sahel may actually be tied to global climate change. The period of drought in the Sahel from 1968 to 1998 represents one of the most sustained dry periods for a region in the global record. This has led to heightened speculation about the relationship of droughts in the Sahel to global climate change. Climate change models show no consensus in predictions for annual rainfall in the region, but do agree that within- and between-year variability will increase. Rainfall during the rainy season will be more likely to occur in large-volume events punctuated by dry periods. This prediction is consistent with what has been observed over the past ten years: very high between-year variation and greater prevalence of local flooding, and "mini-droughts" during the rainy season. Predicting regional shifts in rainfall associated with climate change is notoriously difficult. Still, the empirical record and current predictions raise important environmental justice concerns. Is the 1968–1998 decline, and subsequent increased variability in ecological conditions and agricultural productivity in the Sahel, initially blamed on African mismanagement, linked to the "luxury" greenhouse gas emissions of the industrialized North? Have high living standards in Europe and North America associated with fossil fuel use actually produced, through a changing climate, the economic and ecological impoverishment of the Sahel? And if so, what do these conclusions mean for generating sustainable development for the rural people of the Sahel?

## Human Uses and Environmental Change in the Sahel

In outlining the history of environmental assessment in the Sahel, we have seen how ignorance on the part of outsiders about the Sahelian environment and about the livelihoods of the region's inhabitants has been sustained through the resource management and assessment tools brought to bear on the region's complexities. But what can we say about the relationship between human land use and the environment? We can in fact identify some general points to consider in thinking about improved resource management in the region. First, farming, livestock rearing, fishing, hunting, and foraging activities of rural people do have measurable effects on the environment. Second, significant human-induced environmental degradation is not widespread, but patchily distributed around population centers in the region. Third, there is a remarkable ability for both herbaceous (grasses and forbs) and lignaceous (trees and shrubs) vegetation to recover after overuse and/or drought that points to the importance of soils in land degradation (that is, more persistent declines in land productivity). These are only general points. We should consider the ecology of human land use in more detail to gain insights on how to move forward.

Grazing and farming are the two activities that affect vegetation and soils in the region. The impact of grazing on them is dramatically higher in situations of sustained high grazing pressure during the rainy season. Under open-range conditions, *regional* livestock populations never, except under conditions of significant drought, exceed the capacity of *regional* rangelands to provide forage during the rainy season. Instead, the distribution of livestock with respect to the shifting mosaic of forage availability shapes the sustainability of livestock husbandry. With all else equal, livestock systems that allow more freedom of movement (to redistribute in relation to vegetation) have less impact on soils and vegetation than those that are fixed in one place (see Niamir-Fuller 1999).

### Rangeland Carrying Capacities

The use of carrying capacity of rangeland management, that is, the number of livestock that a rangeland can support, assumes that vegetation is shaped by the grazing of livestock. Deviation from the expected state of the rangeland is seen to require an adjustment in stocking rate. The ecological realities in the Sahel deviate from the inherent assumptions of the carrying capacity model. Given the high variation of rainfall and forage production, it is difficult to define a normal state in dryland areas of Africa such as the Sahel. Except under high grazing pressures, rangeland condition is shaped more by rainfall fluctuations than grazing pressure (see Behnke et al. 1993). At the same time, quality forage is most limiting to livestock production during the dry season when vegetation is actually less vulnerable to grazing (see Turner 1993). This less-than-intuitive situation deviates from standard carrying capacity assumptions.

Farming, however, cannot follow the rain. Therefore, a critical feature of farming in the region is to achieve sufficient harvests during good years to compensate for the deficits that will come during bad years. These surpluses are stored in granaries or in the form of livestock. The environmental effects of the farming of millet, the major grain crop, are variable. Despite the history of agroforestry in some areas of the Sahel, farming generally involves the clearing of trees, and may also involve plowing. Such forms of farming are more environmentally destructive than grazing. In rural areas experiencing rapid population increases, cropland expansion can be quite rapid. In other areas, it occurs because of declining soil fertility and the need to spread out cropped fields to capture spatially heterogeneous rainfall. Even with good rainfall, yields are low. Increased yields result mostly from the application of fertilizers—inorganic fertilizer (which requires a cash source) and/or livestock manure. Given the very limited international demand for millet, programs to increase the productivity of millet fields, largely through improving soil fertility, could (compared to other crops) increase the food security of Sahelian families.

Farming and livestock rearing are strongly interconnected in other ways too. Except for the very rich, most herding families farm. Farmers rely on manure from livestock animals to improve the productivity of fields. They have historically gained access to manure by arrangements with herders that allow them to camp on the farmers' fields. Most farmers likewise choose to store their limited wealth in the form of livestock and utilize livestock for plowing, hauling carts, and other forms of traction. In these ways, livestock rearing and farming in this "agropastoral" region is mutually beneficial. Still, the interests of farming and herding are not always aligned, particularly during the cropping season, when poorly managed livestock may cause crop damage. In addition, the extension of a cropped area may block the movements of livestock, which in turn can lead to crop damage. Thus, the deterioration of farmer and herder relationships can interfere with the mutual benefits of farming and herding in these areas.

For more insight into the practice of transhumance in the Sahel, especially West Africa, see the short video "West Africa: Support Transhumance" published on YouTube by Acting for Life, located at this URL: www.youtube.com/watch?v=yQIMuCNTmYM.

A consideration of the ecology of Sahelian livelihoods reveals some basic features of a more sustainable future where: (1) farmers have access to manure and chemical fertilizer, which will increase food production and limit cropland expansion; (2) the paths, water points, and key rangelands needed to allow livestock herds to move both regionally and locally are protected; and (3) there is greater access to wealth stores (including livestock) for rural inhabitants and a means to properly manage (that is, protect and grow) this wealth. Note that these livelihood strategies require significant coordination, cooperation, and mediation among people. Improvements must build from existing traditional and formal institutions, which vary greatly across the region.

## The Fakara Region of Western Niger

Although we can point to general trends and tendencies in the region, the diversity of the Sahelian region—culturally, economically, and ecologically—precludes a common approach to assessing human–environment dynamics. To provide an on-the-ground view of the constraints on and opportunities for achieving sustainable outcomes for the rural population, we focus in this section on a particular area in western Niger: the Fakara. As mentioned above, improvements in resource management affect the relationship between farming and herding. As this case will show, prospects for improved coordination of different livelihood practices for their mutual benefit depends on the nature of the social relations among groups. In the Fakara, the two social groups of particular importance are Djerma farmers and Fulbe pastoralists. The social relations between these two groups, as mediated by such institutions as land tenure, manuring contracts, and herding contracts, shape the prospects for improved agropastoral coordination. As described in the following section, the sustained fertility of Djerma fields depends on relations of trust and traditional manuring contracts with Fulbe herders. Fulbe herders' ability to properly manage their livestock sustainably requires mobility for their herds, which in turn relies on agreements with the Djerma to delineate and protect livestock corridors. Livestock is the major store of wealth for Djerma and Fulbe alike, and the ability of households to effectively manage their wealth depends not only on the maintenance of livestock corridors but also on the herding labor contracts between them.

The Fakara lies in the southern portion of the bioclimatic Sahel with a long-term mean annual rainfall of 575 mm (23 inches), which falls during a single rainy season from May to October. Overall, soil fertility is very poor even by regional standards, and water tables lie several tens of meters below ground level. Like virtually all areas of the southern Sahel, this is a region where farming, livestock rearing, and labor emigration is practiced by most households. Historically, its resident population was composed largely of Djerma millet farmers with large transhumance herds of Fulbe pastoralists moving through the area at the beginning of the rains and after the harvest during the dry season.

Recurrent drought since the early 1970s encouraged Fulbe from the north and east to settle and farm plots of land loaned to them by Djerma landowners in the Fakara. As a result of these changes, the agrarian landscape consists of Djerma villages with Fulbe homesteads located outside of the villages on their loaned fields, often adjacent to livestock paths. Continued population growth and migration to the area (of Fulbe and others) have resulted in a large increase in cultivated area and growing cropland shortage, especially near the heavily populated fossil valley called the Dallol to the east.

The distribution of wealth and political power among the Djerma is related to landownership, with village chieftaincies that benefit from loaning land to recent immigrants (generally compensated by crop shares). Other Djerma supplement their insufficient crop harvests (which are due to declining soil fertility and recurrent drought) with labor emigration to cities to the south. Emigration is largely seasonal (dry season) but increasingly, young men will remain away throughout the year. Money sent back by migrants is often invested in livestock—a store of wealth that is sold when the family is in need of cash. These livestock are entrusted with local Fulbe during the cropping season.

The Fulbe all farm, and despite lower farming expertise compared to the Djerma, they enjoy good yields due to their greater access to manure. Yet they lack land-tenure security; Djerma landowners occasionally take back fields to enjoy the residual benefits of manuring. There is a wide variation in wealth distribution among the Fulbe, with a small number maintaining large self-owned herds and the rest with herds mainly composed of entrusted livestock for which they receive seasonal payments (per head) for their herding services. Reflecting their relatively recent settlement in the area from diverse places, the Fulbe lack political organization and hold only limited political power to represent their interests.

International donors have pushed political decentralization in Niger, as they have done in other countries in the region. As a result, there has been an uneven shift of effective resource management authority from the central government toward local villages and elected local administrations over the past decade. At the same time, relations between farmers and herders have deteriorated due to increased problems of livestock-induced crop damage and struggles over access to farmland. The increased level of distrust between the two groups has greatly complicated local governance and reduced the effectiveness of resource management.

Increased crop damage is caused by the growing land-use competition between pastures and farmland in the Fakara. Cropped areas have been extended, partly because of a reduction in the fertility of Djerma-owned fields, but most directly because of a growing practice of loaning village land to outsiders by local chieftaincies. In some cases, village lands that were previously left in long-term fallow to serve as important transit points or pastures for local livestock have been loaned to outsiders. The increased presence of livestock in the area during the cropping season is due in part to the partial recovery of Fulɓe livestock wealth after major droughts during the mid-1980s. More important, however, is the reduction of livestock movements out of the Fakara to the north (transhumance) during the cropping season. This reduced mobility has had several causes, including: (1) a shortage of herding labor for smaller Fulɓe-managed herds; this shortage resulting from a shift in labor away from livestock husbandry and a reduction in labor-sharing arrangements among the Fulɓe groups because of conflict as they sided against each other in disputes; (2) Fulɓe-managed herds with Djerma-owned livestock as more Djerma livestock owners require their herders to remain in the area because of concerns that the Fulɓe would secretly sell their livestock while on transhumance; and (3) barriers for all Fulɓe herds along transhumance corridors due to agricultural encroachment as decentralization has progressed. Greater authority given to village and local districts has led to the fragmenting of important regional transhumance corridors since each village or local district does not take into account the importance of its corridor segment for maintaining the continuity of the whole corridor.

Increased distrust and conflict between Fulɓe and Djerma have contributed to the erosion of local manuring arrangements between the two groups. As a result, Fulɓe fields are over-manured while more Djerma fields are not manured. Djerma and Fulɓe have become more prone to revisit and question the rights and responsibilities associated with their social contracts for land and herding services. In most cases, the downward spiral in herder–farmer relations is checked by everyday friendship bonds that have formed over time across the two groups. Still, the relationships have become strained, characterized by growing distrust, an increase in tenure insecurity, and a reduction in livestock mobility.

The likelihood of achieving desirable outcomes in the Fakara rests most realistically on the improved management of the region's meager resources. Such improvements involve institutions (e.g., rules, understandings, social norms, governance bodies) at community and district government levels to mediate the relationships between social groups. These approaches would include:

- The political organization and leadership within the Fulɓe community at the village and local district level need to be developed, so that their interests are represented and agreements made by their leadership can be followed by others. This could also serve to encourage labor sharing among Fulɓe families for herding.
- To avoid the further erosion of relationships between farmers and herders, processes for conflict management need to be improved. These should include transparent methods for determining compensation for crop damage.
- The security of the herding contract should be better ensured so that herders are fairly compensated for their services and owners can trust the herders when their livestock leave the village territory on transhumance. Confidence can be increased with improved farmer–herder relations. The introduction of ownership markings on livestock could also contribute to security.
- The manure contract should be reinvigorated so that the manure produced by resident herds and transhumance herds passing through the area can be more effectively utilized on Djerma fields. This would occur through better compensation of herders for

the manure produced by their herds and a stronger, more positive role played by local Fulɓe as mediators between Djerma and outsider herders.

- The regional government needs to play a larger role in ensuring that interdistrict transhumance corridors are passable, with sufficient pasture and water available for herds moving out of, into, and through the Fakara area. In the current political environment of decentralization, this will require that the national and regional governments take strong leadership roles in protecting these important public goods.

The prospects for a more sustainable future in the Fakara largely rest on governance, politics, and institutions. Since solutions to these governance problems must be comprehensible to and supported by local communities, a grounded, participatory approach to reform is necessary. Although outcomes are unpredictable and require significant local leadership, the initiatives just outlined have all been raised by people in the area. They may be unlikely to lead to a fundamental transformation of the Fakara, yet they show promise as a means to improve farmer–herder relations, increase livestock mobility, lower the chaotic spread of cropping, and increase the productivity of cropped fields. Taken together, the changes could produce measurable and sustained improvements in environmental and social health. Moreover, they would improve the prospects for the successful adoption of appropriate technologies or new forms of technical or financial assistance if they were to become available.

## Sustainability in the Sahel: A Question of Governance

The Fakara case makes clear that improved social relations and strengthened institutions are key to sustainability. We see that to sustain ecologies, we must sustain social relations. Rather than a single cause of mismanagement, we see a complex web of factors that, through their interaction, create environmental stress. Addressing only one of these will not lead to a lasting solution. Introducing a new technology will not transform the ecological system unless governance issues are addressed. In short, there are no easy fixes, but difficult work, involving political leadership, negotiation, and compromise ahead.

These governance issues cannot simply be seen as failures of African institutions. It is true that local capabilities and leadership were eroded in many areas of the Sahel during the colonial and independence periods. Still, the mobility required for more sustainable outcomes in the Sahel call for sophisticated forms of governance—with responsibilities shared across different scales of governance and associated institutions. In the academic literature, this form of governance is called "co-management" and there are few examples of its success in the drylands of the industrial world. The need for innovation in governance is thus a need for a completely new and as yet untried governance form.

This work is not what outside experts and the donor community are best positioned or inclined to do. Not only has the international community brought inappropriate ecological concepts and technologies to the Sahel, they have, in their rush to address the crisis, failed to engage with social and political realities. Moreover, leadership cannot be introduced, but must develop (while supported) from within. The realities vary from one location to the next. Management issues and institutional requirements will differ in the northern areas, where rain-fed agriculture is not practical. Appropriate interventions will also shift depending on the cultural context. Political leadership will vary from one community to the other. While the basic template for the agropastoral Sahel or the pastoral Sahel may be similar, the specifics of how to produce the desired sustainable outcomes will necessarily vary from one location to the next.

## References and Further Research

Bassett, Thomas J., and Koli Bi Zuéli. 2000. "Environmental Discourses and the Ivorian Savanna." *Annals of the Association of American Geographers* 90: 67–95.
Behnke, Roy H., Ian Scoones, and Carol Kerven, eds. 1993. *Range Ecology at Disequilibrium*. London: Overseas Development Institute.
Botkin, Daniel B. 1990. *Discordant Harmonies: A New Ecology of the Twenty-First Century*. New York: Oxford University Press.
Herskovits, Melville J. 1926. "The Cattle Complex in East Africa." *American Anthropologist* 28: 633–644.
Hiernaux, Pierre, and M. D. Turner. 2002. "The Influence of Farmer and Pastoralist Management Practices on Desertification Processes in the Sahel." In *Global Desertification. Do Humans Cause Deserts?*. Edited by James F. Reynolds and M. Stafford Smith, 135–148. Berlin: Dahlem University Press.
Kervan, Carol. 1992. *Customary Commerce: A Historical Reassessment of Pastoral Livestock Marketing in Africa*. London: Overseas Development Institute.
Mangeot, P. 1930. "Le Problème de la Laine en Afrique Occidentale Française." *Comité de l'Afrique Française. Renseignements Coloniaux et Documents*: 311–315.
Middleton, Nick, and David Thomas, eds. 1997. *World Atlas of Desertification*. New York: UNEP.
Mortimore, Michael. 1989. *Adapting to Drought. Farmers, Famines and Desertification in West Africa*. Cambridge: Cambridge University Press.
Niamir-Fuller, Maryam., ed. 1999. *Managing Mobility in African Rangelands*. London: Intermediate Technology Publications.
Scoones, Ian., ed. 1994. *Living with Uncertainty. New Directions in Pastoral Development in Africa*. London: Intermediate Technology Publications.
Stebbing, E. P. 1935. "The Encroaching Sahara: The Threat to the West Africa Colonies." *Geographical Journal* 85: 506–524.
Swift, Jeremy. 1996. "Desertification: Narratives, Winners, and Losers." In *The Lie of the Land*. Edited by Melissa Leach and Robin Mearns, 73–90. London: International African Institute.
Tucker, Compton J., Harold E. Dregne, and Wilbur W. Newcomb. 1991. "Expansion and Contraction of the Sahara Desert." *Science* 253: 299–301.
Turner, Matthew D. 1993. "Overstocking the Range: A Critical Analysis of the Environmental Science of Sahelian Pastoralism." *Economic Geography* 69: 402–421.
———. 2016. "Rethinking Land Endowment and Inequality in Rural Africa: The Importance of Soil Fertility." *World Development*.doi:https://doi.org/10.1016/j.worlddev2016.06.014.
UNEP. 2012. *Sahel Atlas of Changing Landscapes: Tracing Trends in Vegetation Cover and Soil Condition*. Nairobi, Kenya: United Nations Environmental Program.

# About the Editors and Contributors

## Editor

**Andrea J. Nightingale** is a Professor of Human Geography at the University of Oslo, Norway. She was Chair of Rural Development in the Global South at the Swedish University of Agricultural Sciences and Professor II at the Department of International Environment and Development Studies (Noragric), Norwegian University of Life Sciences. Her current research interests include political violence in climate change adaptation programs; climate change adaptation and transformation debates; public authority, collective action, and state formation; and feminist work on emotion and subjectivity in relation to theories of development, collective action, and environmental change. Her recent papers include "Power and Politics in Climate Change Adaptation Efforts: Struggles over Authority and Recognition in the Context of Political Instability" (2017) and "The Socioenvironmental State: Political Authority, Subjects, and Transformative Socionatural Change in an Uncertain World" (2018).

## Associate Editors

**Tom Böhler**, PhD, teaches human ecology at the School of Global Studies at the University of Gothenburg in Sweden. His teaching focuses on what today's mainstream processes of sustainable development in rich consumer societies like Sweden can learn from perspectives such as critical post-humanism (sustainability and trans-species egalitarianism), political ecology (natural resources, power and [in]justice), ecological economy (sustainability and the endurance of natural resources), economic democracy (sustainability and collective ownership), and political ontology (sustainability and pluralism of ontologies).

**Ben Campbell** is Senior Lecturer in the Department of Anthropology, Durham University, UK, and directs the MSc in Sustainability, Culture and Development. He is co-chair of the Low Carbon Energy for Development Network. His research focuses on poverty, livelihoods, and resources in Nepal; on innovation among mountain indigenous communities; on the impacts of environmental conservation on local institutions of common property management; and on energy transitions from fuelwood use to biogas at high altitudes. His publications include *Living between Juniper and Palm: Nature, Culture and Power in the Himalayas* (2013), and "Communities of Energy" (2016).

**Linus Karlsson** is a PhD student in the Department of Urban and Rural Development at the Swedish University of Agricultural Sciences. His research interests include critical agrarian studies, the politics of land use, state formation, and state–citizen relations. In particular, he focuses on how regime shifts in land-use governance generate new conflicts over public authority and rights in resources. In his research, he moves between the fields of geography, anthropology, sociology, and postcolonial theory.

## Contributors

**Sarah Bradshaw** is Professor of Gender and Sustainable Development at Middlesex University, London. She is a feminist scholar-practitioner and has authored numerous articles in the field of gender and development, and engendering disaster response and risk reduction. She has worked with women's groups and movements in Nicaragua for the last twenty years. More recently she has worked with the UN Sustainable Development Solutions Network seeking to engender the Agenda 2030 process, writing the background report on gender for the High Level Panel for the Sustainable Development Goals (SDGs).

**Nancy Langston** is Distinguished Professor of Environmental History at Michigan Technological University who explores the connections between toxics, environmental health, and industrial changes in Lake Superior and other boreal watersheds. She is the author of four books, including the recent *Sustaining Lake Superior: An Extraordinary Lake in a Changing World*. Langston spent seventeen years at the University of Wisconsin-Madison in the Nelson Institute of Environmental Studies and the Department of Forest and Wildlife Ecology. During 2012–2013, Nancy was the King Carl XVI Gustaf Professor of Environmental Science at Umeå University.

**Angus Lyall** is a PhD candidate in Geography at the University of North Carolina at Chapel Hill. He is an economic and cultural geographer who examines the institutions and cultural politics of territorial governance in Ecuador. His research centers on indigenous engagements with oil-driven development and urbanization in the Amazon. He also has ongoing projects on the politics of work in enclaves of cut-flower production and transformations in rural politics related to tourism development in the Andes.

**Phil Macnaghten** is a Professor in the Knowledge, Technology and Innovation (KTI) group at Wageningen University. His PhD is from Exeter and he has held appointments at Lancaster, Durham, and Campinas before joining Wageningen in 2015. His research background is in science and technology studies (STS) and sociology. He has developed in-depth qualitative methodologies for researching controversial technologies which, in turn, have informed policy approaches to dialogue and public engagement. His current research focuses on responsible innovation, gene editing, and gene drives.

**Dinesh Paudel**, a geographer by training, is Associate Professor in the Department of Sustainable Development at Appalachian State University. His research explores the environmental, economic, and political transformations in Nepal and the Himalayas. His recent research programs focused on exploring the pre-history of the Maoist revolution in Nepal. Currently, he is involved in exploring the evolving political and economic transformations in post-earthquake Nepal and the shifting geopolitical dynamics involved in large-scale infrastructure, particularly the Belt and Road initiatives of China, in Nepal, and the Himalayas.

**Pamela Richardson-Ngwenya** is a geographer, currently based at the German Institute for Tropical and Subtropical Agriculture (DITSL) at the University of Kassel, Germany, where she contributes expertise on participatory visual methods, and gender and socio-cultural difference. She undertook action research in Zimbabwe while a postdoctoral fellow at the

Centre for Civil Society, University of KwaZulu-Natal, South Africa, and previous to that, completed her doctorate at the University of Oxford, UK, in 2009. Her research interests cut across agri-food studies, local innovation and knowledge-sharing processes, feminist geographies, critical development studies, and diverse economies. She is committed to developing innovative modes of inquiry, including video methods, which seek to deepen attention to situated knowledges, embodiment, and ecological agencies.

**Laura Saxer** holds a BA in Political Science and a MA in Global Studies. During her training she set a strong focus on environment and sustainable development, graduating with a thesis on the political ecology of changing land tenure systems in pastoral areas in Kenya. Saxer has also conducted an environmental impact analysis of a nongovernmental organization (NGO) project in Nepal and worked as an analyst in a research project on varieties of democracy around the world.

**Matthew D. Turner** is Professor of Geography at the University of Wisconsin-Madison. He has worked in the West African Sahel since the late 1980s. His research covers a number of topics including: politics of environmental governance and assessment; land and property rights; climate change vulnerability; and ecological and land-use change in African drylands. His current work is focused on within- and inter-household variations in nutrition in Burkina Faso; floodplain resiliency to drought in central Mali; and drivers and consequences of trans-Saharan labor emigration from Niger.

**Gabriela Valdivia** is Associate Professor in the Geography Department at the University of North Carolina, Chapel Hill. Her current research interests include: feminist political ecology of extractive industries, environmental mobilization, and environmental justice in Latin America. She is co-author of *Oil, Revolution, and Citizenship in Ecuadorian Amazonia* (2017) and has published extensively on matters of citizenship, social difference, and the governance of natural resources.

# Index

Note: Figures are indicated by page numbers in *italics* and tables are indicated by page numbers in **bold**.

For Product Safety Concerns and Information please contact our EU
representative GPSR@taylorandfrancis.com
Taylor & Francis Verlag GmbH, Kaufingerstraße 24, 80331 München, Germany

www.ingramcontent.com/pod-product-compliance
Ingram Content Group UK Ltd.
Pitfield, Milton Keynes, MK11 3LW, UK
UKHW052220190926
475702UK00006B/1114
*9780765646439*